Springer Series in
MATERIALS SCIENCE 100

Springer Series in
MATERIALS SCIENCE

Editors: R. Hull   R. M. Osgood, Jr.   J. Parisi   H. Warlimont

The Springer Series in Materials Science covers the complete spectrum of materials physics, including fundamental principles, physical properties, materials theory and design. Recognizing the increasing importance of materials science in future device technologies, the book titles in this series reflect the state-of-the-art in understanding and controlling the structure and properties of all important classes of materials.

S. van der Zwaag

Editor

Self Healing Materials

An Alternative Approach to 20 Centuries
of Materials Science

With 221 Figures

 Springer

Prof. Dr. Ir. Sybrand van der Zwaag
Delft Centre for Materials
Technical University Delft
Kluyverweg 1
2629 HS Delft
The Netherlands

Series Editors:

Professor Robert Hull
University of Virginia
Dept. of Materials Science and Engineering
Thornton Hall
Charlottesville, VA 22903-2442, USA

Professor Jürgen Parisi
Universität Oldenburg, Fachbereich Physik
Abt. Energie- und Halbleiterforschung
Carl-von-Ossietzky-Strasse 9–11
26129 Oldenburg, Germany

Professor R. M. Osgood, Jr.
Microelectronics Science Laboratory
Department of Electrical Engineering
Columbia University
Seeley W. Mudd Building
New York, NY 10027, USA

Professor Hans Warlimont
Institut für Festkörper-
und Werkstofforschung,
Helmholtzstrasse 20
01069 Dresden, Germany

The enclosed CD-ROM contains the complete proceedings of the First International Conference on Self Healing Materials, 18ᴶ 20 April 2007, Noordwijk aan Zee, The Netherlands.

A C.I.P. Catalogue record for this book is available from the Library of Congress

ISSN 0933-033x

ISBN 978-1-4020-6249-0 (HB)

ISBN 978-1-4020-6250-6 (e-book)

Published by Springer,
P.O. Box 17, 3300 AA Dordrecht, The Netherlands

www.springer.com

Acknowledgement

The Editor of this volume would like to thank the IOP Self Healing Materials, a subsidiary of the Dutch Ministry of Economical Affairs, for their generous support to the First International Conference on Self Healing Materials, 18–20 April 2007, Noordwijk aan Zee, The Netherlands.

Foreword

"As a general principle natural selection is continually trying to economise every part of the organisation." That was Charles Darwin, writing over 100 years ago about efficiency in nature. Natural materials are remarkably efficient. By efficient we mean that they fulfil the complex requirements posed by the way plants and animals function, and that they do so using as little material as possible. Many of these requirements are mechanical in nature: the need to support static and dynamic loads created by the mass of the organism or by wind loading, the need to store and release elastic energy, the need to flex through large angles, the need to resist buckling and fracture, and to survive damage. Few optimisation algorithms have been more successful than that of "survival of the fittest". The structural materials of nature exemplify this optimisation; even today, few man-made materials do better than those of nature in the function that they fill. And of all the remarkable properties of natural materials, one is truly exceptional – that of the ability for self-repair.

One recurring goal of material development has been to emulate the materials of nature. Among these, the most illusive is that of self-repair. In approaching this it is well to be aware of the nature of the differences that separate the structural materials of man and those of nature. The table itemises some of these: the great differences in chemistry, in mode of synthesis, in structure, and above all in the ability of natural materials for continuous adaptation and replacement, damage-sensing and repair. Man-made materials that achieve it are few: the self healing properties of the lime cement used by the early Roman Empire, the healing of oxide films that provide the corrosion resistance of aluminium, titanium, and stainless steel, the re-crystallisation of metals to restore the pre-deformed properties, and a few others.

A comparison of features of man-made and natural structural materials.

Man-made Structural Materials	Structural Materials of Nature
Based on the entire periodic table	Based on few elements (C, N, O, Ca, Si, etc.)
Thermo-chemistry (high temperature) processing	Ambient temperature processing
Fast production rate	Slow growth rate
Largely monolithic or simple composite structures	Complex, hierarchical structures
Unchanging structure once fabricated	Continuous replacement and renewal
No ability to adapt to change of environment	Ability adapt to an evolving environment
No capacity, in general, for self-repair	Ability to sense damage and self-repair
Thus: requiring "worst-case" design with additional safety factor	Thus: allowing optimal design to match current conditions without penalty of large safety factor

The contributors' initiative to bring together all these interesting concepts and early attempts at developing, describing, and even modelling self healing in materials is visionary and ambitious, and one that takes a broad stance and is the first of its kind. The spread of self healing strategies and of material systems to which they might be applied is broad, including polymers, composites, ceramics, metals, coatings, adhesives, and – significantly – the materials that inspired it all, those of nature. The editor and the contributors are to be congratulated on creating an impressive overture and opening movement to what is a sector of materials research and development with enormous potential.

Mike Ashby
Engineering Department
University of Cambridge
May 2007

Preface

This book *Self healing materials: an alternative to 20 centuries of materials science* is the 100th textbook in the Springer Series on Materials Science and opens the door to a new era in materials science: *self healing materials*.

Over the past 20 centuries the properties of structural materials at our disposal have improved tremendously in every aspect. The early improvements were due to a slow trial and error process, appropriately called "black magic" at the time, guided by just the appraisal of the final material, or more importantly the resulting product performance. The development took place without real knowledge of the material itself, or of the internal changes taking place during materials processing. In the nineteenth, and in particular, in the twentieth centuries, however, the pace of material development accelerated greatly. Two major factors played a role in this development: the notion of a microstructure in combination with physicochemical techniques to quantify it, and the availability of (semiempirical) models linking the microstructure to the final material properties and vice versa. Both factors led to a real acceleration of material development and today material properties can be tuned precisely, and to very high levels, sometimes even approaching the theoretical limits.

With hindsight, all strategies to improve the strength and reliability of materials developed over the past 20 centuries are ultimately based on the paradigm of *damage prevention*, i.e. the materials are designed and prepared in such a way that the formation and extension of damage as a function of load and/or time is postponed as much as possible. Damage is defined here as the presence of micro- or macroscopic cracks not being present initially. Characteristic for the current materials, developed under the damage prevention paradigm, is that the levels of damage in a material can either remain constant or increase, but will never go down spontaneously.

In recent years, however, it has been realized that an alternative strategy can be followed to make materials effectively stronger and more reliable, and that is by *damage management*, i.e. materials have a built-in capability to repair the damage incurred during use. Cracks are allowed to form, but the material itself is capable of repairing the crack and restoring the functionality of the material. Such damage management is encountered in natural materials such as, for example, skin tissue and bone structures. Although the mechanical properties of skin and bone are inferior to those of man-made polymers and ceramics, their ability to repair or heal damage (given the right healing conditions) results in a "lifelong" performance. Of course, given the huge difference in microstructure and chemical species involved, the mechanisms of healing in natural materials cannot be copied exactly. Some forms of self healing have been already been demonstrated, but it is anticipated that many more self healing mechanisms can and will be found with the progression of this novel branch of materials science.

This book aims at bringing together for the first time a more or less complete overview of the developments in the field of self healing materials, not just focused on one material class but covering all important man-made material classes, polymers, composites, metals, concrete, and bituminous materials. To put our human efforts in a wider context, two chapters on skin and bone healing are added as well. As is not uncommon in the field of material science, most of the chapters deal with experimental research, but several chapters also present early theoretical concepts towards modeling self healing phenomena, and are aimed at assisting material developers in creating self healing materials more quickly and more efficiently. As the editor of this first book on self healing materials I am very grateful for all the experts who made their valuable contribution to this book.

The book in its current form should be attractive to the entire community of material scientists working on both structural and functional materials. In addition, the level of subject-specific knowledge is kept to such a level that the book would be very suitable as a textbook in courses on self healing materials at both the undergraduate and graduate level. The CD, which goes with the book, contains the proceedings of the first international conference on self healing materials (18–20 April 2007, Noordwijk, the Netherlands), adding even more aspects of recent works in the field of self healing materials.

Finally, it must be emphasized that the book is written at a relatively early stage of development of the field and that a variety of challenges need yet to be met before we will see a wide application of such materials in daily life, a topic, which will definitely be included in the next edition of this book. The authors, however, are confident that in time we will be able to present a new edition of this book in which commercially and technologically successful applications of self healing materials are presented as well.

Sybrand van der Zwaag
Faculty of Aerospace Engineering
Technical University Delft
July 2007

Contents

An Introduction to Material Design Principles: Damage Prevention versus Damage Management

Sybrand van der Zwaag

Faculty of Aerospace Engineering, Technical University Delft, Delft, The Netherlands
Email: S.vanderZwaag@tudelft.nl

1 Introduction

Materials science is one of the most fascinating and challenging areas of science. It is both an old and a new science. Materials science not only determined our oldest history and even named it (i.e. Bronze Age, Iron Age), but still determines the pace of breakthrough developments in our daily life. Our mobile telephone/camera systems only became reality because of new battery materials, novel light-sensitive detector materials and new developments in silicon IC technology. On the other hand, transport of electrical currents in the national grid system still leads to large energy losses because materials scientists have not been able to create materials showing superconductivity at room temperature. Quantum computing would lead to a quantum jump in computing power, but is still some time away due to the fact that materials with the required properties and stability have not been created yet.

Materials science as a discipline is somewhere in between a multidisciplinary science and a real basic science. In understanding, designing, and creating materials with desired properties we use tools belonging to the basic sciences: physics, chemistry, and engineering. However, all these basic tools together are not sufficient to convert a molecular concept into a material which can be used to fabricate products consumers want and can afford. For this we also need tools exclusively belonging to the realm of materials science, the most important tools being the concept of material microstructure and the relations between microstructure and processing, as well as the relations between microstructure and properties. The poly (p-phenylene terephthalamide) molecule itself is a nice piece of polymer chemistry, but to make the transition from a rigid rod molecule to a high-strength, high-stiffness aramid filament and yarn, the microstructure has to be tuned carefully at several length scales simultaneously. To get these microstructures dedicated spinning routes have to be employed (Young 1989). While the ultimate aramid fibre properties are dictated by the properties of the molecule itself, it is the supermolecular organisation which determines the effective stiffness and strength levels of a polymeric fibre (Staudinger 1931; Northolt 1980, 2002).

Materials science is a science in itself as well as an enabling science. To tune high-strength TRIP steel to the highest levels of strength and ductility, detailed micro-beam X-ray diffraction studies using hard synchrotron radiation are performed to reveal the transformation characteristics of individual austenite grains in a multiphase ferritic–bainitic environment (Jimenez-Melero 2007). However, understanding and optimising the strength of such steels, challenging as it may be, is not a purpose in itself. The TRIP steels derive their real appreciation from the fact that they allow engineers to build lighter cars with a higher crashworthiness (Mahieu et al. 2002; Van Slycken et al. 2006).

As materials are present in all physical devices which are created to satisfy our needs and wishes, and hence cover a tremendous range of properties and underlying molecular concepts, it is now wise to make a restriction as to the materials and topics to be covered in this chapter and this book. From now on we will focus on structural or engineering materials only, i.e. materials which have been made to perform a mechanical function in addition to any other function they have to perform. So, whatever other requirements are formulated, structural materials are judged critically on their strength, stiffness, extendibility, and fracture behaviour. This sets them apart from the functional materials, in which properties other than mechanical properties determine their added value.

As stipulated above, materials and material properties are not an aim in themselves. Materials are there to perform a function and to do so, they are moulded into a shape. It is the combination of material properties and the dimensions and geometry of the product which gives us the desired functionality. Steel itself is considered to be stiff, but when spirally wound, the resulting spring can be rather flexible. Paper is considered to be rather weak and sloppy, but by gluing old scrap paper together in the form of cardboard, rigid and relatively strong structures, suitable as temporary housings in disaster-hit areas, may be created. It is the creation of this functionality in the final consumer product which drives material design and material processing technology. The contribution of material properties and product geometry to the overall performance or functionality of a product has been formulated very elegantly by Ashby (2005) in his so-called performance index for shaped materials:

$$\text{performance index} = (\text{material index}) \cdot (\text{shape efficiency factor}) \qquad (1a)$$

or

$$PI = MI \cdot \phi \qquad (1b)$$

where PI is the performance index, MI is a function of material properties only and ϕ a function depending on the geometry of the shaped material and the loading condition. As will be demonstrated later, this performance index might provide a proper tool to judge the performance of self healing materials against current materials.

2 Current Engineering Material Design Principles: The "Damage Prevention" Concept

In his famous book *The new science of strong materials, or why you do not fall through the floor* the late Professor J.E. Gordon (1968) provided a clear picture of

the principal guidelines behind the development of engineering materials. He opens
the book with a citation from the work of one of the founding fathers of modern
physics, Michael Faraday (1791–1867; "On the various forces of Nature"):

> Now, how curiously our ideas expand by watching those conditions of the attraction of cohe-
> sion! – how many new phenomena it gives to us beyond those of the attraction of gravitation!
> See how it gives us great strength!

Indeed, the cohesion between atoms ultimately limits the properties of any material.
This applies to metals, polymers, ceramics, and composites alike. To make strong
and stiff materials we need to assemble as many atoms with a high bond strength
to neighbouring atoms in as small a volume as possible. Not surprisingly, diamond,
comprised of densely packed small carbon atoms each having a high-covalent bond
strength, is the stiffest and possibly strongest material of all. Carbon nanotubes have
almost the same stiffness as diamond since the arrangement of the carbon atoms and
the spatial density is very comparable to that of diamond. In graphite, the same carbon atoms are present but in the non-basal plane direction the cohesion between the
atomic planes is much weaker, making graphite soft and even plastically deformable!

Materials scientists have used the cohesion between atoms to tune materials to the
desired property level. Generally speaking, a high cohesion leads to a stiff and strong
material, a low cohesion leads to a flexible and weak material. However, stiffness
and strength are rather different properties. Stiffness deals with the reversible material response to a mechanical load. Application of a load leads to an extension or
compression of the material, but after removal of the load the material returns to its
pre-loading configuration. Elastic loading leads to a stretching of atomic bonds but
the atoms retain their mutual position. The resistance of materials against extension,
the stiffness, depends on the strength of the interatomic bonds and their packing density. In the case of crystalline materials these are intrinsic properties and the stiffness
cannot be tuned. Even in the year 4007 the modulus of steel will still be in the range
of 205–210 GPa. In the case of polymers the modulus value seems a less intrinsic
parameter but that is because the modulus of a polymer depends on the polymer
chain configuration too. The modulus of a stretched polyethylene chain measured
along the chain direction is intrinsic and is determined by the chemical and topological structure of the molecule and is also on the order of 200 GPa. This value will not
change but the modulus of polyethylene products may, in the distant future, rise well
above the current value of typically 0.2–1 GPa if we know how to control molecular
alignment better.

Strength on the other hand deals with the ability of a material to sustain a high
load without disintegrating and forming new surfaces. Application of a load just
below the fracture load may or may not cause permanent or plastic deformation of
a material, i.e. atoms are displaced over long distances and relative to their initial
neighbours. The displacement of atoms relative to each other will ultimately lead
to internal defects (nano- and micro-cracks) which by a mechanism of coalescence
might grow into larger cracks and ultimately into cracks that cause the disintegra-
tion of the product. It will be clear that the presence of such defects as a result of an
imperfect positioning of the atoms during the material production phase will have a

negative effect on the strength of a material and can lead to the fracture of the product even without plastic deformation in the material except near the crack tip. The correct description and prediction of the extension of micro defects into larger non-fatal or fatal cracks as a function of applied loading conditions and material microstructure is of such importance that it became a special discipline in the field of materials science, the field of fracture mechanics.

Based on the preceding simplified picture of the material design one can argue that the current design philosophy to create a stronger material is based on creating microstructures which oppose the formation or extension of micro-cracks. In the design of strong materials the following basic guidelines apply:

1. Put all the atoms at the "right" place during the material production cycle (including thermo- and thermo-mechanical post-processing treatments).
2. Arrange the atoms in configurations where they find it hard to move.
3. Tune the material production process such that production defects are avoided.

So, one can argue that the development of new strong materials today, and in the 20 centuries preceding it, is progressing along the paradigm of *Damage Prevention*. The implications of the Damage Prevention concept can be illustrated graphically with the help of Figure 1.

The curve for the reference material (material A) shows very schematically the damage behaviour for any current ductile material exposed to a continuously increasing load or as a function of time for a constant load: at low loads or at early stages of the loading the material is still fully intact and there is no damage. At a particular load or exposure time the first damage is formed but it does not lead to catastrophic failure. Beyond the point of damage initiation, the damage in the material increases continuously, until the level of damage is such that catastrophic failure occurs (damage level = 1).

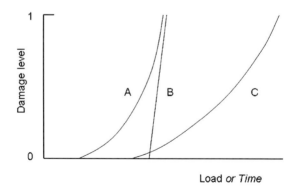

Fig. 1 Schematic diagram of the damage level in classical materials as a function of time or applied load. Material A sets the reference behaviour. Materials B and C are improved material grades. Note that the damage level only increases

The damage curve for the improved material grade (material B) shows a slightly different behaviour. The critical load or loading time for damage initiation is clearly higher than that of the reference material, but once the initial damage has been formed, the damage increases very rapidly to the level of catastrophic failure. This is the characteristic behaviour of brittle materials. Notwithstanding the improvement in the conditions for damage initiation, the short period between the point of damage initiation and the point of fatal damage would greatly reduce the effect of the strength improvement for practical situations, but the material can be called "stronger".

The damage curve for the second improved material grade (material C) shows both an improvement in the point of onset of damage as well as in the rate of damage formation. Clearly, from a materials development perspective, material C is a substantial improvement over the reference material (material A).

At this stage it is important to point out that a critical examination of the damage curves for current materials shows that the damage level as a function of time is either constant or increasing. Even when the load is reduced or removed completely, any damage created will remain present. It will be there "forever". Hence, for current man-made materials the rate of damage formation is either zero or positive at all times:

$$d(\text{damage})/dt \geq 0 \quad (\text{for } 0 < t < \text{life time}) \tag{2}$$

The design concept of *Damage Prevention* has been a very useful and productive concept and will also serve us well in the development of new materials in the future. However, as the formation of damage during use can never be excluded, it also means that structures made out of current materials invariably need periodic inspection to monitor damage development and any damage observed calls for action and costs, sooner or later.

3   Self Healing Material Design Principles: The "Damage Management" Concept

An alternative concept to that of Damage Prevention is Damage Management, which forms the basis of the field of self healing materials. The paradigm is based on the notion that the formation of damage is not problematic as long as it is counteracted by a subsequent autonomous process of "removing" or "healing" the damage.

The effect of such an autonomous healing process on the development of damage as function of time is shown schematically in Figure 2 for a number of cases.

For the material shown in Figure 2a there is a single healing action, which repairs the damage created almost completely. After the healing of the damage the formation of new damage occurs which leads to final fracture, as this material was designed for one healing action only. It should be pointed out that the loading conditions may have been relaxed during the healing cycle to favour the healing process, but this is not always necessary.

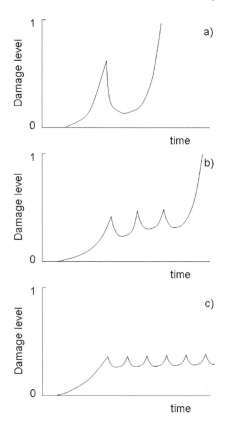

Fig. 2 Schematic diagram of the damage development in three grades of self healing materials (see text). Note that the damage level decreases during certain stages of the life time

The material shown in Figure 2b is capable of showing multiple healing effects in which the damage is partly removed in each cycle but after several healing cycles the material runs out of healing action and catastrophic failure occurs, albeit after a much longer time than when no self healing ability had been built into the material. So, a material can be self healing but still have a finite lifetime.

The material shown in Figure 2c goes a long way towards an ideal self healing material: the damage is healed many times and no accumulation of damage to the level of catastrophic failure occurs. For the loading (and healing) conditions imposed, the material has an infinite lifetime.

A common feature in the three curves shown in Figure 2, and a common feature in all healing materials indeed, is that self healing materials have a negative rate of damage formation at one or more stages of their life:

$$d(\text{damage})/dt < 0 \text{ (for } t_i < t < t_{i+\Delta t \text{ healing}}) \tag{3}$$

The final material performance of a self healing material therefore depends on two time constants: the rate of damage formation and the rate of damage healing! In case the state of reduced loading does not last long enough for healing to occur properly, the life extension will be small and the lifetime extension will fall short of the maximum improvement possible. As stated earlier, for healing to occur properly, it might be necessary to relax the loading conditions during the healing cycle.

Let us now turn to what is necessary at an atomic (or molecular) level to make a material potentially self healing. First of all, like any structural material the self healing material still has to perform a regular mechanical function. So, in our material design we first have to make sure that there are enough atoms and atomic configurations which will give the material these desired mechanical properties. As indicated in the previous paragraphs materials scientists have a huge assortment of tools and methods to do this. It is fair to assume that most of the atoms in a self healing material are there to perform this mechanical function. As a consequence, almost all self healing material grades will look like the classical counterpart designed along the Damage Prevention concept, at least at a first glance.

To make a material self healing, it is necessary that defects disappear more or less spontaneously: the empty space of the defect has to be filled with new matter which has the ability to locally restore the load-bearing capability. Hence, in contrast to current materials, in self healing materials a fraction of the atoms making up the material has to be mobile over shorter or longer distances in order to be able to fill the local damage spots. For materials scientists this is, in general, a strange concept: designing for partial mobility in an otherwise rigid material. The healing agent not only has to move to the damage site, but once it is there, it has to bond the two surfaces together permanently and in doing so lose its ability to move on. This transition of being mobile initially but fixed in space at a later stage can be realised via temperature excursions, chemical reactions but other processes can be envisaged too.

Of course, one can imagine that for a healing process to take place, a damage sensor/healing trigger is required and this feature has to be included in the self healing material design as well. The sensor has to detect the occurrence of damage and to initiate the healing process either by starting the motion of the healing agent, or the rebonding reaction itself. The sensor/trigger might either be identifiable as a separate set of atoms, but it would be much nicer if the damage itself acted as the sensor/trigger.

At this point it is appropriate to note that for self healing to occur it is also crucial that the crack surfaces stay or are brought in close contact and are held in a mutually fixed position while the healing reaction takes place. Any reader having suffered from a broken bone will realise that these requirements also apply to the healing of natural human materials. So, self healing is much easier to realise in situations of partial cracking in which the non-cracked region ahead of the crack tip keeps the fracture surfaces in some registry. When damage results in the full separation of the two fractured pieces, proper healing generally does not occur unless the fracture surfaces are brought together intentionally. This partly defies the notion of autonomous healing.

The requirement of crack surfaces not being too far apart from each other also means that healing is favoured when the applied load is occasionally reduced or removed, as this leads to a reduction of crack opening or even to crack closure. So, self healing materials are more promising for applications under cyclic loading and less promising for structures loaded under a constant load. It has been shown that the fatigue properties of a self healing epoxy, when tested under the right conditions, are exceptional indeed (Brown et al. 2005a, b, 2006; Jones et al. 2007).

In conclusion, it will be quite challenging to make a self healing material, as the various tasks of the atoms in a self healing material are far more complex than those in classical materials. The self healing material is no longer just a material, but has become a system in itself.

The various functions of atoms (and other non-dividing entities such as compounds, monomers and molecules) to be distinguished in a self healing material are summarised in Table 1.

Table 1 Roles to be performed in a self healing material and the corresponding characteristics for behaviour at an atomic level

Role to be Performed	Required Characteristics of the Atoms/Molecules Involved in that Role
Yield the desired performance	Atoms are bonded in such a way they stay in place under the applied load
Repair the damage	Atoms are mobile and but can be transferred to the group of atoms yielding the mechanical properties once done the healing
Sense the damage	Atoms convert information on the state of the material in a response function

At this stage we can continue the "self healing material" definition phase and list the important properties for self healing materials to be technologically and economically attractive. In Table 2 these properties are defined for both an "ideal" self healing material and a "minimal" self healing material.

Table 2 Properties of "ideal" and "minimal" self healing materials

An "Ideal" Self Healing Material:	A "Minimal" Self Healing Material:
Can heal the damage many times	Can heal the damage only once
Can heal the damage completely	Can heal the damage partially
Can heal defects of any size	Can heal small defects only
Performs the healing autonomously	Needs external assistance to heal
Has equal or superior properties to current materials	Has inferior properties to current materials
Is cheaper than current materials	Is terribly expensive

It will be clear that current experimental self healing materials are likely to have characteristics that are close to those of "minimal" self healing materials. The development of "ideal" self healing materials will probably take a very long time, but this should not worry us too much as the alchemist's dream of turning lead into gold also took a long time (and a powerful ion accelerator). Furthermore, even after 20 centuries the ideal classical material has not yet been created either.

Out of the list of characteristics mentioned in Table 2, the requirement of self healing materials having the same or even better properties than the materials developed along the concept of "damage prevention" might intrinsically be the hardest one to meet. The reason is that in a self healing material we have to set aside a certain fraction of the atoms/molecules in the material in order to make it self healing. The mobile "healing" atoms initially do not really contribute to the mechanical properties, or might even reduce the mechanical performance. Hence, a self healing material has some "redundancy" when looking at its mechanical properties from an atomistic/molecular perspective. However, as mentioned in the introduction already, it is not the performance of the material we want to optimise, but the performance of the product we want to make out of it. The performance indicator, which is based on the material properties under ideal conditions and the material being in a pristine state, was given by Equation (1). However, in real-life situations the performance indicator is an upper bound and engineers always apply a safety factor (also called "lack-of-full-knowledge" factor), which takes into account the reduction of performance due to the accumulation of unspecified damage during use. Hence, the effective performance indicator of a shaped material under practical conditions is given by:

$$PI_{practice} = MI \cdot \phi / SF \qquad (4)$$

where SF is the safety factor. In the case of an ideal self healing material any damage is healed fully and perpetually, so the safety factor can be set to 1 and $PI_{practice} = PI$. So, self healing materials are still attractive even if their properties are inferior to those of current materials, provided the reduction in properties is less than the net effect of the application of a safety factor.

4   The Structure of the Book

Let us now turn to the current state of affairs for self healing materials and put the subsequent chapters of this book in a proper and mutual perspective.

In all modesty, it is correct to start by indicating that self healing behaviour is not something very new in materials science. However, up to very recently a self healing capability has been a property which was not imposed intentionally, but one which was there by coincidence. A well-known example of self healing behaviour in inorganic materials is the mortar used by the Romans for the construction of their buildings and large infrastructural works (Riccardi et al. 1998). The mortar holding together the natural stones and man-made bricks, by current standards, does not have very good mechanical properties but nevertheless has held together large civil structures for over 20 centuries. This exceptional durability is the consequence of micro-cracks closing spontaneously due to a chemical reaction between the mortar and the moisture in the air leading to controlled dissolution and reprecipitation (Sanchez-Moral 2004). As the reaction product forms in situ, it shapes itself to the complex crack geometry resulting in a complete stress reduction once the crack is filled. This cracking and refilling of the cracks gave the mortar the ability to fail and adjust its shape while still meeting load-bearing requirements continuously. This example shows that healing not always requires macroscopic stresses to be reduced temporarily. As a consequence of frequent internal healing, the deformation strain in Roman mortar over 20 centuries can be a lot higher than the fracture strain as determined in a standard mechanical test. This discrepancy points to the fact that we will still have to develop proper testing schedules to measure the relevant characteristics of a self healing material. While a standard tensile test, imposing a continuously increasing macroscopic strain level, gives a very good indication of the mechanical properties for a classical non-self healing material, it is clearly insufficient for revealing the attractive properties of self healing materials. Measured on the scale of the parameters listed in Table 2, the Roman mortar is a pretty "ideal" self healing material. The main drawback and the reason it is not being used any longer are that it is not sufficiently resistant to the more aggressive atmosphere of modern society and takes too long to set. Finally, it is important to stress that the presence of old roman buildings held together by a relatively weak mortar demonstrates that self healing can have an enormous impact on the durability of materials and constructions.

In metallurgy, the best-known example of self healing behaviour stems from the field of corrosion protection. Chromate coatings (Zhao et al. 1998; Suda et al. 2002; Kendig and Buchheit 2003) provide an excellent corrosion protection due to the fact that the chromate redeposits as soon as the ferrous substrate is exposed and substrate dissolution has raised the metal ion content in the electrolyte. So, as soon as damage occurs, a chemical reaction is triggered which leads to full coverage of the damaged site with a protective layer. For this healing to occur, it is also not necessary to relax the loading conditions temporarily. So chromate coatings are pretty "ideal" too, but are being banned because of their carcinogenic side effects.

Although it can be argued that human and animal skins, being polymeric in nature, are examples of self healing behaviour in polymers dating back to a distant past, self healing behaviour in man-made polymers necessarily dates from a much more recent time. However, as early as 1966 the first patent of a polymer with intentional self healing characteristics appeared (Craven 1966). Unfortunately, the potential of this route was not appreciated and the field of (intentionally) self healing polymers remained more or less dormant until recently.

The research on creating self healing behaviour in man-made materials probably started with the work of Dry (1994; 1996) but really took off with the landmark paper by White et al. (2001). They embedded brittle microcapsules of hollow fibres filled with a liquid adhesive in a regular matrix material, such as concrete or epoxy respectively. Upon the passage of a crack through the material the microcapsules ruptured and the liquid adhesive flew over the fracture surface, filled the crack, and set. The elegance of their concept is in the notion that liquids are intrinsically mobile (at least relative to solid materials). Hence, embedding a liquid healing agent in a solid material automatically provides the material system with the atomic mobility required to make a material potentially self healing. In this concept there is a complete separation of functions. The matrix provides the desired mechanical function while the healing agent serves no purpose until after the occurrence of damage. The required third functionality, the triggering of the healing process, is provided not by a well-defined damage sensor but by both the combined effects of fracture of the capsule wall, the catalyst in the surrounding matrix and the surface tension of the fracture surfaces, which spreads the liquid healing agent over the fracture surfaces. In their chapter "Self Healing Polymers and Composites" Andersson et al. provide a complete description of the development of the encapsulation route as well as recent results on the healing behaviour under a range of loading conditions.

While the liquid encapsulation technique is very elegant and illustrative, it also has the drawback of allowing only one repair action per broken capsule. Hence, there has been an active interest in systems allowing repeatable healing.

A complete overview of the chemical routes towards repeatable healing in polymeric materials is presented in the chapter "Re-Mendable Polymers" by Bergman and Wudl. Repeatable healing in polymer materials can be realised by fracture and reformation of the molecular chain itself or via fracture and reformation of the intermolecular bonds. Thermally reversible chemical bonds along a covalent polymer backbone via retro-Diels-Alder reactions in furan–maleimide-based polymers have been known for some time (Craven 1966) but the concept was used only recently to create re-mendable polymers (Chen et al. 2002). Sijbesma et al. (1997) used non-covalent, hydrogen-bond interactions to make self-organising polymers which have inherent self healing characteristics. In such systems the spontaneous reformation of principal chemical bonds is due to the dynamic nature of the hydrogen interactions. It should be pointed out that self healing polymers of the bond-reforming type potentially have very interesting recycling characteristics too and should be researched actively not just for their self healing characteristics.

In their chapter "Thermally Induced Self Healing of Thermosetting Resins and Matrices in Smart Composites" Jones et al. present yet another approach leading to repeatable crack-healing behaviour. Their approach focuses on dissolving linear polymer chains into a thermosetting epoxy. Molecules belonging to the thermoset are effectively immobile because of the high crosslink density and, as a consequence, cannot heal any damage. In contrast, well-chosen linear molecules with a carefully tuned solubility in the epoxy can meander through the cross-linked network via a reptation mechanism and can restore significant molecular bonding over the interface provided the two fracture surfaces are brought into intimate contact. The nature of the healing process itself makes it compulsory to heat the material to impart sufficient mobility to the linear molecule and to realise healing on a decent timescale.

Given that most healing processes require some heating of the material, it is interesting to consider damage formation processes at which heat is generated automatically. Ballistic impact might provide appropriate conditions as materials impacted by a projectile locally heat up significantly during the passage of the projectile. This heat can be used to heal the material, provided a second crucial requirement, the crack faces being in contact, is also met. In his chapter "Ionomers as Self Healing Polymers" Varley describes how ionomers show full self healing upon bullet penetration, provided the impact does not remove too much material. It is interesting to point out that in this case the healing is extremely fast (milliseconds), unlike the earlier examples of self healing polymers.

In their chapter "Self Healing Fibre Reinforced Polymer Composites: An Overview" Bond et al. describe their work on self healing carbon–epoxy composites using hollow glass fibres as containers to store two mutually reactive healing agents. For hollow fibre systems the local supply of healing agent can be much larger than in the case of discrete capsules and larger areas of delamination and matrix fracture can be healed. While the initial properties of their self healing composites are below those of regular carbon–epoxy composites, the self healing composites had a much higher residual strength after impact and subsequent healing than the impacted regular carbon fibre composite panels. For many composite applications in aerospace and space applications the post-impact strength is a more relevant design parameter than the initial strength. This work provides good evidence for the earlier statement that the functionality index under practical conditions should be used as the parameter to judge shaped-material performance rather than the pristine functionality index.

In their chapter "Self Healing Polymer Coatings" Van Benthem et al. describe the development of self healing coatings linking both damage and healing processes to the characteristics of coating systems. Coatings are a potentially very attractive application field for self healing materials for several reasons: a small thickness in combination with a large surface area, the presence of a free surface and the presence of a substrate. The small thickness implies that the dimensions of the crack are necessarily small, which makes their repair easier. The large surface area implies that the probability of damage is relatively high per unit of material volume. The presence of a free surface allows the supply of a healing agent or a critical component of the healing agent from the gaseous or liquid medium on the other side of the interface. In case the coating is deposited on a microvascular substrate (White 2006; Toohey et al. 2007), the substrate itself can also be used to supply the interface with healing agent and to restore interfacial and coating cracks. Finally, the substrate will constrain the lateral displacement of cracks in the coating and will keep the crack width relatively small. So, all the ingredients are there to make self healing coatings a reality – and an attractive reality – and some industrial attempts to launch such systems have been made already.

A relatively large fraction of this book is devoted to self healing in polymer-based systems, which should come as no surprise as mobility, the necessary ingredient for self healing, is more inherent to polymers than to ceramics and metals, in which atoms are bonded very strongly in a three-dimensional configuration. However, in ceramics and metals self healing behaviour has been realised too using healing mechanisms tailored to the characteristics of these parent materials.

In their chapter "Self Healing in Concrete Materials" Li and Yang describe self healing in a special type of concrete. Concrete can be made self healing relatively easily by incorporating a sufficient fraction of unreacted cement, which upon ingress of moisture through the micro-cracks reacts further and heals the crack. The main challenge in making proper self healing concrete is to control the crack width as regular concrete is prone to form wide cracks for which healing is not possible. In their engineered cementitious materials the crack width is controlled by the inclusion of polymeric microfibres. The microfibres cause crack bridging and limit the crack opening to less than 40 μm which can be healed well given the right concrete chemistry and a sufficient supply of moisture. Concrete of such a type has already found its first commercial applications in the deck of a suspension bridge in Japan as well as in the central column of a high-rise building in an earthquake-prone city.

The healing in concrete relies on mass transport via diffusion of water through the micro-cracks and the diffusion of ions through the solid material as well as the final precipitation reaction, and is generally rather slow. To speed up the healing process the use of bacteria has been considered. In his chapter "Self Healing Concrete: A Biological Approach" Jonkers describes preliminary results for the healing of concrete containing spore-forming bacteria. The excrements of rejuvenated bacteria are ideal to seal the cracks and to restore their load-bearing ability. The spores can remain dormant but intact over up to 200 years, which means the concrete should demonstrate proper self healing behaviour for a very long time. As before, the research into this type of self healing material is driven primarily by durability considerations.

Another civil engineering material for which self healing behaviour would be very attractive is asphalt, a semi-porous mixture of bitumen and granular filler materials. Little and Bhasin describe the self healing behaviour of asphalt in their chapter "Exploring Mechanisms of Healing in Asphalt Mixtures and Quantifying Its Impact". In self healing of asphalt the healing process takes place at the interface between two dissimilar materials, the bitumen and the (ceramic) aggregate. As expected, the restoration of the interfacial bond is due to the flow of the bitumen only. The flow and the restoration of the interface depend on the polar and non-polar components of the surface energy of the asphalt–aggregate interface as well as the reptation of the polymeric molecules in bitumen. Their model indicates that the healing kinetics should be tailored to the rate of damage formation, which was also a conclusion already drawn on the basis of Figure 2.

As pointed out earlier, self healing behaviour in metals, with the exception of self healing in the semi-liquid state (Bernikowitz 1998), is hard to realise as substantial mobility in a metallic material is limited to temperatures close to the melting point. Hence, self healing in metals will probably be restricted to self healing in corrosive situations involving liquid electrolytes and/or to applications at high temperatures. However, Lumley in his chapter "Self Healing in Aluminium Alloys" presents very interesting work on underaged aluminium alloys in which nano-sized deformation damage is assumed to be healed by the formation of new precipitates at the damage sites. His hypothesis of spontaneous healing of deformation defects by the formation of Cu-rich precipitates is supported by recent positron annihilation measurements (Hautakangas 2007). It is interesting to point out that the effect of healing on the fatigue properties of the aluminium alloy is not unlike that for asphalt mixtures (Little and Bhasin, this vloume).

The topic "Crack and Void Healing in Metals" has been addressed by Wang et al. in their chapter. In their finite element simulations they show that the void instability is a function of the competition between elastic strain energy and interface energies.

The transport mechanism required to move atoms to the defects is a combination of bulk diffusion and grain boundary diffusion. Their work presents the critical conditions for void healing on grain boundaries to occur, as well as the conditions where void growth will occur. The difference in the local energy state of atoms near or away from the voids and cracks provides the driving force for damage restoration. It should be pointed out that in this system there is no chemical difference between the atoms present to perform the mechanical function and the atoms performing the healing action. It is only the location of the atom which determines the difference in mobility and the role to be played. The crack healing described here resembles the healing of two surface melted ice cubes pressed together. Self healing via surface melt phenomena has not been explored actively but has some very attractive features.

Given the fact that the deformation damage in metals which can be healed is of submicron level, dedicated experimental techniques with a high spatial resolution are required. In their chapter "Advances in Transmission Electron Microscopy: Self Healing or is Prevention Better than Cure?"De Hosson and Yaduda demonstrate that a high spatial resolution in itself is not enough to detect and quantify self healing, but that experiments have to be performed in situ to capture the relevant phenomena, greatly increasing the complexity of the experiment. In the same way as current fracture mechanic tests have to be redesigned to capture self healing behaviour in materials, the protocols for material characterisation will have to be adjusted.

The section on self healing in metals ends with the chapter "Self Healing in Coatings at High Temperatures" by Sloof. His chapter deals with self healing in thermal barrier coatings on turbine blades. Given the aggressive combustion environment, the turbine blade surface needs a protective multilayer MCrAlY coating system consisting of carefully tailored oxide-containing layers. Based on thermodynamic and thermokinetic analyses Sloof derived an equation for the conditions for which the degree of spallation damage is equal to the amount of oxide growth. Under these conditions the material stays intact due to a continuous healing of a coating undergoing continuous spallation. It should be realised that a critical component in the healing process, the oxygen, is initially not present in the material itself, but is supplied by the atmosphere surrounding the material. The work highlights the necessity of a chemical reaction as a step in the healing process.

While all the work presented so far has been on man-made materials and man-designed healing concepts only, it is of course interesting to have a look at healing processes in natural materials too. Again a distinction should be made between healing of bulk materials, such as bone, and that of thin film materials, such as skin, in the same way as the healing in bulk polymers is different from that of polymer coatings. It should be realised that these natural materials have a far more developed hierarchical structure than man-made materials and can therefore act as material systems rather than as materials as such. The man-made microvascular systems and the multi-layer thermal barrier coatings only go a little way in that direction but are interesting first steps.

In their chapter "Hierarchical Structure and Repair of Bone: Deformation Remodelling and Healing" Fratzl and Weinkamer point out that bone reconfiguration takes place continuously, not only in the case of bone fracture. The reconfiguration is controlled by the activity of bone-forming cells, osteoblasts, and that of bone-removing cells, osteoclasts, while the actual state of the bone is monitored by the monitoring cells, osteocytes. It is of interest to point out that the bone is also porous, which means that long-distance material transport is very easy. Hence, the ingredients to create additional bone are easily transported to those segments of the tubular structure most heavily loaded. Another lesson for the design of man-made self healing materials is that in bone fibrillar bonding, and rebonding takes place via a viscous displacement of the interfibrillar region involving a volume of only $1 \, nm^3$. This very small volume means that mobility is present but does not lead to macroscopic viscous flow in a material which should give dimensional stability to the human body. The lesson to be learned for materials scientists is that it is indeed possible to create mobility in a material without losing good dimensional stability under load.

The chapter "Modelling of Self Healing of Skin Tissue" by Vermolen et al. does not deal with the biological aspects of skin healing but rather presents a mathematical model of skin healing in which the critical biological processes are captured in rather simplified concepts. Notwithstanding its simplicity the model seems to describe the wound-healing kinetics and wound shape changes rather well, although some further calibration of the model parameters would be beneficial. The main attraction of the model is that its simplicity makes it possible to use it as the nucleus of a generic model for crack healing in man-made materials. Work of such a nature is already ongoing at the TU Delft.

Due to the long history in material design along the lines of fracture prevention, there are many articles and textbooks in the field of fracture mechanics linking microstructures and local stress states to crack formation and crack propagation. In the final chapter of this book entitled "Numerical Modelling of Self Healing Mechanisms" Remmers and De Borst describe a micromechanical model for crack propagation coupled with liquid healing agent transport along the crack and time-dependent rebonding of the crack surfaces due to an unspecified chemical reaction. This multilevel model contains all relevant ingredients for modelling damage management in self healing materials. The degree of mathematical detailing for the three principal processes shows that our current understanding of fracture processes still exceeds our understanding of healing processes by far. This book as a whole aims to make a start at redressing this situation.

5   Concluding Remarks

A well-known concept in materials science is the concept of incubation time. The incubation time is generally defined as the time between the formal start of a process and the time at which a measurable fraction of the material has undergone a transformation or chemical reaction. During this incubation time, net productivity is rather small but the many barely detectable reactions taking place on a very small scale everywhere in the material provide the nuclei for the final reaction and the final success. It is safe to assume that the field of self healing materials science is now in the incubation stage. At a large number of academic and industrial centres new concepts are being explored for all sorts of materials, possibly even wider than the range covered in this book. However, the amount of self healing material actually being sold today commercially is minimal.

It is hard to predict where and when the first real large-scale commercialisation of a self healing material in a particular application will take place. Very recently automotive paint systems were introduced which reportedly show some self healing behaviour; superficial surface scratches due to car-washing actions disappear spontaneously under the action of solar heating. Similar action was reported for ski goggles becoming scratch-free after storing them overnight above the radiator of the central heating system. Also in the case of concrete structures the real introduction of self healing concrete may not be far, but so far has been limited to isolated cases. It is expected that ultimately the drivers for the development of self healing materials will come from three or four societal needs:

1. All applications for which repair is intrinsically very costly but the demands on reliability are very high (wind farms at sea, underground pipe systems, etc.)
2. All applications where a longer than 40-year performance has to be guaranteed (tunnels and large infrastructural works, etc.)
3. All applications where a very high reliability is required (aircraft, nuclear storage systems, etc.)
4. All applications where a high surface quality is highly appreciated (cars, optical systems, windows in public transport, etc.)

While working very hard towards the early commercialisation one should also be a little realistic and aware of historical facts. All recent breakthroughs of novel structural materials, such as aramid and HM polyethylene fibres, TRIP steel, and GLARE fibre metal laminates, took typically 20 years to grow from the conceptual phase to the industrial production phase. I am sure that the great benefits self healing materials hold for making products safer and more durable and our society more sustainable will lead to the development time for commercial self healing materials to be less than the historic time constant of 20 years. In fact, it is our expectation that the second edition of this book on self healing materials will already have several chapters on the application of self healing materials.

Acknowledgements For any new field to be opened up successfully a constructive cooperation between "funding fathers" and researchers in the field is required. Hence, it is a pleasure to thank the College van Bestuur of the Technical University Delft, the ministry of Economic Affairs, the board members of the Delft Centre for Materials and all academic colleagues of DCMat for their financial and intellectual support in establishing a multi-material research program on Self Healing Materials in the Netherlands.

I also thank all the contributors to the book for their fantastic work and most notably for the minimum amount of energy required to persuade them to make a contribution.

I take this opportunity to thank the many leading scientists who at various stages of my professional career introduced me to their favourite branch of materials science: the late Professor Peter Jongenburger, Professor John E. Field, Professor Arie van den Beukel, Dr. John J. van Aartsen and Dr. Maurits. G. Northolt, Professor Steve J. Picken, Professor Mike F. Ashby, Professor Sieb Radelaar, Drs. Jilt Sietsma and Niels van Dijk, Professor Kees Vuik en Dr. Fred J. Vermolen, Professor Klaas van Breugel, and Dr. Tom Scarpas.

Finally, I thank my staff members Dr. Theo Dingemans and Dr. Pedro Rivera for the memorable brainstorming dinner during which we selected self healing material concepts as *the* topic for our future research.

References

Ashby M.F. (2005) Materials selection in mechanical design. Elsevier Amsterdam, The Netherlands

Bernikowitz P. (1998) Design of tin based biomimetic self healing alloy tensile specimens. TMS outstanding student paper contest. www.tms.org/students/winners

Brown E.N., White S.R., Sottos N.R. (2005a) Retardation and repair of fatigue cracks in a microcapsule toughened epoxy composite – Part I: Manual infiltration. Compos. Sci Technol 65: 2466–2473

Brown E.N., White S.R., Sottos N.R. (2005b) Retardation and repair of fatigue cracks in a microcapsule toughened epoxy composite – Part II: In situ self-healing. Compos Sci Technol 65: 2474–2480

Brown E.N., Sottos N.R., White S.R. (2006) Fatigue crack propagation in microcapsule-toughened epoxy. J Mater Sci 41:6266–6273

Chen X., Dam M.A., Ono K., Mal A., Shen H., Nutt S.R., Sheran K., Wudl F. (2002) A thermally re-mendable cross-linked polymeric material. Science 295:1698–1702

Craven J.M. (1966). Cross linked thermally reversible polymers produced from condensation polymers with pendant furan groups cross-linked with maleimides. US patent 3,3435,003

Dry C. (1994) Matrix cracking, repair and filling using active and passive modes for smart timed releases of internal chemicals. Smart Mater Struct 3:118–123

Dry C. (1996) Procedures developed for self-repair of polymer matrix composite materials. Compos Struct 35(3):263–269

Gordon J.E. (1968) The new science of strong materials, or why you don't fall through the floor. Princeton University Press, Princeton, USA

Hautakangas S., Schut H., van Dijk N.H., Rivera del Castillo P., van der Zwaag S. (2007) Self healing of deformation damage in underaged Al-Cu-Mg alloys. Submitted to Scripta Materialia

Jimenez-Melero E., van Dijk N.H., Zhao L., Sietsma J., van der Zwaag S. (2007) Martensitic transformation of individual grains in low-alloyed TRIP steels. Scripta Materialia 56(5):421–424

Jones A.S., Rule J.D., Moore J.S., Sottos N.R., White S.R. (2007) Life extension of self-healing polymers with rapidly growing fatigue cracks. J Roy Soc Interf 4:395–403

Kendig M.W., Buchheit R.G. (2003) Corrosion inhibition of aluminum and aluminum alloys by soluble chromates, chromate coatings, and chromate-free coatings. Corrosion 59(5):379–400

Mahieu J., Maki J., De Cooman B.C. (2002) Phase transformation and mechanical properties of Si-free CMnAl transformation-induced plasticity-aided steel. Metall Mater Trans A 33(8): 2573–2580

Northolt M.G. (1980) Tensile deformation of poly (p-phenylene terephthalamide) fibres, an experimental and theoretical analysis. Polymer 21(10):1199–1204

Northolt M.G., Baltussen J.J.M. (2002) The tensile and compressive deformation of polymer and carbon fibers. J Appl Polym Sci 83(3):508–538

Riccardi M.P., Duminuco P., Tomasi C., Ferloni P. (1998) Thermal, microscopic and X-ray diffraction studies on some ancient mortars. Thermochim Acta 321(1–2):207–214

Sanchez-Moral S., Garcia-Guinea J., Luque L. (2004). Carbonation kinetics in roman-like lime mortars. Materiales de Construction 54(275):23–37

Van Slycken J., Verleysen P., Degrieck J., Bouquerel J., De Cooman B.C. (2006) Crashworthiness characterization and modelling of high-strength steels for automotive applications. Proceedings of the Institution of Mechanical Engineers Part D–J of Automobile Engineering 220(D4): 391–400

Staudinger H. (1931) Die Hochmolekularen Organische Verbindungen. Springer, Berlin, p 111

Suda A., Shinohara T. (2002) Effects of colloidal silica addition on the self-healing function of chromate coatings. ISIJ Int 42(5):540–544

Sijbesma R.P., Beijer F.H., Brunsveld L., Folmer B.J.B., Hirschberg J.H.K., Lange R.F.M., Lowe J.K.L., Meijer E.W. (1997) Reversible polymers formed from self-complementary monomers using quadrupole hydrogen bonding. Science 278:1601–1604

Toohey K.S., White S.R., Lewis J.A., Moore J.S., Sottos N.S. (2007) Self-healing materials with microvascular networks. Nature Materials (in press) (doi:10.1038/nmat1934)

White S.R., Sottos N.R., Geubelle P.H., Moore J.S., Kessler M.R., Sriram S.R., Brown E.N., Viswanathan S. (2001) Autonomic healing of polymer composites. Nature 409:794–797

White S.R. (2006) Microvascular autonomic composites. MURI Annual Report, AFOSR Grant # FA9550–05–1–0346

Young H.H. (1989) Aromatic high strength fibres. Wiley, New York

Zhao J., Frankel G., McCreery R.L. (1998) Corrosion protection of untreated AA-2024-T3 in chloride solution by a chromate conversion coating monitored with Raman spectroscopy. J Electrochem Soc 145(7):2258–2264

Self Healing Polymers and Composites

HM Andersson[1], MW Keller[2], JS Moore[3], NR Sottos[4], and SR White[5]

[1] *Beckman Institute, University of Illinois Urbana-Champaign, Urbana, USA*
Email: magand@uiuc.edu
[2] *Department of Mechanical Science and Engineering, University of Illinois Urbana-Champaign, Urbana, USA*
[3] *Department of Chemistry, University of Illinois Urbana-Champaign, Urbana, USA*
[4] *Department of Materials Science and Engineering, University of Illinois Urbana-Champaign, Urbana, USA*
[5] *Department of Aerospace Engineering, University of Illinois Urbana-Champaign, Urbana, USA*

1 Introduction

Structural polymers are susceptible to damage in the form of cracks, which form deep within the structure where detection is difficult and repair is almost impossible. Damage in polymeric coatings, adhesives, microelectronic components, and structural composites can span many length scales. Structural composites subject to impact loading can sustain significant damage on centimeter length scales, which in turn can lead to subsurface millimeter-scale delaminations and micron-scale matrix cracking (Figure 1).

Coatings and microelectronic packaging components have cracks that initiate on even smaller scales. Regardless of the application, once cracks have formed within polymeric materials, the integrity of the structure is significantly compromised.

1.1 *Autonomic Healing of Polymer Composites*

Inspired by biological systems in which damage triggers a healing response, structural polymeric materials with the ability to autonomically heal cracks have been developed at the University of Illinois (White et al. 2001). Figure 2 illustrates the autonomic healing concept. Healing is accomplished by incorporating a microencapsulated healing agent and a catalytic chemical trigger within an epoxy matrix. An approaching crack ruptures embedded microcapsules, releasing healing agent into the crack plane through capillary action. Polymerization is triggered by contact with the embedded catalyst, bonding the crack faces.

1.2 *Materials Systems*

Successful completion of the self healing process presents a complex set of requirements on storage, rupture, transport, and healing: quiescent, stable storage of liquid-healing agent, mechanical triggering, release and transport of the healing agent,

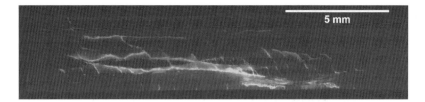

Fig. 1 Delaminations and transverse matrix cracks in a fiberglass-reinforced epoxy laminate after impact damage (UV illumination of cross-section of impact site with fluorescently highlighted damage). (From Patel and White 2006)

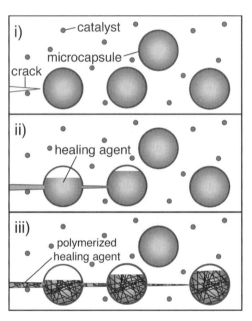

Fig. 2 The autonomic healing concept: A microencapsulated healing agent is embedded in a structural composite matrix containing a catalyst capable of polymerizing the liquid healing agent. (i) Cracks form in the matrix wherever damage occurs. (ii) The crack ruptures the microcapsules, releasing the healing agent into the crack plane through capillary action. (iii) The healing agent contacts the catalyst, triggering polymerization that bonds the crack faces closed. (From White et al. 2001)

chemical triggering, polymerization, and recovery of mechanical toughness – all must occur without significantly impacting the inherent properties of the material. To date, the microencapsulated approach to self healing has yielded high-healing efficiencies in several materials systems. The initial focus of this research was brittle thermoset materials, e.g. epoxy and vinyl ester. In epoxy, healing has been achieved with two different catalysts, Grubbs' catalyst (Brown et al. 2002, 2004, 2005b; Jones et al. 2006, 2007; Mauldin et al. 2007; Rule et al. 2005; White et al. 2001) and tungsten hexachloride (Rule 2005). The former approach has also been successful in a vinyl ester matrix (Wilson et al. 2007). Another healing system that has been successfully integrated in vinyl ester is based on an encapsulated organotin catalyst and a phase-separated healing agent (Cho et al. 2006).

For structural composite materials, the presence of fiber reinforcement increases the number of damage modes and the complexity of the healing process. Self healing of fiber-reinforced composites has been demonstrated for both glass fiber-reinforced epoxy (Patel et al. 2007) and carbon fiber-reinforced epoxy (Kessler et al. 2003).

Unlike thermoset materials, elastomers do not fail in a brittle manner. However, the tear failure of elastomers generally does not involve large, permanent deformations. As such, a self healing elastomeric system based on the encapsulation of a two-component PDMS has recently been developed (Keller et al. 2007).

1.3   Scope

Based on the microencapsulated approach to self healing, this chapter describes the chemistry, components and mechanical characterization necessary for a successful autonomic healing system. Critical issues for self healing fiber-reinforced composites and elastomeric materials are also addressed. Although the microencapsulated approach has yielded high-healing efficiencies in several materials systems, the number of possible healing events is limited by the supply of healing agent in the capsules. To address this issue, research has commenced on a second generation of self healing materials that utilize microvascular networks to store and transport healing agent to damaged areas in a continuous fashion (Toohey et al. 2007).

2   Self Healing Chemistry

The self healing process relies on a suitable chemistry to polymerize the healing agent in the fracture plane. Among the requirements of a self healing system are long shelf life, low monomer viscosity and volatility, rapid polymerization at ambient conditions, and low shrinkage upon polymerization.

2.1   Ring-Opening Metathesis Polymerization (ROMP) of Dicyclopentadiene

2.1.1   Grubbs' Catalyst

Poly(dicyclopentadiene) (polyDCPD) is a highly cross-linked polymer of high toughness formed by a ring-opening metathesis polymerization (ROMP). ROMP of DCPD is highly exothermic because of the relief of ring strain energy and can be initiated by transition-metal/alkylidene complexes. Bis(tricyclohexylphosphine)benzylideneruthenium(IV)di-chloride (Grubbs' catalyst) shows high metathesis activity with the

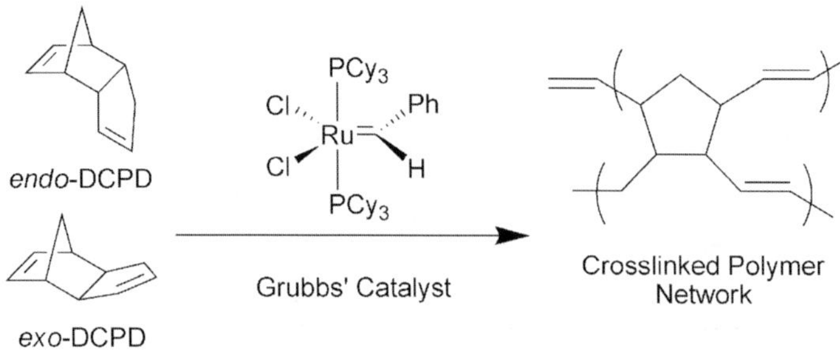

Fig. 3 ROMP of *endo*- and *exo*-DCPD with Grubbs' catalyst

DCPD monomer (Figure 3) coupled with extreme chemical stability (Grubbs and Tumas 1989; Schwab et al. 1996; Sanford et al. 1998).

Grubbs' catalyst can exist in different crystal morphologies, each with a distinct crystal shape (Figure 4), thermal stability and dissolution kinetics.

Dissolution kinetics and catalyst concentration have a large effect on cure kinetics (Kessler and White 2002) and the more rapidly dissolving polymorph shows superior healing efficiency (Jones et al. 2006). Complete coverage of the crack plane with polymerized healing agent is required for optimal recovery of mechanical integrity and is achieved when the kinetics of dissolution and curing are compatible (Figure 5).

The extent to which successful self healing is achieved for a given time and temperature is determined by the interplay between the mechanical and chemical kinetics of the system.

2.1.2 Stereoisomers of Dicyclopentadiene

DCPD can exist as an *endo* or an *exo* isomer (Figure 3). While the initial work on self healing systems used the commercially available *endo*-isomer (commercially available DCPD is >95% *endo*), the *exo*-stereoisomer is known to have much faster olefin metathesis reaction rates with first-generation Grubbs' catalyst (Rule and Moore 2002). For many self healing applications, it is desirable to have the fastest healing kinetics possible so long as the quality of the repair is not compromised. However, self healing is a complex problem that involves monomer transport, mixing, catalyst dissolution, and catalyst transport in addition to healing polymerization. The fast-reacting *exo*-DCPD leads to rapid gelation and insufficient time to completely dissolve the embedded Grubbs' catalyst (Mauldin et al. 2006). Jones et al. (2006) showed that complete dissolution of Grubbs' catalyst occurs in the range of 5–10 min. This dissolution rate is acceptable for self healing with *endo*-DCPD, which gels in approximately 20 min at room temperature. For *exo*-DCPD, which gels in seconds, much of the catalyst remains undissolved and a largely heterogenous polyDCPD film is formed on the crack plane (see Figure 6).

Fig. 4 ESEM images of Grubbs' catalyst for two different crystal morphologies as received from (**a**) Strem Chemicals (Newburyport, MA, USA) and (**b**) Sigma-Aldrich (St. Louis, MO, USA). (From Jones et al. 2006)

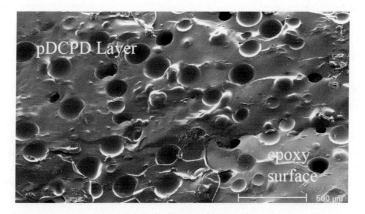

Fig. 5 ESEM image of a fracture surface showing a continuous polyDCPD layer on a specimen with a rapidly dissolving catalyst polymorph, see Figure 4a. (From Jones et al. 2006)

Fig. 6 ESEM image of an epoxy matrix fracture plane for exo-DCPD-healing agent in a self-activated sample after healing and subsequent fracture. Scale bar = 200 μm. (From Mauldin et al. 2007)

However, the desired combination of fast kinetics and high-healing efficiency was demonstrated by appropriate blending of *exo/endo*-DCPD healing agents and by adjusting catalyst loadings to optimal levels (Mauldin et al. 2006). In addition to the accelerated polymerization, *exo*-DCPD has significantly different phase transition properties compared to *endo*-DCPD. The *exo*-DCPD monomer does not solidify until temperatures below −50°C (compared to 15°C for *endo*-DCPD), making it a potential healing agent for cryogenic applications. Initial fracture tests for *exo*-DCPD performed in a cold room environment (0–4°C) showed no degradation in healing capability as compared to the same tests performed at ambient conditions (Mauldin et al. 2006).

2.1.3 Tungsten Hexachloride Catalyst

While Grubbs' catalyst possesses many of the primary attributes required of a self healing catalyst phase, high cost and limited availability coupled with moderate temperature stability motivate the search for alternative ROMP catalysts for self healing polymers. Alternative catalysts for ROMP healing have been surveyed and tungsten (VI) chloride (WCl_6) and a coactivator were determined to have the greatest potential to address many of the limitations of Grubbs' catalyst. The WCl_6 system offers a cost-effective alternative that is widely available with a melting point of $T_m = 275°C$ that is significantly higher than Grubbs' catalyst ($T_m = 153°C$). Moreover, the commercial potential of WCl_6 has been demonstrated in the Metton (Metton America) liquid molding two part resin system based on the polymerization of *endo*-DCPD with WCl$_6$ and a coactivator for reaction injection molding of high volume parts. Preliminary studies of self healing epoxy system using WCl$_6$ catalyst and *exo*-DCPD healing agent have demonstrated high levels of fracture toughness recovery (Kamphaus et al. 2007).

2.2 Poly(dimethyl siloxane) Healing Chemistry

2.2.1 Tin-Catalyzed Polycondensation

Based on a tin-catalyzed polycondensation of hydroxy end-functionalized poly(dimethyl siloxane) (HOPDMS) and poly(diethoxy siloxane) (PDES) using a di-n-butyltin dilaurate (DBTL) catalyst (Figure 7), Cho et al. (2006) has recently demonstrated a self healing chemistry where the siloxane-based healing agent is phase-separated in the matrix while the catalyst is encapsulated (Figure 8).

The polycondensation of HOPDMS with PDES occurs rapidly at room temperature in the presence of amine and carboxylic acid organotin catalysts and because side reactions are limited, organotin catalysts are highly desirable for curing PDMS-based systems, even in open air. The PDMS-based self healing chemistry remains stable at elevated temperatures ($> 100°C$) as well as in humid or wet environments. The components are also widely available and comparatively low in cost.

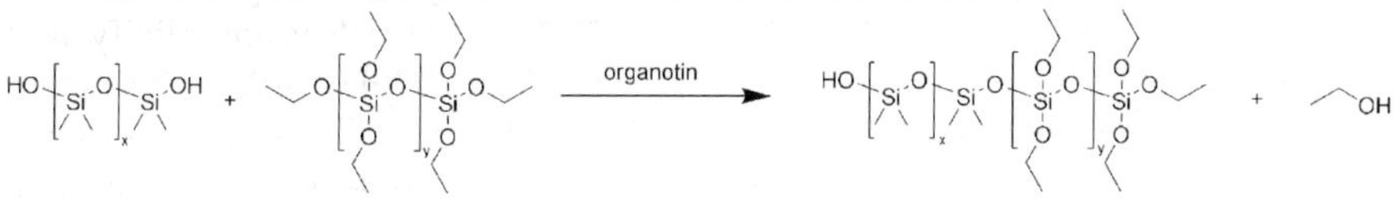

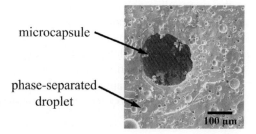

Fig. 7 Tin-catalyzed polycondensation of silanol-functionalized PDMS. A hydroxyl-terminated PDMS chain (**a**) is linked to an ethylsilicate (PDES) (**b**) in the presence of an organo-tin catalyst. This reaction produces a cross-link (**c**) and releases an ethanol condensation product (**d**)

Fig. 8 Scanning electron microscopy (SEM) image of a fracture surface, showing an empty catalyst microcapsule and phase-separated healing agent regions. (From Cho et al. 2006)

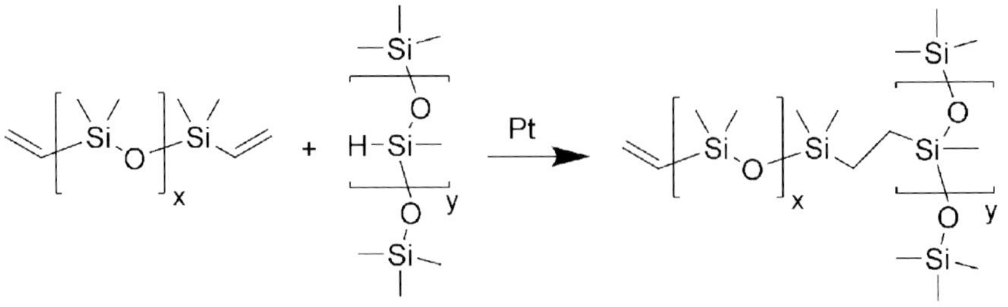

Fig. 9 Hydrosilylation of vinyl-terminated PDMS. A vinyl-terminated PDMS chain (**a**) is linked to an active methylhydrosiloxane copolymer (**b**) through the action of a platinum catalyst producing a crosslink site (**c**)

2.2.2 Platinum Catalyzed Hydrosilylation

An alternative route to a PDMS-based chemistry is the platinum-catalyzed hydrosilylation of vinyl terminated PDMS with a methylhydrosiloxane copolymer. This chemistry has been demonstrated to provide an efficient healing mechanism for PDMS elastomers (Keller et al. 2007). Hydrosilylation avoids the use of potentially environmentally hazardous materials such as tin and produces no condensation products (Figure 9).

Formation of a cross-linked PDMS network occurs very quickly in this system. The chemical kinetics can be modified through the addition of inhibitors, which allows for the chemistry to be tailored to the anticipated mechanical kinetics.

3   Microencapsulation and Catalyst Protection

Microencapsulation of the liquid healing agent provides a mechanical trigger for the self healing process when a crack ruptures the capsule. High interfacial bond strength to the surrounding polymer matrix combined with moderate strength capsules are required to ensure the capsules survive processing of the host polymer, yet rupture when the polymer is damaged. The capsules must also be impervious to leakage and diffusion of the encapsulated liquid healing agent for secure storage during quiescent times. Microcapsule diameter and surface morphology significantly influence capsule rupture behavior and healing agent release in self healing polymers. The addition of microcapsules to an epoxy matrix also provides a unique toughening mechanism for the composite system.

3.1   Poly(urea-formaldehyde) Microcapsules

The required characteristics of an encapsulated healing agent are achieved by microencapsulation of DCPD utilizing acid-catalyzed in situ polymerization of urea with formaldehyde to form the capsule wall (White et al. 2001; Brown et al. 2003). High-quality poly(urea-formaldehyde) (UF) microcapsules with a DCPD core are prepared by following the in situ processing route in an oil-in-water emulsion (Figure 10).

The microcapsules are spherical and free flowing after drying (Figure 11). Average microcapsule diameter is controlled over a wide range of sizes by agitation rate as shown in Figure 12. As the agitation rate is increased, a finer emulsion is obtained and the average microcapsule diameter decreases. Current methods of encapsulation via a macroemulsion interfacial polymerization can produce capsules ranging from 1, 000 μm down to a limit of 10 μm (Brown et al. 2003). The use of ultrasonication combined with miniemulsion techniques has significant potential to produce capsules as small as 300 nm in diameter (Blaiszik et al. 2006).

In addition to DCPD core materials, the in situ encapsulation technique can be used to encapsulate a number of other reactive materials. Recent work has utilized UF microcapsules to sequester functionalized poly(dimethyl siloxane) resins and

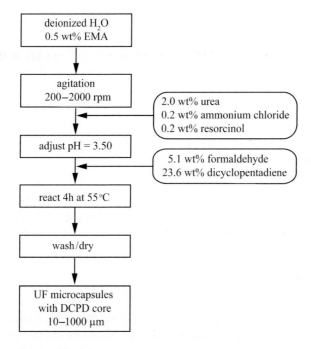

Fig. 10 Microencapsulation of DCPD by acid-catalyzed in situ polymerization of urea with formaldehyde (wt% relative to H_2O)

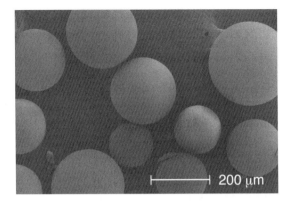

Fig. 11 ESEM image of UF microcapsules (DCPD core) prepared at 550 rpm agitation rate. (From Brown et al. 2003)

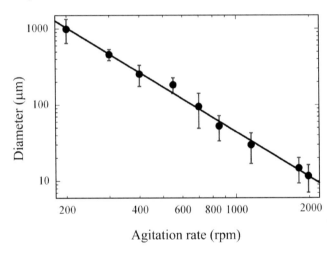

Fig. 12 Mean UF microcapsule diameter vs. agitation rate. The solid line corresponds to a linear fit on a log–log scale. (Fom Brown et al. 2003)

copolymers such as silanol- and vinyl-terminated PDMS resins for use in elastomeric self healing matrices (Keller et al. 2007).

3.2 Polyurethane Microcapsules

In a PDMS-based self healing system reported by Cho et al. (2006) the healing agent is phase-separated in the matrix, while the organo-tin catalyst phase is encapsulated. The low solubility of siloxane-based polymers enables the healing agent to be directly blended with the vinyl ester matrix prepolymer, forming a distribution of stable phase-separated droplets. The microcapsules containing the catalyst are

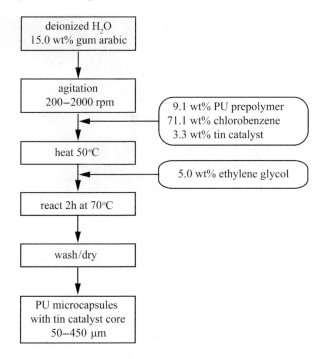

Fig. 13 Microencapsulation of tin-catalyst by interfacial polymerization of urethane prepolymer (wt% relative to H_2O)

formed through interfacial polymerization and consist of a polyurethane shell surrounding a mixture of di-n-butyltin di-laurate (DBTL) and chlorobenzene. In this procedure, a hydrophobic monomer or prepolymer is dissolved in the organic phase (e.g. toluene diisocyanate in chlorobenzene) which is then emulsified in an aqueous medium. A hydrophilic chain extender (e.g. ethylene glycol) is then added to the aqueous phase. The hydrophilic monomer and the hydrophobic chain extender react at the aqueous-organic interface producing the polymeric shell wall of the capsule (Figure 13).

High-quality polyurethane microcapsules with a smooth surface are obtained by interfacial polymerization (Figure 14). Just like UF microcapsules, the average diameter of polyurethane microcapsules is a function of agitation rate and for this self healing system typically ranging from 450 μm down to 50 μm in diameter (Cho et al. 2006).

3.3   Wax-protected Microspheres

The Grubbs' catalyst based self healing system reported by White et al. (2001) performs well with a relatively large loading (2.5 wt%) of catalyst. By embedding

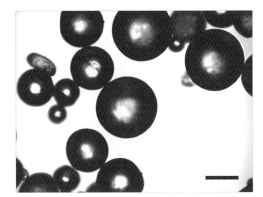

Fig. 14 Optical microscopy image of tin-catalyst containing polyurethane micro-capsules prepared at 1,000 rpm agitation rate. Scale bar = 100 μm. (Fom Cho et al. 2006)

Grubbs' catalyst into wax microspheres, Rule et al. (2005) achieved effective healing at dramatically lower catalyst loadings (down to 0.25 wt%). The wax microspheres containing Grubbs' catalyst were formed through a hydrophobic congealable disperse phase encapsulation (Rule et al. 2005). In this procedure, a mixture of molten wax and catalyst is poured into a hot, rapidly stirred, aqueous solution of poly(ethylene-co-maleic anhydride) (Figure 15).

Wax microspheres obtained from rapid cooling of the suspension of molten wax droplets are shown in Figure 16. The wax microspheres serve two purposes for improved healing at lower catalyst loadings. Firstly, by improving the dispersion of the catalyst in the epoxy matrix, regions of incomplete healing due to nonuniform catalyst availability is minimized. Secondly, the wax protects the catalyst from deactivation upon exposure to the epoxy curing agent, diethylenetriamine (DETA), effectively increasing the amount of active catalyst available for the promotion of healing.

3.4 Smaller Size Scales for Efficient Healing

While the repair of large-scale cracks and delaminations have been shown to be feasible with current materials systems, self healing may be best applied at the micron and nanometer length scale when damage is small. The development of smaller microcapsules through ultrasonication in miniemulsions is an important step forward in self healing research (Blaiszik et al. 2006). Fiber-reinforced composite materials, for example, will benefit from nanometer-scaled healing system components so that submicron capsules are able to reside in the matrix zones between individual fibers. By repairing small-scale microcracking that frequently initiates near fibers, large-scale critical damage such as delamination could be delayed or potentially prevented. Extending self healing to the nanoscale presents several technical challenges (Rule

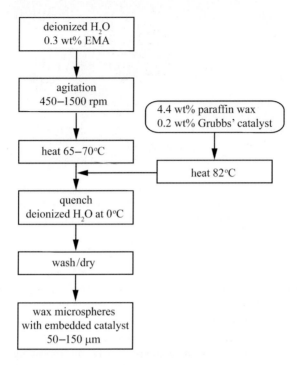

Fig. 15 Protection of Grubbs' catalyst in microspheres by hydrophobic congealable disperse phase encapsulation of paraffin wax (wt% relative to H_2O)

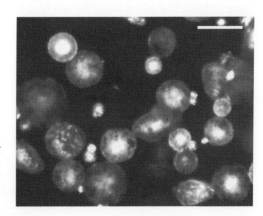

Fig. 16 Optical microscopy image of Grubbs' catalyst particles suspended in wax microspheres prepared at 1,000 rpm agitation rate. Scale bar = 100 μm. (From Patel and White 2006)

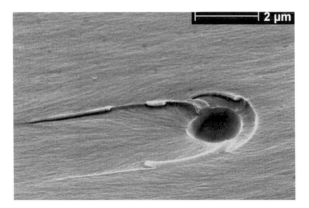

Fig. 17 SEM image of a ruptured UF capsule (diameter = 1.5 μm, shell wall thickness = 80 nm) in an epoxy matrix. The distinct tail structure on the fracture surface indicates crack-pinning and crack-deflection toughening mechanisms. (From Blaiszik BJ, University of Illinois, unpublished results)

et al. 2007). In addition to producing smaller capsules, interfacial bond strength between the healing agent and the fiber reinforcement becomes a critical parameter, as well as developing tools for mechanical assessment of the rupture, release, and healing mechanisms at this scale (Figure 17).

4 Mechanical Characterization

For self healing materials the ultimate goal is to demonstrate functional recovery in some fashion. The initial target for self healing materials has encompassed primarily a mechanics-based definition of healing and recovery. Significant effort has been made to develop a set of experimental protocols that may be used to evaluate self healing materials in both static and dynamic fracture conditions.

4.1 Static Fracture Testing

For quasi-static fracture conditions healing efficiency is defined in terms of the recovery of fracture toughness (White et al. 2001)

$$\eta_s = \frac{K_{IC}^{healed}}{K_{IC}^{virgin}}. \tag{1}$$

A tapered double-cantilever beam (TDCB) geometry (see inset Figure 18) is utilized to determine the fracture toughness (K_{IC}) of the virgin and healed states. This

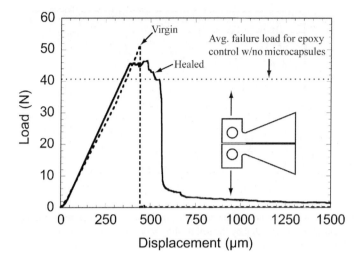

Fig. 18 Load-displacement data for a virgin and healed TDCB self healing epoxy sample with 2.5wt% Grubbs' catalyst and 5 wt% UF microcapsules (DCPD core). Healing efficiency η_s = 90.3% after 48 h healing at room temperature. Inset: Schematic of TDCB sample. (From Brown et al. 2002)

geometry provides a crack-length-independent measure of fracture toughness and allows the fracture toughness to be related to the peak load by a simple geometric coefficient.

Healing evaluation begins with a virgin fracture test of an undamaged TDCB sample. A precrack is introduced to sharpen the crack-tip and loading of the specimen is increased until the crack propagates along the centerline of the sample until failure. The crack is then closed and allowed to heal at room temperature with no external intervention. After healing, the sample is loaded again until failure. Figure 18 shows a representative load–displacement data for virgin and healed tests for a self healing epoxy. The dashed line indicates the average peak load for a neat epoxy specimen. Based on this test protocol, conclusive demonstration of self healing for an epoxy matrix containing DCPD-filled microcapsules and a solid phase Grubbs' catalyst was achieved (White et al. 2001). With optimized concentrations of microcapsules and catalyst, healing efficiencies of 90% were reported (Brown et al. 2002).

4.2 Fatigue Testing

For dynamic fracture conditions, healing efficiency based on static fracture toughness recovery is no longer meaningful. Instead, we define healing in terms of the life extension factor (Brown et al. 2005a)

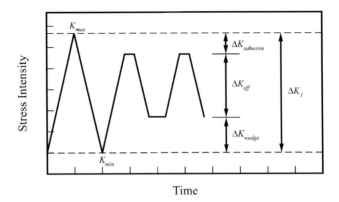

Fig. 19 Schematic of a cyclic-loading profile showing the stress intensity-reducing effects of crack-tip shielding mechanisms activated by the self healing process

$$\eta_d = \frac{N_{healed} - N_{control}}{N_{control}} \qquad (2)$$

where N is the number of fatigue cycles to failure. The fatigue of microcapsule-based self healing epoxy composites has been studied in a number of recent papers (Brown et al. 2005a, b, 2006; Jones et al. 2007). These studies have shown that self healing significantly extends the fatigue life when compared to the neat resin material.

Fatigue performance of self healing epoxy was evaluated using the TDCB sample geometry. Load was applied using a triangular waveform varying between maximum crack tip stress intensity, K_{max}, and minimum stress intensity, K_{min}, as shown schematically in Figure 19. The crack-length independent nature of the TDCB specimen ensures a constant crack growth rate over the majority of the specimen, in the absence of any crack shielding mechanisms.

4.2.1 Crack-Tip Shielding Mechanisms

There are several mechanisms which combine to improve the fatigue behavior of a self healing material by reducing the crack tip stress intensity. In virgin, undamaged material a crack tip is subjected to the global stress intensity, ΔK_I, as dictated by the far-field loading conditions. Once self healing occurs, a wedge of polymerized healing agent forms at the crack tip. The stress intensity driving the crack tip forward is significantly reduced (ΔK_{eff}) since the wedge prevents full loading ($\Delta K_{adhesive}$) and unloading (ΔK_{wedge}) of the crack tip. These mechanisms are active when the released healing agent has sufficiently polymerized, however, the presence of a liquid-healing agent on the crack plane also introduces hydrodynamic crack tip shielding but is less effective at slowing crack growth (Brown et al. 2005a).

4.2.2 Fatigue Performance

Fatigue of a self healing epoxy system highlights the important role of the interplay between chemical and mechanical kinetics in this system. Attempting to retard and ultimately arrest fatigue cracks hinges on the ability of the healing chemistry to pro- duce polymer at rates that are comparable to the crack propagation speed (Jones et al. 2007). In a sample cycled at high frequency or high stress levels, the healing kinet- ics are not fast enough to significantly retard crack growth. However, if rest periods are incorporated into the loading cycle, then sufficient time for healing is maintained and significant life extension can be achieved (Figure 20a). By modifying the kinet- ics of the healing chemistry, the effect of polymerization rate on healing performance is shown in Figure 20b. For this regime of fatigue, faster healing kinetics produces greater life extension. If a sample is fatigued at low frequency or low stress levels, healing chemical kinetics dominates and the crack is effectively arrested (Figure 20c).

4.3 Tear Testing

For elastomeric self healing materials, the TDCB-based fracture toughness proto- col to evaluate healing performance is inappropriate. Instead, the recovery of tear strength using a tear specimen is used to define healing efficiency, where

$$\eta_e = \frac{T_{healed}}{T_{virgin}}. \tag{3}$$

A tear test utilizes a rectangular coupon of material with a large axial precut that produces two loading arms. These arms are loaded in tension until the tear propagates through the rest of the specimen (see inset of Figure 21). Healing evaluation begins with a virgin tear test of an undamaged sample. After failure, the sample loading arms are reregistered and healing occurs at room temperature with no external intervention. After healing, the tear sample is loaded again to failure.

Using this testing protocol a self healing poly(dimethyl siloxane) (PDMS) elastomer has been shown to routinely recover more than 70% of the original tear strength (Keller et al. 2007). Full recovery of the original tear strength is also possible and is demonstrated when the healed tear propagates into virgin material as shown in Figure 22.

For this type of system, the healing chemistry and the matrix chemistry are iden- tical. The closely matched material properties of the healed polymer and the sur- rounding matrix can allow the tear to deviate from the original healed tear path and propagate into virgin material, yielding essentially complete recovery of virgin tear strength.

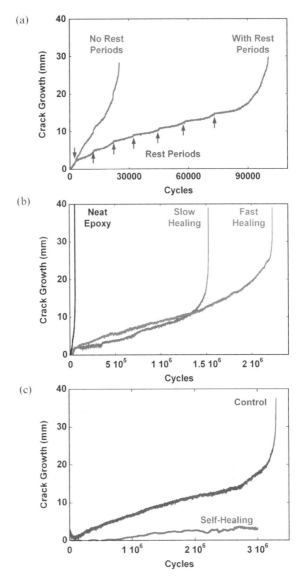

Fig. 20 Fatigue testing of TDCB self healing epoxy samples with 20 wt% UF microcapsules (DCPD core) and 5 wt% wax microspheres containing 5wt% Grubbs' catalyst. Control samples contain 20 wt% UF microcapsules and no catalyst. All samples are tested at 5 Hz. (**a**) Rapid crack growth at high stress levels ($\Delta K = 0.72\,\text{MPa}\sqrt{\text{m}}$) where the rate of mechanical damage exceeds the kinetics of the polymerization reaction. By incorporating rest periods of 10–12 h each, the crack is retarded and fatigue life is extended (life extension factor $\eta_d \approx 300\%$). (**b**) Intermediate fatigue loading ($\Delta K = 0.61\,\text{MPa}\sqrt{\text{m}}$) where faster chemical kinetics due to a rapidly dissolving catalyst polymorph enables significant life extension (life extension factor for slow healing $\eta_d \approx 1300\%$ and fast healing $\eta_d \approx 2000\%$). (**c**) Healing chemical kinetics dominates at low stress levels ($\Delta K = 0.50\,\text{MPa}\sqrt{\text{m}}$) and complete crack arrest is achieved (life extension factor $\eta_d = \infty$). (From Jones et al. 2007)

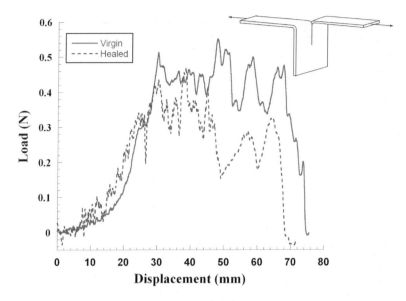

Fig. 21 Representative load–displacement data for a virgin (solid line) and healed (dashed line) tear test of a self healing PDMS sample with 10 wt% UF microcapsules (resin core) and 5 wt% UF microcapsules (initiator core). Healing efficiency $\eta_e = 76\%$ after 48 h healing at 25°C. Inset: Schematic of a tear sample. (From Keller et al. 2007)

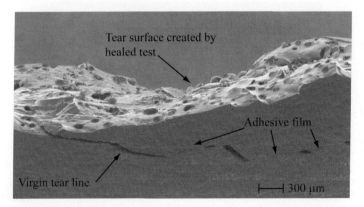

Fig. 22 SEM image of a healed PDMS tear specimen with 20 wt% UF microcapsules (resin core) and 5 wt% UF microcapsules (initiator core) showing adhesive film along the virgin tear line. The healed tear path deviates from the original virgin tear path. The healing efficiency for this sample was $\eta_e = 115\%$ after 48 h healing at 25°C. (From Keller et al. 2007)

4.4 Microcapsule-induced Toughening

Microcapsules not only provide convenient and stable storage of healing monomer, but also significantly improve the fracture toughness of the matrix polymer. The addition of microcapsules can toughen an epoxy matrix by as much as 127% and

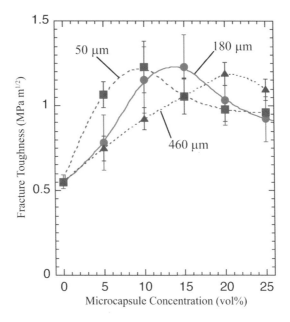

Fig. 23 Effect of UF microcapsule (DCPD core) concentration and diameter on virgin fracture toughness for epoxy TDCB samples. (From Brown et al. 2004)

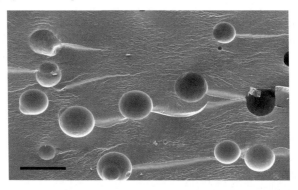

Fig. 24 SEM image of a fracture plane in an epoxy matrix with 10 wt% UF microcapsules (DCPD core) showing tails emanating from the trailing edges of the microcapsules. This image also shows the extensive hackle marking of the fracture surface. Crack propagation is from left to right, scale bar = 200 µm. (From Brown et al. 2004)

is a strong function of capsule concentration (Figure 23). Microcapsules produce marked morphological changes on the fracture plane (Figure 24). Tail structures on the trailing edges of microcapsules indicate energy adsorbing mechanisms such as crack pinning. Furthermore, the surface of the fracture plane is covered in hackle-marking which is indicative of significant subsurface microcracking and further energy absorption (Brown et al. 2002, 2004).

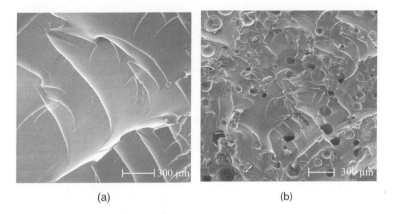

(a)   (b)

Fig. 25 Comparison of tear surfaces in PDMS samples. (**a**) Neat resin tear sample and (**b**) tear sample with 20 wt% UF microcapsules (resin core) and 5 wt% UF microcapsules (initiator core). Tear propagation is from bottom to top. (From Keller et al. 2007)

Microcapsule-induced toughening is not limited to brittle, thermosetting materials such as epoxy. The inclusion of microcapsules in a self healing PDMS improved the tear strength of the elastomeric matrix by 25% at 25 wt% microcapsules (Keller et al. 2007). Figure 25 shows a comparison of a neat PDMS tear surface and a tear surface of a sample that contained 25 wt% microcapsules. As in self healing epoxy, the addition of microcapsules induces significant morphological changes on the failure plane. Microcapsules induce many small tear ridges, analogous to the capsule tails in Figure 24, and generally cause the tear surface to be rougher than in the neat case.

5 Self Healing Fiber-reinforced Composites

Brittle polymers and the fiber-reinforced structural composites made from them are susceptible to microcracking when subjected to thermomechanical loading. Transverse cracks and delaminations can also occur in laminated composites even for modest levels of impact. Whatever the mode of damage, severely damaged structural components are typically replaced entirely while repairs are generally attempted for less extensive damage. Common repair methods include injecting resin via an access hole and bonding or bolting of reinforcing patches to the failed areas, all requiring time-consuming and costly manual intervention by a trained technician. Healing of microcracks and delaminations could potentially delay or prevent large-scale damage in structural composites.

5.1 Static Fracture Testing

The self healing polymer developed by White et al. (2001) offers significantly extended life of polymeric components by autonomically healing microcracks

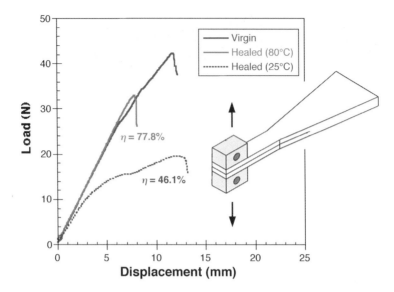

Fig. 26 Typical loading curves for virgin and healed self healing carbon fiber-reinforced epoxy WTDCB specimens with 5 wt% Grubbs' catalyst and 20 wt% UF microcapsules (DCPD core). Healing conditions are 48 h at room temperature (green curve) and 48 h at 80°C (red curve). Inset: Schematic of WTDCB specimen. (From Kessler et al. 2003)

whenever and wherever they occur. For structural composite materials, the presence of fiber reinforcement increases the number of damage modes and the complexity of the healing process. Although resin microcracks can be healed similarly, the development of a self healing composite is fundamentally more difficult (Kessler and White 2001). For fiber-reinforced composites a width-tapered double cantilever beam (WTDCB) testing protocol was developed for healing evaluation of delamination damage. The WTDCB geometry (Figure 26 inset) provides a crack length independent measure of fracture toughness and healing efficiency (η_s) is calculated based on the ratio of peak fracture loads for healed and virgin specimens.

Repair of large-scale delaminations in a self healing fiber-reinforced structural polymer composite material was achieved by incorporating DCPD containing UF microcapsules and Grubbs' catalyst in a carbon-fiber-reinforced epoxy matrix (Kessler et al. 2003). With no manual intervention, a recovery of nearly 50% was achieved after 48 h healing at room temperature, and upon elevating the healing temperature to 80°C for 48 h, healing efficiency was dramatically increased to over 75% (Figure 26).

5.2 Impact Testing

Out-of-plane impact events with fiber-reinforced polymer composites can lead to damage that significantly compromises the integrity of the structure. In addition to the complicated methods of repair, this type of damage is often hidden or barely visible,

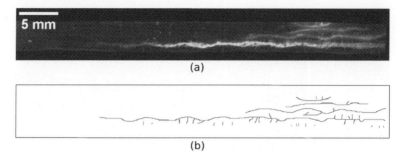

Fig. 27 (**a**) UV illumination of fluorescently marked delaminations and transverse matrix cracks in a fiberglass reinforced epoxy after impact damage. (**b**) Skeletonization of marked cracks. (From Patel et al. 2007)

making it difficult to detect. Damage from impact can also include significant fiber breakage, matrix cracking and gross delaminations. While self healing materials are unlikely to address all these damage modes, specific damage modes can be targeted by a careful choice of correct size scale of the self healing components.

In a study by Patel et al. (2007), glass fiber-reinforced epoxy composites with DCPD-filled UF microcapsules and wax-protected Grubbs' catalyst were investigated for mitigation of impact damage (Figure 27). A dual capsule approach (microcapsules of 35 and 180 μm diameter) was used to provide both high dispersion and sufficient healing agent delivery. Using a drop weight impact tester for low-velocity impact testing, delaminations caused by low-velocity damage were reduced by 51% (quantified by cross-sectional crack length reduction). Ongoing work is focused on postimpact mechanical assessment of structural recovery.

6 Biomimetic Microvascular Autonomic Composites

Self healing polymers and composites are inspired by biological systems in which healing occurs in an autonomic fashion with site-specific control. In self healing polymers, healing is accomplished by incorporating a microencapsulated healing agent and a chemical catalyst within a polymer matrix. Damage in the form of a crack serves as the triggering mechanism for self healing. As with biological systems, this activation scheme provides site-specific control of healing in an autonomic fashion. Even so, this concept yields materials with limited healing lifetimes since rupture of the entire microcapsule population depletes the supply of healing agent and healing functionality is no longer maintained. Looking to biological models for guidance, the solution to limited resources is a circulatory system that continually replenishes and perfuses the host material with the chemical building blocks of healing (Figure 28). More importantly, a circulatory network provides a mechanism to deliver not only healing agents, but also molecular triggers and sensors that impart additional functionality such as self-diagnosis and temperature regulation.

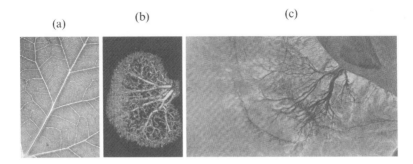

Fig. 28 Natural vascular structures and systems. (**a**) dendritic design of the vasculature of a leaf (www.istockphoto.com), (**b**) dichotomous branching in the arteries of the rabbit kidney (from Sernetz et al. 1995), and (**c**) aerial view of a drainage basin. (From US Navy Photo)

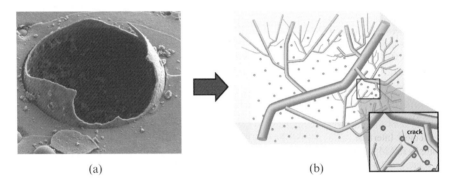

Fig. 29 Future development of self healing materials. (**a**) The first generation of autonomic healing is based on compartmentalization (via microcapsules) of the healing agent (from Kessler 2002). (**b**) In the future, the healing agent will be supplied via a microvascular network as shown in this conceptual image of a 3D network embedded in a structural matrix (From White 2006)

In moving from a compartmental (microcapsule) to a circulatory (network) approach, the primary technical challenge is to design and fabricate a pervasive, interconnected, three-dimensional (3D) vascular network across multiple-length scales (Figure 29). While the level of vascular interconnectivity and complexity in natural systems is formidable, simplified networks have recently been fabricated and tested (Therriault et al. 2003; Toohey et al. 2007). Preliminary results indicate that resupply of the healing agent is feasible and extended life of polymers can be achieved in response to repeated damage (Toohey et al. 2007).

References

Blaiszik B.J., White S.R., Sottos N.R. (2006) Nanocapsules for self-healing composites. Proceedings of the 2006 SEM Annual Conference and Exposition on Experimental and Applied Mechanics 1:391–396

Brown E.N., Sottos N.R., White S.R. (2002) Fracture testing of a self-healing polymer composite. Exp Mech 42:372–379

Brown E.N., Kessler M.R., Sottos N.R., White S.R. (2003) In situ poly(urea-formaldehyde) microencapsulation of dicyclopentadiene. J Microencapsul 20:719–730

Brown E.N., Sottos N.R., White S.R. (2004) Microcapsule induced toughening in a self-healing polymer composite. J Mater Sci 39:1703–1710

Brown E.N., White S.R., Sottos N.R. (2005a) Retardation and repair of fatigue cracks in a microcapsule toughened epoxy composite – Part I: Manual infiltration. Compo Sci Technol 65:2466–2473

Brown E.N., White S.R., Sottos N.R. (2005b) Retardation and repair of fatigue cracks in a microcapsule toughened epoxy composite – Part II: In situ self-healing. Compos Sci Technol 65: 2474–2480

Brown E.N., Sottos N.R., White S.R. (2006) Fatigue crack propagation in microcapsule-toughened epoxy. J Mater Sci 41:6266–6273

Cho S.H., Andersson H.M., White S.R., Sottos N.R., Braun P.V. (2006) Polydimethylsiloxane-based self-healing materials. Adv Mater 18:997–1000

Grubbs R.H., Tumas W. (1989) Polymer synthesis and organotransition metal chemistry. Science 243:907–915

Jones A.S., Rule J.D., Moore J.S., White S.R., Sottos N.R. (2006) Catalyst morphology and dissolution kinetics of self-healing polymers. Chem Mater 18:1312–1317

Jones A.S., Rule J.D., Moore J.S., Sottos N.R., White S.R. (2007) Life extension of self-healing polymers with rapidly growing fatigue cracks. Accepted for publication in Journal of the Royal Society Interface

Kamphaus J.M., Rule J.D., Delafuente D.A., Moore J.S., Sottos N.R., White S.R. (2007) Development and evaluation of an alternative catalyst for self-healing polymers. Proceedings of the 1st International Conference on Self-Healing Materials, Noordwijk, The Netherlands

Keller M.W., White S.R., Sottos N.R. (2007) A self-healing poly(dimethyl siloxane) elastomer. Submitted to Advanced Functional Materials

Kessler M.R., White S.R. (2001) Self-activated healing of delamination damage in woven composites. Composites: Part A 32:683–699

Kessler M.R. (2002) Characterization and performance of a self-healing composite material. PhD Thesis in Theoretical and Applied Mechanics, Graduate College of the University of Illinois at Urbana-Champaign

Kessler M.R., White S.R. (2002) Cure kinetics of the ring-opening metathesis of dicyclopentadiene. J Polym Sci: Part A: Polym Chem 40:2373–2383

Kessler M.R., Sottos N.R., White S.R. (2003) Self-healing structural composite materials. Composites: Part A 34:743–753

Mauldin T.C., Rule J.D., Sottos N.R., White S.R., Moore J.S. (2007) Self-healing kinetics and the stereoisomers of dicyclopentadiene. Accepted for publication in Journal of the Royal Society Interface

Patel A.J., White S.R. (2006) Self-healing composite armor. ARL Progress Report

Patel A.J., Sottos N.R., White S.R. (2007) Self-healing composites for mitigation of low-velocity impact damage. Proceedings of the 1st International Conference on Self-Healing Materials, Noordwijk, The Netherlands

Rule J.D., Moore J.S. (2002) ROMP reactivity of endo- and exo-dicyclopentadiene. Macromolecules 35:7878–7882

Rule J.D., Brown E.N., Sottos N.R., White S.R., Moore J.S. (2005) Wax-protected catalyst microspheres for efficient self-healing materials. Adv Mater 17:205–208

Rule J.D. (2005) Polymer chemistry for improved self-healing composite materials. PhD Thesis in Chemistry, Graduate College of the University of Illinois at Urbana-Champaign

Rule J.D., Sottos N.R., White S.R. (2007) Effect of microcapsule size on the performance of self-healing polymers. Submitted to Polymer

Sanford M.S., Henling L.M., Grubbs R.H. (1998) Synthesis and reactivity of neutral and cationic ruthenium(II) tri(pyrazolyl)borate alkylidenes. Organometallics 17:5384–5389

Schwab P., Grubbs R.H., Ziller J.W. (1996) Synthesis and applications of $RuCl_2(= CHR')(PR_3)_2$: the influence of the alkylidene moiety on metathesis activity. J Am Chem Soc 118:100–110

Sernetz M., Justen M., Jestczemski F. (1995) Disperse characterization of kidney arteries by three dimensional mass-radius-analysis. Fractals 3:879–891

Therriault D., White S.R., Lewis J.A. (2003) Chaotic mixing in three-dimensional microvascular networks fabricated by direct-write assembly. Nat Mater 2:265–271

Toohey K.S., White S.R., Lewis J.A., Moore J.S., Sottos N.S. (2007) Self-healing materials with microvascular networks. Submitted to Nature Materials

White S.R., Sottos N.R., Geubelle P.H., Moore J.S., Kessler M.R., Sriram S.R., Brown E.N., Viswanathan S. (2001) Autonomic healing of polymer composites. Nature 409:794–797

White S.R. (2006) Microvascular autonomic composites. MURI Annual Report, AFOSR Grant # FA9550-05-1-0346

Wilson G.O., Andersson H.M., Sottos N.R., White S.R., Moore J.S. (2007) Autonomic healing in epoxy vinyl esters via ring opening metathesis polymerization (ROMP). Submitted to Advanced Functional Materials

Re-Mendable Polymers

Sheba D. Bergman and Fred Wudl

Department of Chemistry and Biochemistry, University of California, Santa Barbara
Email: wudl@chem.ucsb.edu

1 Introduction

Polymers have become an indispensable material resource, representing billions of dollars worth of material consumption every year. The rising prices and exhaust of natural resources such as petroleum, combined with rising environmental concerns, have prompted the development of recyclable and degradable polymers. Polymers that can be reverted back to their monomers or to shorter repolymerizable oligomers, hence, reversible polymers are particularly enticing in this respect because they essentially prevent any material loss with multiple recycling. While reversible polymers have been known for a long time, there has been recent renewed interest in such polymers, since their reversibility can be exploited for repair at the molecular level.

Polymers have a finite lifetime, their inherent properties degrade with age due to accumulated stresses and strains. The cracking or breaking of a material starts at the microscopic level and usually goes unnoticed until it reaches the macroscopic level, where the material fails and requires mending. Mending at the macroscopic level can be done with adhesives, but does not refurbish the original properties of the polymer. Mending at the microscopic level, as enabled by a reversible polymerization process, fully restores the original properties of the material and this process can be repeated many times. Historically, such reversible polymerizations would constitute the formation and cleavage of covalent bonds. The recent advent of Supramolecular Chemistry has brought about a new type of such polymerizations that are based on formation and severing of noncovalent interactions. Due to space constraints, this chapter will emphasize polymer systems that have been shown to undergo crack-healing by external intervention that exploits the reversible nature of the covalent bonds. For convenience, these are labeled "re-mendable", in contrast to self healing or "smart" materials that are endowed with self-repair; i.e. autonomous mending. The latter are covered in Chapter 2 of this book and will be mentioned only briefly here.

2 Externally Mendable Polymers

Externally mendable polymers are polymers that upon crack formation stay in their failed state until healed by an external intervention. This can be in the form of thermal-, photo-, or chemical-induced healing.

S. van der Zwaag (ed.), *Self Healing Materials. An Alternative Approach to 20 Centuries of Materials Science*, 45–68.
© 2007 *Springer*.

2.1 Diels–Alder (DA)-based Polymers

One of the most relevant aspects of the DA reaction for re-mendable polymers is its thermal reversibility, a process known as the retro-Diels–Alder (RDA) reaction. While a substantial amount of work has been published concerning fabrication of reversible DA-based polymers, only recently has the healing ability of such polymers been demonstrated.

2.1.1 Furan–Maleimide-based Polymers

Furan–maleimide-based polymers have recently been reviewed [1]. Probably the earliest report of incorporation of furan–maleimide moieties into polymers for the purpose of achieving thermal reversibility was reported in 1969 by Craven [2]. Since then several patents and journal articles have been published, all concerning the fabrication of a thermally reversible polymer network bearing DA-reactive furan and maleimide units, either as pendant groups (for reversible cross-linking) [3–14], or as part of the polymer backbone (for reversible polymerization) [15–20]. Noteworthy is the work of McElhanon et al. [21–24], who have successfully exploited furan–maleimide reversible DA chemistry for the design of debondable epoxy resins. The built-in reversibility of the material enabled debonding at 90°C.

Despite these and other earlier studies, Wudl and co-workers [25–27] were the first to employ the DA–RDA strategy to prepare thermally re-mendable polymers [28–30]. Multifunctional furan- and maleimide-based monomers were used to form highly cross-linked polymeric networks. The first system that was developed is shown in Scheme 1, where a tris-maleimide (3M) and a tetra-furan (4F) were allowed to react to afford a clear solid DA-step-growth polymer (3M4F). These polymers were submitted to heating/cooling cycles and their structural changes followed by solid-state ^{13}C NMR spectroscopy studies. These measurements clearly demonstrated the occurrence of the RDA reaction at ~120°C. Samples of this polymer were stressed to complete failure and subsequently healed by heating to ~90–120°C, followed by cooling to room temperature (Figure 1). The healed polymer exhibited about 57% of the original polymer strength. Subsequently, an improved system was designed which employed 4F in conjunction with either 2ME or 2MEP (Figure 2). The polymer 2MEP4F exhibited crack-healing with as much as 83% recovery of the polymer's original strength (Figure 3). Thus, it was shown that the DA/RDA principle provides a simple and efficient way to prepare re-mendable polymers, which can go through repeated cycles of cracking and re-mending at the same site. The limitations of this system are the working temperature of the materials, which is too low (<120°C) for many applications and the lengthy and costly synthesis of the monomers that is problematic for large-scale production.

Since the publication of this work, a number of related polymer systems have been reported in the literature. Liu et al. [31] reported a similar cross-linked polymer constructed from the tris-maleimide and tris-furan monomers TMI and TF, respectively

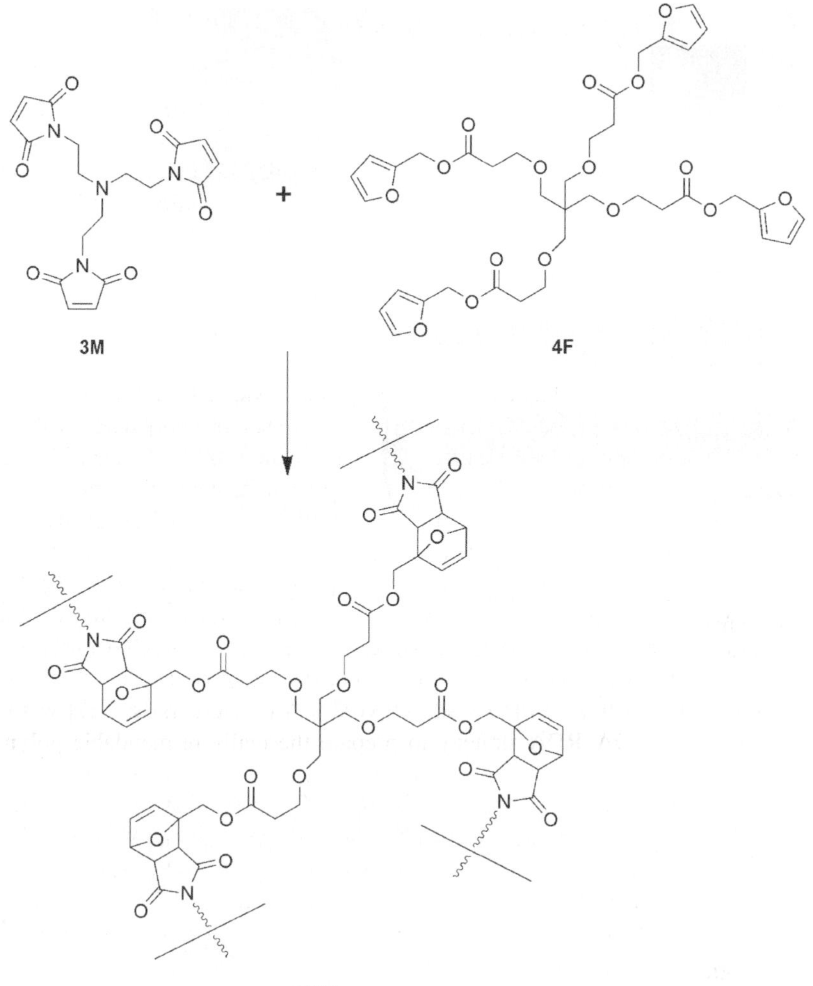

Scheme 1 Preparation of a highly cross-linked furan–maleimide-based polymer network

(Figure 4). The polymer exhibited thermal re-mendability and removability through DA and retro-DA reactions (Figure 5). These materials were shown to be applicable as advanced encapsulants and structural materials.

Watanabe et al. [32] reported the synthesis and recycling of polymers made up of bis-furan terminated poly(ethylene adipate) (PEA2F) and multi-maleimide linkers (3M and 2M) (Figure 6). They did not study the healing of the polymer.

Liu et al. [33] designed polyamides containing various amounts of maleimide and furan pendant groups that exhibited thermally reversible cross-linking behavior via DA and RDA reactions. Films of these polymers were cut and then subjected to thermal mending. For these polymers, only partial healing was observed (Figure 7).

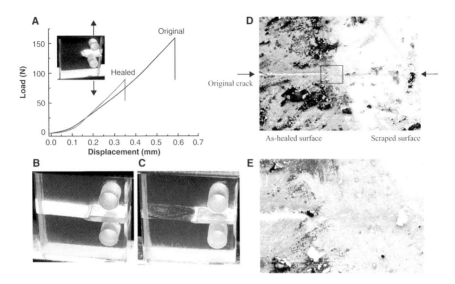

Fig. 1 Thermal mending of the polymer. (**A**) Mending efficiency obtained by fracture toughness testing of compact tension test specimens. Values for the original and healed fracture toughness were determined by the propagation of the starter crack along the middle plane of the specimen at the critical load. (**B**) Image of a broken specimen before thermal treatment. (**C**) Image of the specimen after thermal treatment. (**D**) SEM image of the surface of a healed sample: the left side is the as-healed surface and the right side is the scraped surface. (**E**) Enlarged image of the boxed area in (**D**). (Reproduced from Chen et al. [28])

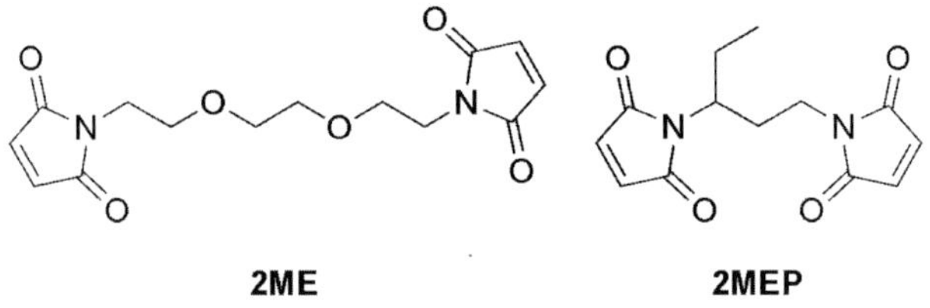

Fig. 2 Structure of 2ME and 2MEP

An exciting application of these polymer systems has been reported by Gostmann et al. [34]. A novel furan–maleimide polymer was designed that allows switching between two different states: a rigid, highly cross-linked, low-temperature state, and a deformable, fragmented, high-temperature state. The switching between these two states was exploited in this work for data-storage and lithographic applications.

2.1.2 Dicyclopentadiene-based Polymers

Cyclopentadiene (CP) and its derivatives have been employed for some time now as DA-active units in polymers, the major advantage being that they function as both diene and dienenophile, self-reacting to produce dicyclopentadiene (DCP) and

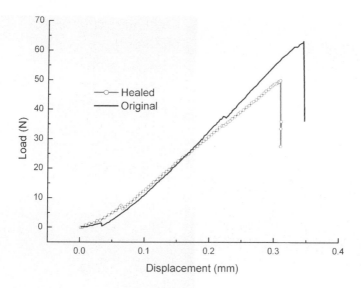

Fig. 3 Load vs. displacement diagram of fracture toughness testing of specimens of original and healed polymer 2MEP4F. (Reproduced from Chen et al. [29])

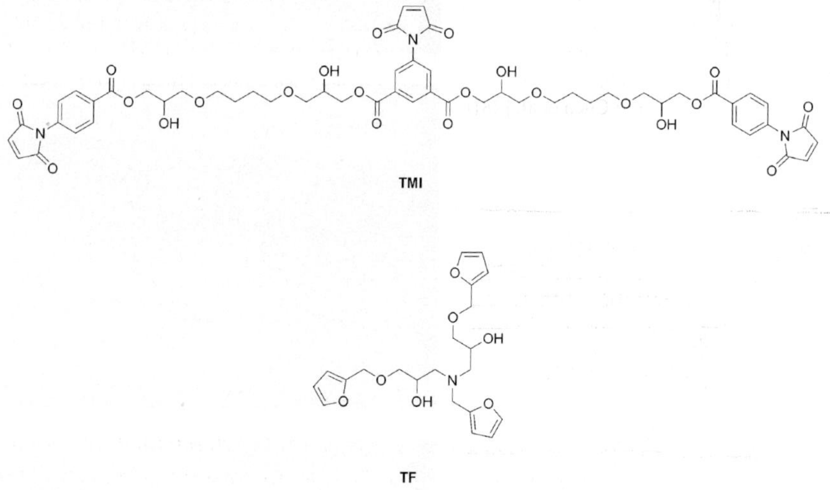

Fig. 4 Structure of the tris-maleimide and tris-furan TMI and TF, respectively

DCP-derived materials. Thus, only a single component monomer is required to achieve polymerization and cross-linking.

In 1961, Stille and Plummer [35] reported the utilization of 1,n-alkane (biscyclopentadienes) as monomers for the production of high molecular weight DA-based polymers. They prepared homo-polymers from 1,6-bis(cyclopentadienyl) hexane, bis(cyclopentadienyl)nonane, and α,α'-bis (cyclopentadieny1)-p-xylene

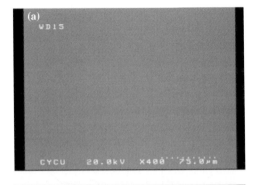

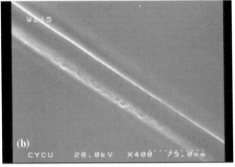

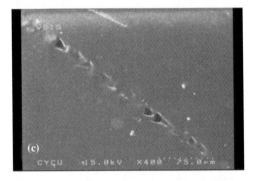

Fig. 5 SEM micrographs of: (**a**) pristine crosslinked polymer; (**b**) knife-cut sample; (**c**) thermally mended sample (50°C; 12 h); and (**d**) thermally mended sample (50°C; 24 h). (Reproduced Liu and Hsieh [31])

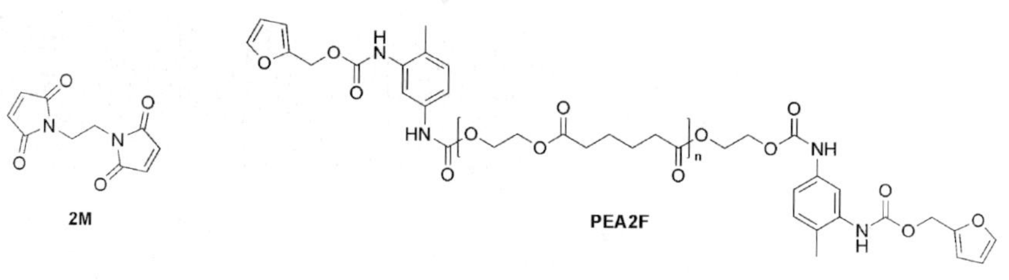

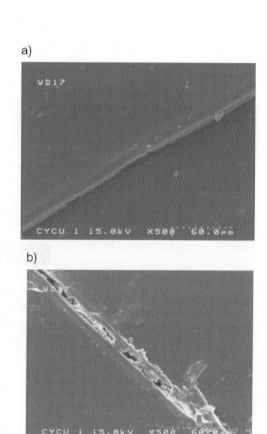

Fig. 6 Structure of PEA2F and 2M

a)

b)

Fig. 7 Test of healing ability: (**a**) knife cut cross-linked film surface; (**b**) thermally self-repaired crosslinked film surface: 120°C at 3 h and 50°C for 5 days. (Reproduced from Liu and Chen [33])

(Figure 8), as well as copolymers of these monomers with the bis-dienophiles *p*-benzoquinone and N,N′-hexamethylenebismaleimide. However, the reversibility of these polymerizations was not studied.

While Takeshita et al. [36] published an early patent concerning the use of CP for reversible cross-linking of polymers, Kennedy and Castner [37, 38] were the first to actually demonstrate the thermal reversibility of such cross-linked systems. They

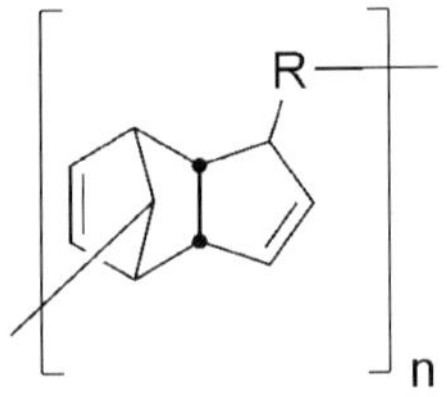

Fig. 8 DCP-based polymers prepared from 1,6-bis(cyclopentadienyl)hexane, bis(cyclopentadienyl)nonane, and α,α′-bis (cyclopentadienyl)-*p*-xylene

$R = -(CH_2)_6- / -(CH_2)_9- / -CH_2-C_6H_4-CH_2-$

reported the preparation of polymers bearing pendant CP groups by substitution of active-halogen-containing polymers. They were able to obtain DA cross-linked polymer networks, and demonstrated the reversibility of the cross-linking by thermally reversing the reaction at 215°C in the absence and presence of maleic anhydride. In addition, samples of the polymers were cut into small pieces and subjected to "curing" at 170°C (high enough temperature to effect cracking of DCP), this process was repeated three times [38]. After the first and second curing cycles, the cyclopentadienylated polymer behaved as a cross-linked elastomer. The cyclopentadienylated polymer assumed the shape of the mold cavity and was continuous and smooth. A stretching experiment revealed that on releasing the tension the polymer snapped back to its original shape. In the third curing cycle the derivatized polymer did not flow in the mold, and the pieces did not coalesce. No explanation was offered for this behavior, though one can speculate that the reversal temperatures were sufficiently high to cause undesirable decomposition reactions.

Despite these and several later reports, [39–43] where DCP-based systems were employed to obtain reversible cross-linking, they have not been reported in the context of re-mendable polymers. Wudl and co-workers [44] presented preliminary results of a DCP-based monomer that upon heating yields the DA-polycondensation polymer.

2.1.3 Other DA-based Polymers

Few reports have appeared in the literature concerning thermally reversible polymers based on anthracene-maleimide DA chemistry [45–47]. Jones et al. [47] reported the reversible cross-linking of polymers bearing pendent anthracene moieties with maleimides (Figure 9). The healing capabilities of these systems have not been studied.

An interesting report by Brand and Klapper [48] reported the thermally reversible DA polymerization of α,ω-bis(3-furylmethyl)-pentaethylene glycol and α,ω-bis(trans-4,4,4-trifluorocrotonylethyl)-polyehtylene glycol (Figure 10). Under inert conditions, the authors were able to cycle the polymerization/de-polymerization

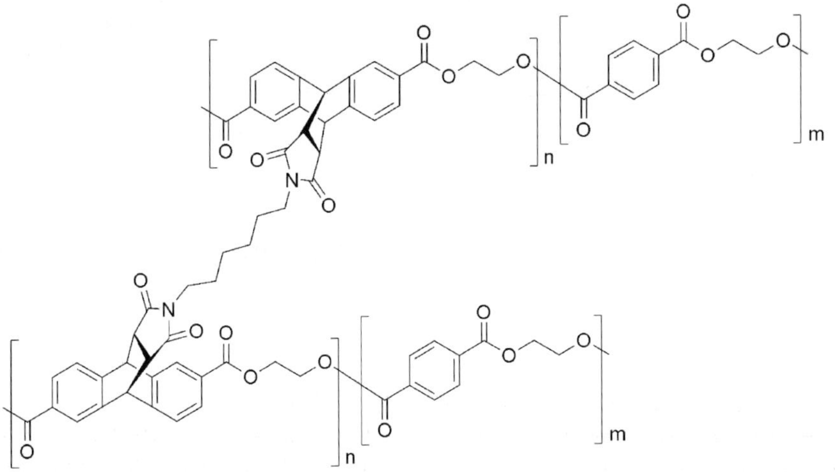

Fig. 9 Structure of the reversibly cross-linked anthracene-maleimide based polymer

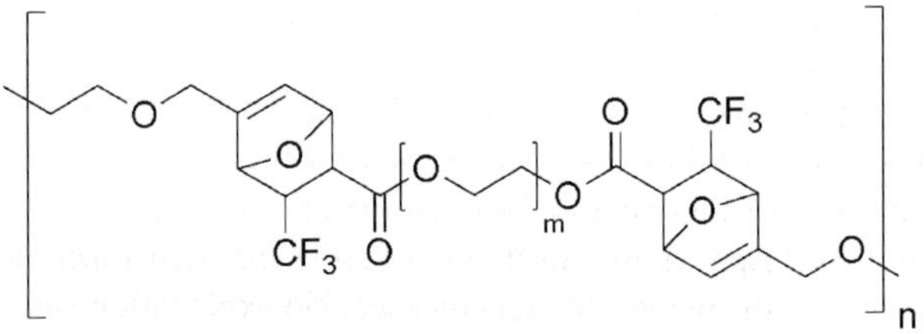

Fig. 10 Structure of the polymer obtained by DA reaction of α,ω-bis(3-furylmethyl)-pentaethylene glycol and α,ω-bis(trans-4,4,4-trifluorocrotonyl-ethyl)-polyehtylene glycol

multiple times. This system was studied in the context of variable viscosity, and did not demonstrated re-mendability in the solid state.

2.2 *Photodimerization-based Polymers*

A number of publications have appeared concerning reversible photo-induced cross-linking of polymers, based on photodimerization of anthracene, sulfides, and coumarins [49–55]. Chung et al. [56] presented a 1,1,1-tris-(cinnamoyloxymethyl) ethane (TCE)-based polymer (Figure 11), that undergoes photodimerizaion to produce a highly cross-linked matrix. The authors claimed that these networks, once cracked, can undergo photo-induced healing by irradiating with light of $\lambda > 280$ nm. Flexural strength measurements indicated a drastic decrease in strength of the healed vs. the original polymer specimens, raising doubts as to the claimed re-mendability of this polymer.

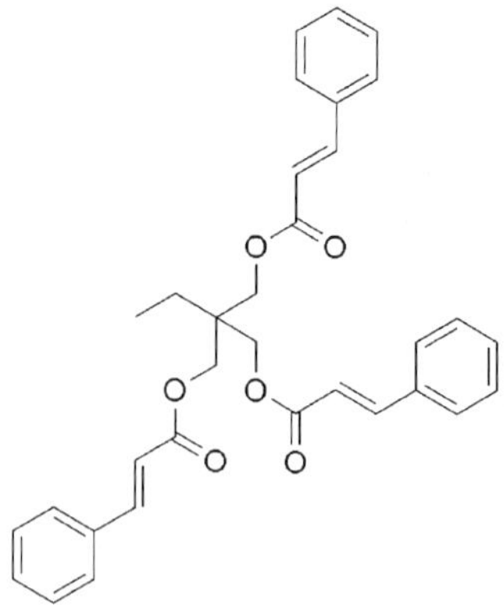

Fig. 11 Structure of 1,1,1-tris-(cinna-
moyloxymethyl) ethane (TCE)   **TCE**

2.3 Supramolecular-based Polymers

Supramolecular polymers are defined as materials that result from association of
monomers via noncovalent interactions. The reversibility of noncovalent interactions
renders these polymers with dynamic features such as the ability to change in length,
constitution, and structure. As a result, one of their inherent traits is the ability to
self-repair and self-heal. While the concept has been known for some time, useful
materials have been made only recently. Since supramolecular polymers have been
extensively reviewed, the interested reader is kindly referred to the recent literature
on the subject [57–63]. Here we will briefly present some of the latest interesting
and potentially useful systems that have been developed by Meijer and co-workers
[64–81].

Meijer and co-workers [64–81] have introduced supramolecular polymers that
utilize the cooperative effect and directionality of quadruple hydrogen bonds.
These polymers were fabricated by employing 2-ureido-4-pyrimidone (UPy) end
groups that form dimers held together by self-complementary DDAA (donor–donor–
acceptor–acceptor) hydrogen bonds. Monomers containing two- and three-binding
sites led to the formation of linear and cross-linked polymers, respectively. The
high dimerization constant ($> 10^6\,M^{-1}$ in CHCl$_3$ [65, 66]) leads to a high degree of
polymerization. The polymeric networks generated by this method, above $\sim$90°C,
dissociate and melt, and behave much like thermoplastic elastomers (Figure 12). As
a consequence, these polymeric systems can undergo thermal mending.

The ureido-pyrimidone-based materials represent the first example of a truly
reversible supramolecular polymer network that can be easily synthesized from
commercially available starting materials, where the ureido-pyrimidone dimerization is strong enough to construct supramolecular materials possessing acceptable
mechanical properties. Meijer et al. [77] have recently reviewed their progress in
the field.

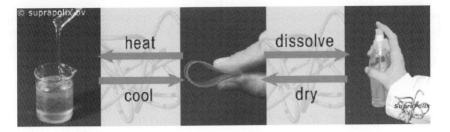

Fig. 12 Illustration of the phase changes in a rubbery material based on Kraton, (poly-ethylene-co-butylene), modified with UPy, exemplifying the diversity in phase behavior because of the supramolecular interactions. (Reproduced from Bosman et al. [77])

Interest from industry in this reversible polymeric system led to the inauguration of SupraPolix, whose purpose is the manufacture of materials based on supramolecular polymers. It is claimed that incorporation of even a small amount of UPy in existing plastics dramatically simplifies the processing of the material. It is further claimed that, production is cheaper and faster, though there should be additional costs incurred in the modification of conventional polymers. The materials should be easily recycled and be consumer-friendly. (For further information: www.suprapolix.com.)

2.4 Other Externally Mendable Polymer System

Bleay et al. [82] employed hollow glass fiber composites filled with an x-ray opaque dye penetrant and one and two-part curing resin systems. After impact damage, the healing resin could be effectively drawn out of the fibers by simultaneous application of heat and vacuum. The damage and repair processes could be monitored by x-ray spectroscopy, the repaired samples exhibiting a 10% improvement in compression strength.

Zako and Takano [83] introduced a method of impregnating small particles (50 μm) of thermoplastic adhesive in a glass/epoxy composite laminate. The cure temperature of the epoxy matrix was 110°C. The embedded thermoplastic particles melted when damaged composites were subsequently heated to 120°C for 10 min on a hot plate. In subsequent three point bend testing, the load-displacement curve indicated that stiffness was recovered in the repaired specimen.

3 Autonomous Mendable Polymers

Self healing materials are inspired by biological systems where damage initiates an autonomous healing process, i.e. in contrast to conventional repair, no external action (e.g. heating) is required. Such materials respond autonomously to damage, and posses the capability to restore the undamaged material's properties without affecting the overall properties of the system. The benefits of such materials are increased safety and reliability, prolonged work-effective lifetimes, minimal maintenance, and thus reduced costs.

3.1 Autonomous Repair by Chemical Catalysis

One of the most efficient ways to produce an autonomously self healing material is to store healing agents inside composite materials that partially restore the mechanical properties after damage. Healing agent storage methods have been developed based on the use of hollow tubes and fibers, particles and microcapsules, all of which can provide an integral healing agent storage capacity.

Dry and coworkers [84–87] embedded methyl methacrylate (MMA) inside hollow porous polypropylene fibers within concrete. Bleeding of MMA from these fibers was shown to reduce concrete permeability. In addition, it was shown that the release of crack-adhering adhesive from hollow glass pipettes into the concrete increased the ability by 20% to carry load under a subsequent flexural test. Similarly, Li et al. [88] employed Superglue (ethyl cyanoacrylate) as a healing agent within 500 μm diameter hollow glass tubes embedded in cementitious composites.

Dry et al. [89–94] have investigated damage-associated matrix microcracking. A single repair fiber was embedded in a polymer–matrix and tests performed to visually verify the release of the healing agent. Motuku et al. [95] further developed this concept by considering different critical parameters, such as method of storage (glass, copper, and aluminium tubing), and healing agents (vinyl ester 411-C50 and EPON-862 epoxy). In their report, glass tubing was found to be an effective encapsulation medium for the healing agent, allowing its release into the matrix upon fiber breakage. In both works [94, 95], dye release accompanied the healing agent; however, this combination resulted in an inability to cure and thus no improvement in mechanical properties was reported.

White and coworkers [96–102] (see Chapter 2) have achieved autonomous mending of cracks by embedding microcapsules of monomer healing agent and an appropriate catalyst throughout a polymer matrix (Figures 13 and 14). Recent work by Liu et al. [103] revealed that the healing process could be improved by using a blend of DCP with 5-ethylidene-2-norbornene (ENB) in a ratio of 1:3. Recently, a computational model of fatigue crack retardation in White's system, taking into account the reaction kinetics of the polymeric material, has been presented [104].

While autonomous self-mending, this system could have limitations such as healing being limited to small cracks that enable capillary action, inability to heal multiple recurring cracks and incompatibility of different polymeric matrices.

Braun and co-workers [105] improved on this concept by designing self healing materials based on the tin-catalyzed polycondensation of phase-separated droplets

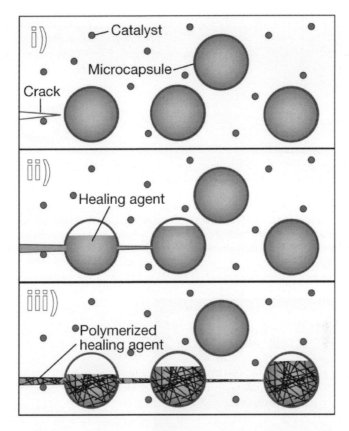

Fig. 13 The self healing concept where a microencapsulated healing agent is embedded in a structural composite matrix containing a catalyst capable of polymerizing the healing agent. (i) Cracks form in the matrix. (ii) The crack ruptures the microcapsules, releasing the healing agent into the crack plane through capillary action. (iii) The healing agent contacts the catalyst, triggering polymerization that bonds the crack faces closed. (Reproduced from White et al. [96])

containing hydroxyl end-functionalized polydimethylsiloxane (HOPDMS) and polydiethoxysiloxane (PDES) (Figure 15). While this system is stable in ambient and working (>100°C) conditions and is relatively cheap as well as easy to fabricate, the healing efficiency is low and the polymer itself is not likely to find applications as a structural engineering material but as one where soft matter is required.

Bond and coworkers [106] have introduced a novel fiber reinforced plastic which also utilizes a bleeding mechanism to achieve self-repair and visualization of damage. The fibers are filled with a two part epoxy healing and fluorescent indicator system (Figure 16), enabling the monitoring of bleeding by spectroscopic methods. This

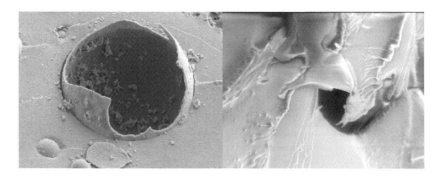

Fig. 14 A scanning electron micrograph shows the fracture plane of a self healing epoxy with a ruptured urea-formaldehyde microcapsule in the center of the image. A scanning electron micrograph
shows the fracture plane of a self healing epoxy and the polymerized healing agent coating the original fracture plane. A broken (emptied) microcapsule appears in the background. (Reproduced from
White et al. [96])

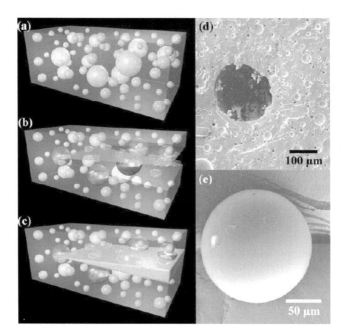

Fig. 15 Schematic of self healing process: (**a**) self healing composite consisting of microencapsulated catalyst (yellow) and phase-separated healing-agent droplets (white) dispersed in a matrix
(green); (**b**) crack propagating into the matrix releasing catalyst and healing agent into the crack
plane; (**c**) a crack healed by polymerized PDMS (crack width exaggerated). Scanning electron
microscopy (SEM) images of (**d**) the fracture surface, showing an empty microcapsule and voids
left by the phase-separated healing agent, and (**e**) a representative microcapsule showing its smooth,
uniform surface. (Reproduced from Cho et al. [105])

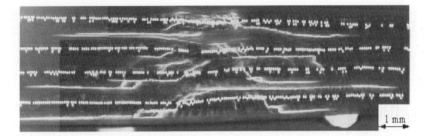

Fig. 16 Impact damaged cross section of composite laminate containing healing agent and UV dye filled filaments. (Reproduced from Pang and Bond [106])

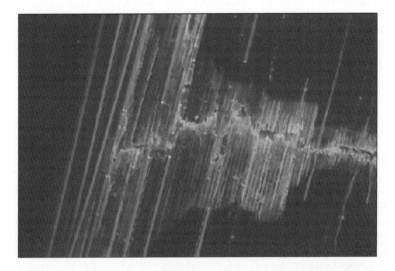

Fig. 17 Damaged hollow fiber composite material viewed under UV light shows "bleeding" of the fluorescent dye mixed with the healing resin. (Reproduced from Pang and Bond [106])

system exhibited up to 97% recovery of-the original flexural strength. The release and infiltration of a UV fluorescent dye from fractured hollow fibers (Figure 17) into damage sites was shown to be an effective method of quickly and easily highlighting damage. In a later study by the same group [107] a fiber-reinforced epoxy matrix was prepared, where the fibers were filled with dilute epoxy healing agent mixed with fluorescent dye (Figure 18). This composite exhibited similar self healing as well as convenient spectroscopic monitoring of the process and featured up to 93% restoration of flexural strength. Despite the gains in material longevity, this healing system ultimately decreases flexural stiffness of a structure when incorporated into laminate structures and it should also be quite difficult to fabricate.

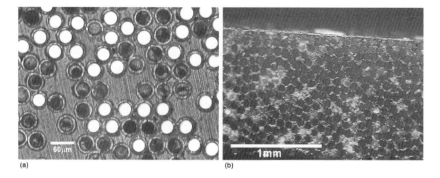

Fig. 18 Optical micrographs of the manufactured fibers and composites: (**a**) hollow glass fibers of 60 μm external diameter with a hollowness of 50% and (**b**) the same fibers within a Hexcel 913 epoxy matrix. (Reproduced from Pang et al. [107])

3.2 Microbiology in Autonomous Repair

Bang et al. [108–112] of South Dakota School of Mines and Technology presented a novel approach for crack healing using microorganisms. The chemistry of the process was based on the function of microbial urease of *Bacillus pasteurii*, which induces $CaCO_3$ (calcite) precipitation as a by-product of the hydrolysis of urea to produce ammonia and carbon dioxide. The ammonia produced increases the pH, leading to accumulation of insoluble $CaCO_3$. These microorganisms were subsequently microencapsulated in polyurethane (PU), to protect them from the extremely alkaline environment of concrete, while serving as additional nucleation sites for calcite crystals [110]. Upon cracking, which ruptured these micelles as well, healing was achieved by calcite precipitation (Figure 19). While the calcite precipitation did not affect the tensile strength and the modulus of elasticity of the PU polymer, the healed cement exhibited a significant increase in compressive strength. Although *B. pasteurii* is a common soil microbe, its urease enzyme is being sought as a more attractive sealing catalyst. This system holds much promise.

3.3 Ionomers

Ionomers are thermoplastic ionic polymers, e.g. hydrocarbon polymers bearing pendant carboxylic acid groups that are either partially or completely neutralized with metal or quaternary ammonium ions [113]. These polymers were originally developed by DuPont via partial neutralization of the random copolymer poly(ethylene-co-methacrylic acid) with alkali metals or zinc hydroxides. DuPont commercialized the sodium and zinc ionomers under the trade name Surlyn, while the acid form is

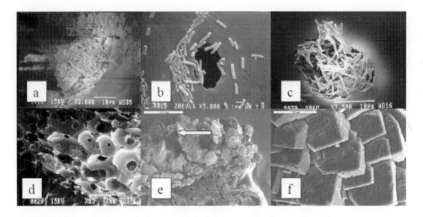

Fig. 19 Scanning electron micrographs of calcite precipitation induced by *B. pasteurii* immobilized in PU. (**a**) Porous PU matrix without microbial cells showing open-cell structures. Bar: 1 mm. (**b**) Distribution of microorganisms on the PU surface. Bar: 1 μm. (**c**) Microorganisms densely packed in a pore of the PU matrix. Bar: 10 μm. (**d**) Calcite crystals grown in the pore (shown in **c**) of the PU matrix. Bar: 10 μm. (**e**) Calcite crystals grown extensively over the PU polymer. Bar: 500 μm. (**f**) Magnified section pointed with an arrow in (**e**) shows crystals embedded with microorganisms. Bar: 20 μm. (Reproduced from website [112])

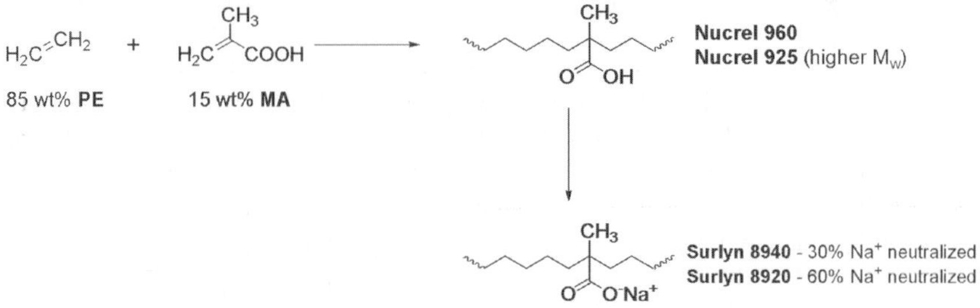

$H_2C \overset{CH_2}{=}$ + $H_2C \overset{CH_3}{\underset{COOH}{=}}$ $\longrightarrow$ structure with CH_3 and $O \overset{}{\underset{OH}{=}}$ groups

85 wt% **PE**   15 wt% **MA**

Nucrel 960
Nucrel 925 (higher M_w)

$\downarrow$

structure with CH_3 and $O \overset{}{\underset{O^-Na^+}{=}}$ groups

Surlyn 8940 - 30% Na$^+$ neutralized
Surlyn 8920 - 60% Na$^+$ neutralized

Scheme 2 Preparation and structure of ionomers investigated for self healing properties

sold under the trade name of Nucrel [114]. Their impact toughness and abrasion and chemical resistance have made them marketable as materials for food packaging, sporting goods, and coatings [115–117].

Several ionomers exhibit instantaneous and autonomous self healing in response to projectile puncturing. Of the ionomers manufactured by DuPont, only Surlyn 8920, Surlyn 8940, Nucrel 960, and Nucrel 925 (Scheme 2) have been examined in depth for their self healing properties. Surlyn 8940 is licensed by Reactive Target Systems and sold as React-A-Seal; it is currently being used at some shooting ranges as target backings to stop or slow stray rounds [118–120].

Initially, the self healing observed in these ionomers was largely attributed to the thermally responsive ionic cross-linking present in the Surlyn polymers [121, 122]. However, later studies [118] revealed that the healing mechanism was more sensitive to temperature and projectile shape than to ionic content. This, combined with the

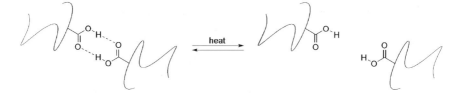

Scheme 3 Thermally controlled hydrogen bond cross-linking in Nucrel

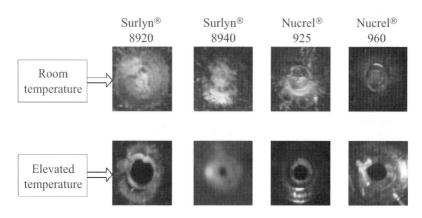

Fig. 20 Temperature-dependant studies of ionomer healing. (Reproduced from Kalista [118])

fact that the hydrogen bond cross-linked Nucrel polymers performed similarly to Surlyn, indicated that the healing action occurs via thermally controlled reversible hydrogen bonding (Scheme 3).

Although the healing mechanism is activated by a sharp increase in temperature after projectile impact, elevated temperatures do not lead to healing of the damaged sample (Figure 20). The healing response was found to be dependent on the shape of the projectile, both Surlyn and Nucrel fairing better in puncture tests with pointed, rather than blunt, projectiles. Preliminary work has been performed on carbon nanotube-reinforced ionomer composites with positive results [123]. Of all the self healing systems, self healing ionomers are able to autonomously recover from the most devastating damage in a very short period of time, and are by far the least expensive to manufacture.

4 Summary and Outlook

The emphasis in this chapter has been the use of weak covalent bonds and weak intermolecular interactions, either as reversible covalent C–C bonds or multiple hydrogen-bonds, to effect repair of polymeric materials. A number of polymer systems have been presented that exhibit externally induced healing. Autonomous healing of cracks was only briefly mentioned, and the interested reader is referred to Chapter 2 in this book dedicated to such systems. The use of the Diels–Alder reaction is, so far, the most successful entry into re-mendable polymers for fully covalent systems, and quadruple hydrogen-bonded links as well as inter-carboxylate hydrogen bonds are, so far, the most successful for noncovalently bound polymers. The recent developments in this field of "smart materials" provide exciting prospects for the future.

Acknowledgements We thank the Air Force for support through a MURI grant administered by University of Illinois.

References

1. Gandini A. (2005) The application of the Diels-Alder reaction to polymer syntheses based on furan/maleimide reversible couplings. Polímeros: Ciência e Tecnologia 15:95–101
2. Craven J.M. (1969) Cross-linked thermally reversible polymers produced from condensation polymers with pendant furan groups cross-linked with maleimides. US Patent 3,435,003
3. Stevens M., Jenkins A. (1979) Crosslinking of polystyrene via pendant maleimide groups. J Polym Sci 17:3675–3685
4. Chujo Y., Sada, K., Saegusa T. (1990) Reversible gelation of polyoxazoline by means of Diels-Alder reaction. Macromolecules 23:2636–2641
5. Imai Y., Itoh H., Naka K., Chujo Y. (2000) Thermally reversible IPN organic-inorganic polymer hybrids utilizing the Diels-Alder reaction. Macromolecules 33:4343–4346
6. Canary S.A., Stevens M.P. (1992) Thermally reversible crosslinking of polystyrene via the furan-maleimide Diels-Alder reaction. J Polym Sci: Polym Chem 30:1755–1760
7. Laita H., Boufi S., Gandini A. (1997) The application of the Diels-Alder reaction to polymers bearing furan moieties. 1. Reactions with maleimides. Eur Polym J 33:1203–1211
8. Goussé C., Gandini A., Hodge P. (1998) Application of the Diels-Alder reaction to polymers bearing furan moieties. 2. Diels-Alder and retro-Diels-Alder reactions involving furan rings in some styrene copolymers. Macromolecules 31:314–321
9. Gheneim R., Perez-Berumen C., Gandini A. (2002) Diels-Alder reactions with novel polymeric dienes and dienophiles: synthesis of reversibly cross-linked elastomers. Macromolecules 35:7246–7253
10. Goiti E., Huglin M.B., Rego J.M. (2001) Some observations on the copolymerization of styrene with furfuryl methacrylate. Polymer 42:10187–10193
11. Goiti E., Huglin M.B., Rego J.M. (2003) Thermal breakdown by the retro Diels-Alder reaction of crosslinking in poly[styrene-co-(furfuryl methacrylate)]. Macromol Rapid Commun 24:692–696
12. Goiti E., Huglin M.B., Rego J.M. (2004) Some properties of networks produced by the Diels-Alder reaction between poly(styrene-co-furfuryl methacrylate) and bismaleimide. Eur Polym J 40:219–226
13. Goiti E., Huglin M.B., Rego J.M. (2004) Kinetic aspects of the Diels-Alder reaction between poly(styrene-co-furfuryl methacrylate) and bismaleimide. Eur Polym J 40:1451–1460
14. Gheneim R., Perez-Berumen C., Gandini A. (2002) Diels-Alder reactions with novel polymeric dienes and dienophiles: synthesis of reversibly cross-linked elastomers. Macromolecules 35:7246–7253
15. Kuramoto N., Hayashi K., Nagai K. (1994) Thermoreversible reaction of Diels-Alder polymer composed of difurfuryl adipate with bismaleimidodiphenylmethane. J Polym Sci: Polym Chem 32:2501–2504

16. Tesoro G.C., Sastri V.R. (1986) Synthesis of siloxane-containing bis(furans) and polymerization with bis(maleimides). Ind Eng Chem: Prod Res Dev 25:444–448
17. Diakoumakos C.D., Mikroyannidis J.A. (1992) Polyimides derived from Diels-Alder polymerization of furfuryl-substituted maleamic acids or from the reaction of bismaleamic with bisfurfurylpyromellitamic acids. J Polym Sci Polym Chem 30:2559–2567
18. Diakoumakos C.D., Mikroyannidis J.A. (1994) Heat-resistant resins derived from cyano-substituted Diels-Alder polymers. Eur Polym J 30:465–472
19. Goussé C., Gandini A. (1999) Diels-Alder polymerization of difurans with bismaleimides. Polym Intern 48:723–731
20. McElhanon J.R., Wheeler D.R. (2001) Thermally responsive dendrons and dendrimers based on reversible furan-maleimide Diels-Alder adducts. Org Lett 3:2681–2683
21. McElhanon J.R., Russick E.M., Wheeler D.R., Loy D.A., Aubert J.H. (2002) Removable foams based on an epoxy resin incorporating reversible Diels-Alder adducts. J Appl Polym Sci 85:1496–1502
22. Loy D.A., Wheeler D.R., McElhanon J.R., Durbin-Voss M.L. (2002) Method of making thermally removable Diels-Alder adduct-containing polyurethanes. US Patent 6403753
23. Loy D.A., Wheeler D.R., Russick E.M., McElhanon J.R., Sanders R.S. (2002) Method of making thermally removable epoxy resins. US Patent 6403753
24. Small J.H., Loy D.A., Wheeler D.R., McElhanon J.R., Saunders R.S. (2001) Method of making thermally removable polymeric encapsulants from maleimide resins. US Patent 6271335
25. Loy D.A., Wheeler D.R., Saunders R.S., McElhanon J.R., Small J.H. (2001) Method of making thermally removable polymeric encapsulants US Patent 6271335
26. Kamahori K., Tada S., Ito K., Itsuno S. (1999) Optically active polymer synthesis by Diels-Alder polymerization with chirally modified lewis acid catalyst. Macromolecules 32:541–547
27. Goussé C., Gandini A. (1998) Synthesis of 2-furfurylmaleimide and preliminary study of its Diels-Alder polycondensation. Polym Bull 40:389–394
28. Chen X., Dam M.A., Ono K., Mal A.K., Shen H., Nutt S.R., Sheran K., Wudl F. (2002) A thermally re-mendable cross-linked polymeric material. Science 295:1698–1702
29. Chen X., Wudl F., Mal A.K., Shen H., Nutt S.R. (2003) New thermally remendable highly cross-linked polymeric materials. Macromolecules 36:1802–1807
30. Wudl F., Chen X. (2004) Thermally re-mendable crosslinked polymers, monomers, their manufacture, and thermally mending. US Patent 2004014933
31. Liu Y.L., Hsieh C.Y. (2006) Crosslinked epoxy materials exhibiting thermal remendablility and removability from multifunctional maleimide and furan compounds. J Polym Sci: Polym Chem 44:905–913
32. Watanabe M., Yoshie N. (2006) Synthesis and properties of readily recyclable polymers from bisfuranic terminated poly(ethylene adipate) and multi-maleimide linkers. Polymer 47:4946–4952
33. Liu Y.L., Chen Y.W. (2007) Thermally reversible cross-linked polyamides with high toughness and self-repairing ability from maleimide- and furan-functionalized aromatic polyamides. Macromol Chem Phys 208:224–232
34. Gotsmann B., Duerig U., Frommer J., Hawker C.J. (2006) Exploiting chemical switching in a Diels-Alder polymer for nanoscale probe lithography and data storage. Adv Funct Mater 16:1499–1505
35. Stille J.K., Plummer L. (1961) Polymerization by the Diels-Alder reaction. J Org Chem 26:4026–4029
36. Takeshita Y., Uoi M., Hirai Y., Uchiyama M. (1972) Polymers crosslinked through dicyclopentadiene rings. German Patent 2164022
37. Kennedy J.P., Castner K.F. (1979) Thermally reversible polymer systems by cyclopentadienylation. I. A model for termination by cyclopentadienylation of olefin polymerization. J Polym Sci: Polym Chem Ed 17:2039–2054
38. Kennedy J.P., Castner K.F. (1979) Thermally reversible polymer systems by cyclopentadienylation. II. The synthesis of cyclopentadiene-containing polymers. J Polym Sci: Polym Chem Ed 17:2055–2070

39. Miura M., Akutsu F., Usui T., Ikebukuro Y., Nagakubo K. (1985) Soluble cyclopentadienylated polymers. Makromol Chem 186:473–481
40. Salamone J.C., Chung Y., Clough S.B., Watterson A.C. (1988) Thermally reversible, covalently crosslinked polyphosphazenes. J Polym Sci: Polym Chem Ed 26:2923–2939
41. Chen X.N., Ruckenstein E.J. (1999) Thermally reversible linking of halide-containing polymers by potassium dicyclopentadienedicarboxylate. J Polym Sci: Polym Chem 37:4390–4401
42. Ruckenstein E.J., Chen X.N. (2000) Crosslinking of chlorine-containing polymers by dicyclopentadiene dicarboxylic salts. J Polym Sci: Polym Chem 38:818–825
43. Chen X.N., Ruckenstein E.J. (2000) Thermally reversible covalently bonded linear polymers prepared from a dihalide monomer and a salt of dicyclopentadiene dicarboxylic acid. J Polym Sci: Polym Chem 38:1662–1672
44. Murphy E., Bolanos E., Wudl F. (2007) Manuscript in preparation.
45. Stevens M.P. (1984) Diels-Alder polymer of N-(2-anthryl)maleimide. J Polym Sci: Polym Lett Ed 22:467–471
46. Grigoras M., Colotin G. (2001) Copolymerization of a bisanthracene compound with bismaleimides by Diels–Alder cycloaddition. Polym Int 50:1375–1378
47. Jones J.R., Liotta C.L., Collard D.M., Schiraldi D.A. (1999) Cross-linking and modification of poly(ethylene terephthalate-co-2,6-anthracenedicarboxylate) by Diels-Alder reactions with maleimides. Macromolecules 32:5786–5792
48. Brand T., Klapper M. (1999) Control of viscosity through reversible addition of telechelics via repetitive Diels-Alder reaction in bulk. Des Monomers Polym 2:287–309
49. Muehlebach A., Hafner A., Van Der Schaaf P.A. (1996) Crosslinkable polymers and process for crosslinking them. Int Patent 9623829
50. Chen Y., Geh J.L. (1996) Copolymers derived from 7-acryloyloxy-4-methylcoumarin and acrylates: 2. Reversible photocrosslinking and photocleavage. Polymer 37:4481–4486
51. Matsui J., Ochi Y., Tamaki K. (2006) Photodimerization of anthryl moieties in a poly(methacrylic acid) derivative as reversible cross-linking step in molecular imprinting. Chem Lett 35:80–81
52. Zheng Y., Micic M., Mello S.V., Mabrouki M., Andreopoulos F.M., Konka V., Pham S.M., Leblanc R.M. (2002) PEG-based hydrogel synthesis via the photodimerization of anthracene groups. Macromolecules 35:5228–5234
53. Torii T., Ushiki H., Horie K. (1993) Fluorescence study on dynamics and structure of polymer chain. II. Dependence of intramacromolecular photocrosslinking reaction of polymers with anthryl and eosinyl moieties on number of crosslinkable side chains. Polym J 25:173–183
54. Chujo Y., Sada K., Nomura R., Naka A., Saegusa T. (1993) Photogelation and redox properties of anthracene-disulfide-modified polyoxazolines. Macromolecules 26:5611–5614
55. Chang J.M., Aklonis J.J. (1983) A photothermal reversibly crosslinkable polymer system. J Polym Sci: Polym Lett Ed 21:999–1004
56. Chung C.M., Roh Y.S., Cho S.Y., Kim J.G. (2004) Crack healing in polymeric materials via photochemical [2+2] cycloaddition. Chem Mater 16:3982–3984
57. Serpe M.J., Craig S.L. (2007) Physical organic chemistry of supramolecular polymers. Langmuir 23:1626–1634
58. Harada A., Hashidzume A., Takashima Y. (2006) Cyclodextrin-based supramolecular polymers. Adv Polym Sci 201:1–43
59. Lehn J.M. (2005) Dynamers: dynamic molecular and supramolecular polymers. Prog Polym Sci 30:814–831
60. Ciferri A. (2005) Supramolecular polymers, 2nd edn. Taylor & Francis, Boca Raton, FL
61. Armstrong G., Buggy M. (2005) Hydrogen-bonded supramolecular polymers: a literature review. J Mater Sci 40:547–559
62. Brunsveld L., Folmer B.J.B., Meijer E.W., Sijbesma R.P. (2001) Supramolecular polymers. Chem Rev 101:4071–4097
63. Moore J.S. (1999) Supramolecular polymers. Curr Opin Coll Int Sci 4:108–116

64. Sijbesma R.P., Beijer F.H., Brunsveld L., Folmer B.J.B., Ky Hirschberg J.H.K., Lange R.F.M., Lowe J.K.L., Meijer E.W. (1997) Reversible polymers formed from self-complementary monomers using quadruple hydrogen bonding. Science 278:1601–1604

65. Beijer F.H., Sijbesma R.P., Kooijman H., Spek A.L., Meijer E.W. (1998) Strong dimerization of ureidopyrimidones via quadruple hydrogen bonding. J Am Chem Soc 120:6761–6769

66. Söntjens S.H.M., Sijbesma R.P., van Genderen M.H.P., Meijer E.W. (2000) Stability and lifetime of quadruply hydrogen bonded 2-ureido-4[1H]-pyrimidinone dimers. J Am Chem Soc 122: 7487–7493

67. Ky Hirschberg J.H.K., Beijer F.H., van Aert H.A., Magusin P.C.M.M., Sijbesma R.P., Meijer E.W. (1999) Supramolecular polymers from linear telechelic siloxanes with quadruple-hydrogen-bonded units. Macromolecules 32:2696–2705

68. Lange R.F.M., Van Gurp M., Meijer E.W. (1999) Hydrogen-bonded supramolecular polymer networks. J Polym Sci: Polym Chem 37:3657–3670

69. Folmer B.J.B., Sijbesma R.P., Versteegen R.M., van der Rijt J.A.J., Meijer E.W. (2000) Supramolecular polymer materials: chain extension of telechelic polymers using a reactive hydrogen-bonding synthon. Adv Mater 12:874–878

70. El-ghayoury A., Peeters E., Schenning A.P.H.J., Meijer E.W. (2000) Quadruple hydrogen bonded oligo(p-phenylene vinylene) dimers. Chem Commun 19:1969–1970

71. Ky Hirschberg J.H.K., Brunsveld L., Ramzi A., Vekemans J.A.J.M., Sijbesma R.P., Meijer E.W. (2000) Helical self-assembled polymers from cooperative stacking of hydrogen-bonded pairs. Nature 407:167–170

72. Schenning A.P.H.J., Jonkheijm P., Peeters E., Meijer E.W. (2001) Hierarchical order in supramolecular assemblies of hydrogen-bonded oligo(p-phenylene vinylene)s. J Am Chem Soc 123: 409–416

73. El-Ghayoury A., Schenning A.P.H.J., Van Hal P.A., Van Duren J.K.J., Janssen R.A.J., Meijer E.W. (2001) Supramolecular hydrogen-bonded oligo(p-phenylene vinylene) poly-mers. Angew Chem Int Ed 40:3660–3663

74. Keizer H.M., Sijbesma R.P., Jansen J.F.G.A., Pasternack G., Meijer E.W. (2003) Polymerization-induced phase separation using hydrogen-bonded supramolecular polymers. Macromolecules 36:5602–5606

75. Keizer H.M., van Kessel R., Sijbesma R.P., Meijer E.W. (2003) Scale-up of the synthe-sis of ureidopyrimidinone functionalized telechelic poly(ethylenebutylene). Polymer 44: 5505–5511

76. Bosman A.W., Brunsveld L., Folmer B.J.B., Sijbesma R.P., Meijer E.W. (2003) Supramolecu-lar polymers: From scientific curiosity to technological reality. Macromol Symp 201:143–154

77. Bosman A.W., Sijbesma R.P., Meijer E.W. (2004) Supramolecular polymers at work. Mater Today 7:34–39

78. Ligthart G.B.W.L., Ohkawa H., Sijbesma R.P., Meijer E.W. (2005) Complementary quadru-ple hydrogen bonding in supramolecular copolymers. J Am Chem Soc 127:810–811

79. Scherman O.A., Ligthart G.B.W.L., Sijbesma R.P., Meijer E.W. (2006) A selectivity-driven supramolecular polymerization of an AB monomer. Angew Chem Int Ed 45:2072–2076

80. Kautz H., van Beek D.J.M., Sijbesma R.P., Meijer E.W. (2006) Cooperative end-to-end and lateral hydrogen-bonding motifs in supramolecular thermoplastic elastomers. Macromole-cules 39:4265–4267

81. Ohkawa H., Ligthart G.B.W.L., Sijbesma R.P., Meijer E.W. (2007) Supramolecular graft copolymers based on 2,7-diamido-1,8-naphthyridines. Macromolecules 40:1453–1459

82. Bleay S.M., Loader C.B., Hawyes V.J., Humberstone L., Curtis P.T. (2001) A smart repair system for polymer matrix composites. Composites A 32:1767–1776

83. Zako M., Takano N. (1999) Intelligent material systems using epoxy particles to repair micro-cracks and delamination damage in GFRP. J Int Mater Sys Struct 10:836–841

84. Dry C. (1991) Alteration of matrix permeability, pore and crack structure by the time release of internal chemicals. Ceramic Trans: Adv Cem Mater 16:729–768

85. Dry C. (1994) Matrix cracking repair and filling using active and passive modes for smart timed release of chemicals from fibres into cement matrices. Smart Mater Struct 3:118–123

86. Dry C., McMillan W. (1996) Three-part methylmethacrylate adhesive system as an internal delivery system for smart responsive concrete. Smart Mater Struct 5:297–300

87. Dry C. (2001) Three designs for the internal release of sealants, adhesives and waterproofing chemical into concrete to release. Cement Concrete Res 30:1969–1977

88. Li V.C., Lim Y.M., Chan Y.W. (1998) Feasibility of a passive smart self-healing cementitious composite. Composites B 29B:819–827

89. Dry C., Corsaw M., Bayer E. (2003) A comparison of internal self-repair with resin injection in repair of concrete. J Adh Sci Tech 17:79–89

90. Dry C., Corsaw M. (2003) A comparison of bending strength between adhesive and steel reinforced concrete with steel only reinforced concrete. Cem Conc Res 33:1723–1727

91. Dry C. (2007) Multiple function, self-repairing laminated composites with special adhesives. Int Patent 2007005657

92. Dry C. (1996) Procedure developed for self-repair of polymer matrix composite materials. Comp Struct 35:263–269

93. Dry C., Sottos N. (1996) Passive smart self-repair in polymer matrix composite materials. Proc SPIE: Smart Mater Struct 1916:438–444

94. Dry C., McMillan W. (1997) A novel method to detect crack location and volume in opaque and semi-opaque brittle materials. Smart Mater Struct 6:35–39

95. Motuku M., Vaidya U.K., Janowski G.M. (1999) Parametric studies on self-repairing approaches for resin infused composites subjected to low velocity impact. Smart Mater Struct 8:623–638

96. White S.R., Sottos N.R., Geubelle P.H., Moore J.S., Kessler M.R., Sriram S.R., Brown E.N., Viswanathan S. (2001) Autonomic healing of polymer composites. Nature 409:794–797

97. Kessler M.R., White S.R. (2001) Self-activated healing of delamination damage in woven composites. Composites A 32:683–699

98. Kessler M.R., Sottos N.R., White S.R. (2003) Self-healing structural composite materials. Composites A 34:743–753

99. Brown E.N., Sottos N.R., White S.R. (2002) Fracture testing of a self-healing polymer composite. Exp Mech 42:372–379

100. Brown E.N., Kessler M.R., Sottos N.R., White S.R. (2003) In situ poly (ureaformaldehyde) microencapsulation of dicyclopentadiene. J Microencapsul 20:719–730

101. Brown E.N., White S.R., Sottos N.R. (2004) Microcapsule induced toughening in a self-healing polymer composite. J Mater Sci 39:1703–1710

102. Jones A.S., Rule J.D., Moore J.S., White S.R., Sottos N.R. (2006) Catalyst morphology and dissolution kinetics of self-healing polymers. Chem Mater 18:1312–1317

103. Liu X., Lee J.K., Yoon S.H., Kessler M.R. (2006) Characterization of diene monomers as healing agents for autonomic damage repair. J Appl Polym Sci 101:1266–1272

104. Maiti S., Shankar C., Geubelle P.H., Kieffer J. (2006) Continuum and molecular-level modeling of fatigue crack retardation in self-healing polymers. J Eng Mat Tech 128:595–602

105. Cho S.H., Andersson H.M., White S.R., Sottos N.R., Braun P.V. (2006) Polydimethylsiloxane-based self-healing materials. Adv Mater 18:997–1000

106. Pang J.W.C., Bond I.P. (2004) Bleeding composites – damage detection and self-repair using a biomimetic approach. Composites A 36:183–188

107. Pang J.W.C., Jody W.C., Bond I.P. (2005) A hollow fibre reinforced polymer composite encompassing self-healing and enhanced damage visibility. Comp Sci Tech 65:1791–1799

108. Gollapudi U.K., Knutson C.L., Bang S.S., Islam M.R. (1995) A new method for controlling leaching through permeable channels. Chemosphere 30:695–705

109. Stocks-Fischer S., Galinat J.K., Bang S.S. (1999) Microbiological precipitation of $CaCO_3$. Soil Bio Biochem 31:1563–1571

110. Bang S.S., Galinat J.K., Ramakrishnan V. (2001) Calcite precipitation induced by polyurethane-immobilized *Bacillus pasteurii*. Enz Microb Tech 28:404–409

111. Bachmeier K., Williams A.E., Warmington J., Bang S.S. (2002) Urease activity in microbiologically-induced calcite precipitation. J Biotech 93:171–181
112. http://ssbang.sdsmt.edu/PDF%20Files/MECR.pdf
113. Holliday L. (1975) Ionic polymers. Wiley, New York
114. http://www2.dupont.com/Surlyn/en_US/ and http://www2.dupont.com/Products/en_RU/Nucrel_en.html
115. Eisenberg A., Kim J.S. (1998) Introduction to ionomers. Wiley, New York
116. Tant M.R., Mauritz K.A., Wilkes G.L. (1997) Ionomers: Synthesis, structure, properties, and applications. Chapman & Hall, New York
117. Wilson A.D., Prosser H.J. (1983) Developments in ionic polymers. Applied Science, New York
118. Kalista S.J. (2003) Self-healing of thermoplastic poly(ethylene-co-methacrylic acid) copolymers following projectile puncture. M.Sc. Thesis, Virginia Tech http://scholar.lib.vt.edu/theses/available/etd-12162003–103411/
119. Kalista S.J., Ward T.C., Oyetunji Z. (2003) Self-healing behavior of ethylene-based ionomers. Proc Ann Meet Adhes Soc 26:176–178
120. Kalista S.J., Ward T.C. (2004) A quantitative comparison of puncture-healing in a series of ethylene based ionomers. Proc Ann Meet Adhes Soc 27:212–214
121. Fall R. (2001) Puncture reversal of ethylene ionomers-mechanistic studies. M.Sc. Thesis, Virginia Tech http://scholar.lib.vt.edu/theses/available/etd-08312001–084412/
122. Huber A., Hinkley J.A. (2005) Impression testing of self-healing polymers. NASA Tech Man 2005–213532
123. Kalista S.J., Ward T.C. (2006) Self-healing in carbon nanotube filled thermoplastic poly(ethylene-co-methacrylic acid) ionomer composites. Proc Ann Meet Adhes Soc 29: 244–246

Thermally Induced Self Healing of Thermosetting Resins and Matrices in Smart Composites

FR Jones, W Zhang, and SA Hayes

Ceramics and Composites Laboratory, Department of Engineering Materials, University of Sheffield, UK
E-mail: F.R.Jones@sheffield.ac.uk

1   Introduction

Advanced composite materials combine high-performance-reinforcing fibres with matrix resins. Most high-performance applications currently utilise thermosetting resins because they provide stable high-modulus matrix systems. There is some interest in thermoplastic polymer matrices. These are often found in short-fibre-reinforced thermoplastic materials and occasionally in high-performance composites. The latter are likely to receive more interest in future because the weldability of thermoplastics can be advantageous in tape placement fabrication. In recent times there has been an increased interest in the development of durable systems in which the polymer can self heal. There are many opportunities for self healing because the micromechanical mechanism of first ply failure of an advanced composite is transverse or matrix cracking. Transverse cracking of a laminated composite arises because the ply failure strain at 90° to the fibre is normally less than that of the polymer alone. Furthermore, these cracks propagate within the interphase material adjacent to the fibres. In addition, the fibres are configured at angles to each other so that these transverse cracks are pinned at the higher modulus ply interface and do not have a large-crack displacement. There is, therefore, ample scope for the repair of first ply failure cracks. The micromechanics of laminates often involves the initiation of delamination at transverse cracks. Thus, the damage accumulation mechanisms involve fracture of the polymeric matrix which provides the opportunities for self healing of fibre composites.

Four techniques for self healing or repair of cracks in composite materials have been proposed. The first approach described by Dry [1] used glass capillaries which contained a liquid repair resin, embedded in the polymer matrix. A crack running in the material would sever the capillary releasing the liquid resin which would then polymerise and effectively cause the fracture sites to heal. This principle was extended by Bleay et al. [2] and, recently, by Pang and Bond [3, 4], who used hollow glass fibres to encapsulate the liquid resin. A similar, but related system, developed at the University of Illinois, Urbana-Champaign [5–7] employs a liquid monomer which has been micro-encapsulated in a polymeric shell. In a similar manner, a matrix crack propagates into the shells releasing the healing agent. The polymerisation catalyst is dispersed in a soluble wax which dissolves in the liquid monomer

S. van der Zwaag (ed.), *Self Healing Materials. An Alternative Approach to 20 Centuries of Materials Science*, 69–93.

© 2007 *Springer*.

69

or healing agent. Healing occurs through polymerisation of the monomeric agent which has diffused into matrix cracks. Wudl et al. [8, 9] have designed a remendable polymer which uses a reverse Diels–Alder mechanism for the reformation of the cross-linked molecular bonds using a thermal intervention, with a heating cycle at 115°C. Hayes et al. [10, 11] have reported a solid-state healing system in which a linear polymer was incorporated into a thermoset epoxy resin, which diffuses to a crack for closure and healing. This was part of a smart composite system whereby the reinforcing carbon fibres can be used to detect damage in a self-sensing approach. In addition, the self-sensing carbon fibres can be used to supply resistance heating to the localised damaged area for healing. This paper reviews the requirements for solid-state healing of fibre composite materials.

2 Equilibria in Polymerisations

Two main polymerisation mechanisms exist: (i) addition polymerisation, which is a chain reaction with initiation, propagation and termination stages; and (ii) step-growth polymerisation, which is statistically controlled.

2.1 Chain Addition Polymerisation

Addition polymerisation is a chain reaction involving initiation, propagation, transfer, and termination stages.

$$I \rightarrow R^*$$
$$R^* + M \rightarrow P_1^* \text{ Initiation} \tag{1}$$
$$P_1^* + (n-1)M \rightleftharpoons P_n^* \text{ Propagation} \tag{2}$$
$$P_n^* \xrightarrow[\text{termination}]{\text{transfer}} \text{Polymer} \tag{3}$$

where P^* is an active site which may be free radical, ionic, or covalent (in certain cases). Thermodynamically, propagation and depropagation (stage 2) are in equilibrium providing termination (stage 3) is absent or at best much slower than propagation. The degree of polymerisation (n) achieved in the propagation stage is determined by the equilibrium constant (K_p) for stage 2.

$$K_p = \frac{[P_n^*]_e}{[P_n^*]_e[M]_e} = \frac{1}{[M]_e} \tag{4}$$

where $[P_n^*]_e$ is the equilibrium concentration of polymer and $[M]_e$ is the equilibrium concentration of monomer. Since ΔG_p°, the standard state Gibbs free energy for propagation, is equal to $-RT \ln K_p$, where R is the gas constant and T is the temperature, we can define a ceiling temperature, T_c, as

$$T_c = \frac{\Delta H_p^o}{\Delta S_p^o + R \ln[M]_o} \tag{5}$$

where ΔH_p^o and ΔS_p^o are the standard state enthalpy and entropy of propagation. $[M]_o$ is the concentration of monomer in the reaction volume, because at T_c the polymer is thermodynamically unstable and $[M]_e = [M]_o$.

Realistically, the equilibrium between monomer and polymer can be achieved with living polymers. Thus, the anionic polymerisation α-methyl styrene shows a T_c of 334 K and the cationic polymerisation of tetrahydrofuran also has a low ceiling temperature. This topic has been reviewed by Szwarc [12] and Dainton and Ivin [13].

Equation (5) shows that there will be a ceiling temperature at which the monomer is more stable than the polymer. In other words, polymerisation is no longer thermodynamically possible above T_c. Also, from Equation (5) a floor concentration of monomer for polymerisation can also be defined.. A classic example of a depolymerisation/polymerisation equilibrium exists for the commercial acetal homopolymer, which needs to be stabilised by end-capping so that it can be injection moulded above T_c. Confirmation that depolymerisation is a chain depropagation is shown by the observation that acetal can also be stabilised by copolymerisation in which the comonomer units (such 1, 3-dioxolan) prevent the depropagation of the chain. Wudl [8, 9] has applied these concepts to a reverse Diels–Alder cross-linked polymer where thermal healing of cracks can be achieved. For example, a four functional furan was reacted with a difunctional bismaleimide in a cycloaddition curing reaction. Metathetical polymerisations, which are used in the Illinois system also exhibit polymerisation and depolymerisation equilibria. While the early catalysts were tungsten coordination compounds [14, 15] the Grubbs catalysts [16] have proved more consistent.

2.2   Step-Growth Polymerisation

Step-growth or condensation polymerisation occurs through statistical reaction of multifunctional monomers containing at least two reactive functional groups. For bifunctional monomers linear polymers form, whose average degree of polymerisation $(\overline{P}_n)$ is determined by the extent of reaction of the functional groups (p). The Carothers equation shows the relationship between the two important variables.

$$\overline{P}_n = \frac{1}{(1 - p)} \tag{6}$$

Flory [17] demonstrated that with trifunctional and higher functional monomers, highly branched polymers form which eventually gel into cross-linked polymers.

Equation 6 shows that a high value of $\overline{P}_n$ can only be achieved at high extents of polymerisation, p. Furthermore, the molar concentration of the two functional groups need to be exactly equal for a high molecular weight polymer. When the concentration of the two functional groups is not equal, the polymers and oligomers will be terminated with the functional group in excess and further polymerisation is effectively terminated. Commercially, for polyamide 66, the monomers are cocrystallised to form a "nylon salt" in which the amine and carboxylic acid concentrations are equal. For polyesters, transesterifications can be used for chain extension. This also shows that there is potential for end-groups reacting with the skeletal bonds. Thus intermolecular chain–chain reactions lead to equilibriation of the molecular weight. While intramolecular ring–chain equilibration leads to cyclic formation, R′ and R are polyester chains of differing length.

$$
\underset{\displaystyle \sim\!\!\sim\!\!\sim C}{\overset{\displaystyle \overset{O}{\|}}{}} {}^-O \sim\!\!\sim\!\!\sim R' + HO \sim\!\!\sim\!\!\sim R
$$

$$
\updownarrow
$$

$$
\underset{\displaystyle \sim\!\!\sim\!\!\sim C}{\overset{\displaystyle \overset{O}{\|}}{}} {}^-O \sim\!\!\sim\!\!\sim R + HO \sim\!\!\sim\!\!\sim R' \tag{7}
$$

(i) Intermolecular chain–chain equilibriation

$$
\underset{\displaystyle \sim\!\!\sim C}{\overset{\displaystyle \overset{O}{\|}}{}} -O \sim\!\!\sim\!\!\sim \underset{\displaystyle C}{\overset{\displaystyle \overset{O}{\|}}{}} -OH \rightleftharpoons \sim\!\!\sim \underset{\displaystyle C}{\overset{\displaystyle \overset{O}{\|}}{}} O H + \underset{\displaystyle C}{\overset{\displaystyle \overset{O}{\|}}{}} -O \tag{8}
$$

(ii) Intramolecular ring–chain equilibration

The concentration of cyclics at equilibrium is a function of the conformation of the polymer chain, which determines the statistics of reaction 8 [18]. Jacobson and Stockmayer [19] provided an equation for the equilibrium constant (K_x) of a cyclic of degree of polymerisation (x);

$$
K_x = (3/2\pi < r_x^2 >)^{3/2}(1/N_A \, \sigma_{Rx}) \tag{9}
$$

where $< r_x^2 >$ is the mean square end-to-end distance of x-meric chains, σ_{Rx} is the number of skeletal bonds of an x-meric ring that can open in the reverse reaction, and is usually taken as $1x$ or $2x$.

Thus, for many linear polymers equilibration of molecular structure can provide mechanisms for thermal healing. For example, linear siloxane polymers, which form the basis of silicone materials readily equilibrate [20]. By using multifunctional monomers in an analagous reverse Diels–Alder mechanism a cross-linked polymer can also be provided with healability [8, 9]. The healing temperature can be identified in a similar analysis to Equations 4 and 5.

2.3   Polymers with Reversible Skeletal Bonds

One class of polymers which employ reactive bonds within the skeleton is the ionomer systems. As shown below, ionomer bonds are formed between inorganic elements and a carboxyl group. Thus, low molecular weight polymers can react together to form longer chains with ionic reversible bonds.

$$HOOC–R–COOH + 3MgO + HOOC–R–COOH$$
$$\downarrow -H_2O \tag{10}$$
$$-R–COO^-Mg^{2+-}OOC–R–COO^-Mg^{2+}\ {}^-OOC–R–COO^-\ Mg^{2+-}OOC–R–$$

where R is a polymeric chain.

Equation (10) illustrates the chain extension which can cause a polymer solution to exhibit an increase in viscosity. This chemistry is used to thicken unsaturated polyester resins (in styrene solution) required for the manufacture of a handleable sheet moulding compound (SMC). The DuPont Ionomer resins are linear polymers with fusible ionomer bonds using analogous chemistry. Here the anion can be either Na^+ or Zn^{2+} and with the relatively non-polar ethylene-co-methacrylic acid matrix anions and cations are agglomerated into ion clusters which act as thermally fusible crosslinks. Therefore the ionomer systems also have potential for providing healing in linear polymers by a thermal equilibration of the ionomer bonds [21].

where M^+ is Mg^{2+}, Zn^{2+} or Na^+ (11).

Crack healing will require both diffusion of the anion-terminated polymer and ionomer bond reformation.

3   Diffusional Solid State Healing

Hayes et al. [10, 11] have reported that the incorporation of a soluble linear polymer into an epoxy resin can lead to the recovery of up to 70% in fracture toughness. Epoxy resins are usually made from the condensation of epichlorohydrin (iii) with a diol. A typical aromatic diol used in the manufacture of an epoxy resin is bisphenol A (iv). As shown below, the epoxy resin (v) produced can have varying degrees of polymerisation, n, depending on the stoichiometry of the reactants. With an excess of epichlorohydrin, a diglycidyl ether of bisphenol A (DGEBA) forms. This polymer is available commonly under the trade names of Araldite MY750 or Epikote 828. This resin is not completely monomeric, because it contains other oligomers, and $n \approx 0.1$.

Table 1 shows that a range of epoxy resins can be synthesised from these two monomers to meet the requirements of differing applications. A variety of alternative multifunctional hydroxyl or amine compounds can be used to prepare other specialist resins.

3.1   *Requirements for Solid State Self Healing of Epoxy Resins*

According to Hayes et al. [23], for successful healing the healing agent should have the following properties:

(i) The healing agent should be reversibly bonded to the cross-linked epoxy resin network through intermolecular bonds such as hydrogen bonding.
(ii) The healing agent should become mobile above the minimum healing temperature so that diffusional bridging of a crack can occur.
(iii) The addition of the linear chain molecule should not significantly reduce the thermomechanical properties of the matrix resin.

Examination of the structure of the epoxy resin (v) shows that a resin with a large value of n can meet the above requirements; Hydroxyl groups can provide hydrogen-bonding capability. Thus, a linear polymer with a large value of n (average molecular weight of $44{,}000$ g mol^{-1}) has the potential for healing an epoxy resin of similar

Table 1 Different epoxy resins from Bisphenol A, BA, and Epichlorohydrin, ECH (Diglycidyl Ethers of Bisphenol A). (From Jones [22])

n (Structure iii)	ECH:BA (Mole Ratio)	Ideal M (g mol^{-1})	Observed M (g mol^{-1})	Epoxy Groups (mol^{-1})	[a]EPEW (g mol-1)	[b]S pt (°C)	Viscosity (cps)	[c]Epikote Designation
0	4	340	380	2	170	$<RT$	4,000–6,000	825
0.07	–	360	–	2	180	$<RT$	7,000–9,000	826
0.14	–	380	–	2	190	$<RT$	10,000–16,000	828
	2		451	1.39	314	43	–	–
1	1.4	624	791	1.34	592	84	–	–
2	1.33	908	802	1.10	730	90	–	–
2.3	–	993	–	1.99	500	65 – 75	–	1,001
3	1.25	1,192	1,132	1.32	862	100	–	–
4	1.20	1,476	1,420	1.21	1,176	112	–	–
34	–	9,996	–	1.99	5,000	155–165	–	1,010

[a] Epoxy-equivalent weight or epoxy molar mass
[b] Softening point
[c] Hexion Epikote or Epon resin

structure. Therefore, we have studied the healability a cured epoxy resin based on a DGEBA resin containing a high molecular weight linear polymer of similar chemical structure. Nadic Methylene anhydride (NMA) was chosen as the curing agent.

3.2   Estimation of Compatibility of a Healing Agent

The solubility parameter (δ) is a good way of identifying the compatibility of a potential linear-healing agent with a cured resin.

For a polymer solution, the Gibbs equation provides the thermodynamic estimate of free energy of mixing ΔG_m.

$$\Delta G_m = \Delta H_m - T \Delta S_m \tag{13}$$

where the subscript m refers to mixing. This shows that for a polymer in solution in a solvent, the entropy of mixing ΔS_m will be positive as a result of an increase in disorder of the system, encouraging dissolution. However, for the mixing of polymers, ΔS_m is very small and close to zero because the degree of disorder does not change significantly. Thus the important requirement for blending of two polymers is that ΔH_m is less than zero. Since

$$\Delta H_m = v_1 v_2 (\delta_1 - \delta_2)^2 V \tag{14}$$

where $\delta_1\delta_2$ are the solubility parameters of components 1 and 2. V is the total volume and v_1v_2 are the volume fractions of components 1 and 2.

The solubility parameter δ is defined as square root of cohesive energy density:

$$\delta = (\Delta U_E/V_m)^{1/2} = [(\Delta H_E - RT)/V_m]^{1/2} \tag{15}$$

where ΔU_E and ΔH_E are the internal energy and enthalpy of evaporation, V_m is the molar volume.

For mixing, $\Delta H_m = 0$, which is satisfied when $\delta_1 = \delta_2$ Equation (14); thus, the statement "like dissolves like". δ can be calculated from Equation (15) where ΔH_E is determinable from the pressure dependence of the boiling point of a solvent. The solubility parameter for a polymer can be determined by finding a solvent (of known solubility parameter) into which it dissolves in all proportions, or causes maximum swelling of a lightly cross-linked polymer.

δ can also be estimated from the molar interaction parameters of the molecular components within the polymer. Using the Hoy method [24], the solubility parameters of a diglycidyl ether of bisphenol F (DGEBF) and bisphenol A (DGEBA) have been calculated. The structure of DGEBF (vi) is similar to DGEBA (v) except that the linking group between the phenylene groups is –CH$_2$– instead of –C(CH$_3$)$_2$–.

(vi)

Where R' is CH$_2$ ——— CH — CH$_2$ —

R is CH$_2$ ——— CH — CH$_2$ — O — ⟨◯⟩ — CH$_2$ — ⟨◯⟩ — O —

Figure 1 compares the effect of the degree of polymerisation, n, on the solubility parameters of the DGEBA and DGEBF. For both polymers δ increases with n tending to plateau above 20.

For DGEBF, a polymer with 20 repeat units has a molecular weight of 5438 g mol^{-1}. However, the solubility parameter can be more easily calculated from the simple repeat unit (vi) as 22.36 MPa$^{1/2}$. This can be seen to provide a good approximation for the linear polymer. This is compared to a value of 21.26 MPa$^{1/2}$ calculated from the repeat unit for the polymer from epichlorohydrin and bisphenol A(v).

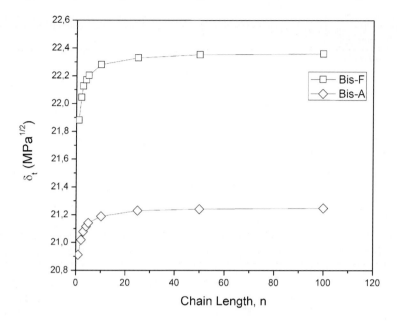

Fig. 1 Comparison of the chain length dependence of the solubility parameters of linear copolymers of epichlorohydrin with diglycidyl ether of bisphenol A (Bis-A) and Bisphenol F (Bis-F)

3.2.1 The Polar Contribution to the Solubility Parameter

The exact value of the solubility parameter is determined by a number of molecular interaction components which contribute to the cohesive energy. The intermolecular forces consist of dipole, polar, and hydrogen bonding. These contribute to the total solubility parameter (δ_t) which can be calculated according to Equation 16:

$$\delta_t = \left(\delta_d^2 + \delta_p^2 + \delta_h^2\right)^{1/2} \tag{16}$$

where δ_d (dipole δ), δ_p (polar δ), δ_h (hydrogen δ) are the dipole, polar, and hydrogen bonding contributions to the solubility parameter.

Table 2 illustrates how these components can be estimated.

Each molecular contribution is identified and assigned a known value. The hydrogen-bonding component is relatively low for DGEBA high molecular weight polymer. Figure 2 shows how the hydrogen-bonding component of DGEBF increases with the degree of polymerisation. This arises from the additional hydroxyl groups in polymer skeleton at high values of n. This occurs at the expense of the dipole component (Figure 3).

Table 2 The dipole, polar, and hydrogen-bonding components of the solubility parameter of DGEBA high polymer

Groups Present in Repeat Unit	No. of Groups in the Repeat Unit
$- CH_2 -$	2
$- CH -$	1
Phenyl group	2
$- O -$	2
$- OH$ Hydrogen Bonded	1
$- C -$	1
$- CH_3$	2

Δ	δ in $(cal/cm^3)^{1/2}$	δ in $(J/cm^3)^{1/2}$
δ_t	10.39	21.26
δ_d	7.93	16.22
δ_p	5.41	11.07
δ_h	3.98	8.13

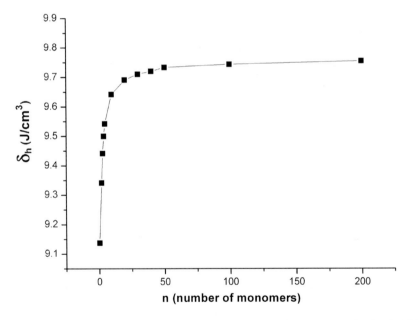

Fig. 2 The change in hydrogen-bonding component of the solubility parameter (δ_h) for the high polymer of DGEBF

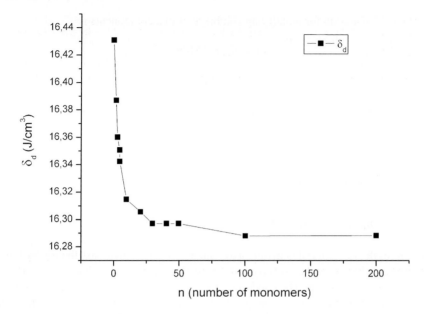

Fig. 3 The change in the dipole component of the solubility parameter (δ_d) for the high polymer of DGEBF

Table 3 Calculated solubility parameters (δ_t) and the hydrogen-bonding component (δ_h) for a range of epoxy resin segments in linear and cured networks

Epoxy Monomeric Segment	$\delta_t(MPa^{1/2})$	$\delta_h(MPa^{1/2})$
DGEBF	22.36	9.75
DGEBA	21.26	8.13
TGAP	22.92	10.34
TGDDM	21.84	8.88
DDS	27.18	9.18
TGAP/DDS(2:1)	24.06	10.25
TGDDM/DDS(5:1)	22.30	8.95
TGDDM/TGAP/DDS(19:13:10)	22.89	9.4
DGEBA/NMA(1:2)	24.00	–

TGAP = Triglycidyl p-amino phenol
TGDDM = Tetraglycidyl diamino diphenyl methane
DDS = p-diamino diphenyl sulphone

3.3   The Solubility Parameter of Cured Epoxy Resins

Epoxy resins used for aerospace composites are often blended to provide optimum processing conditions. The solubility parameters of a range of epoxy network segments in cured resins have been calculated. Table 3 illustrates how δ_t differs for the differing epoxy monomers and on curing.

The rule-of-thumb for identifying solubility of two components is

$$\delta_1 = \delta_2 \pm 2\,\text{MPa}^{1/2}$$

Therefore, we can assume that a high molecular weight poly bisphenol A-co-epichlorohydrin (DGEBA with $n > 20$) would be soluble in most of the cured epoxy resins. In which case, the polymer can diffuse throughout the cured epoxy resin network and promote healing.

3.4 Assessment of Healing in Resins

3.4.1 Compact Tension Test

The strength of a material is mainly associated with a statistical distribution by flaws of differing size. A fracture mechanics test, which can be used to determine a critical strain energy release rate (G_c) and/or a critical stress intensity factor (K_c) is a more appropriate method of determining a material property representative of strength. Thus, self healing efficiency is best quantified in this way. The compact tension test is preferred for cast resins because of the more manageable dimensions of the specimens. Even these are difficult to cast and machine into defect free coupons of standard dimensions. Thermosetting resins have exothermic curing reactions, and heat dissipation is slow because of their insulating properties. Most thermosets are applied in a "thin" film form or as a matrix in a fibre composite, where the local volume of resin is significantly lower than in a resin casting. The compact tension test coupon is described in Figure 4 where the British Standard test dimensions are given [23].

To ensure that the coupons of the more reactive curing systems were bubble and defect-free, it has proved necessary to reduce the thickness of the casting from 12 to 8 mm. However, the conditions for plain strain in the plain of the crack are still maintained.

For healing, the fractured specimens need to be precisely aligned with fracture surfaces in perfect contact. While this can be achieved using a healing system where light pressure can be applied, it is better to use a mechanism which prevents complete separation of the specimen. This can be achieved by introducing a 3 mm diameter hole in the plane of the growing crack at 8 mm from the tip of the notch. This approach was introduced in ceramics studies and also used by Chen et al. [8, 9].

3.4.2 Charpy Impact Test

The preparation of compact tension test specimens is time consuming so that Hayes et al. [10, 11, 23] have examined the Charpy Impact Test as a potential method for rapid examination of the variables; healing agent concentration, healing temperature, healing time and repeated healability.

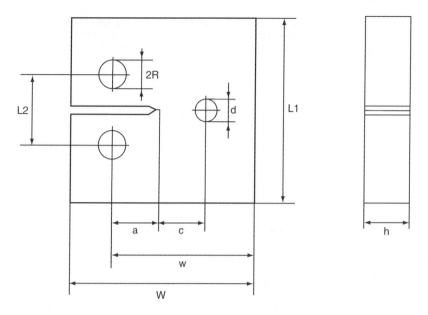

Fig. 4 Compact tension sample geometry and dimensions (mm) W $=$ 30, w $=$ 24, L1 $=$ 28.8, L2 $=$ 8, R $=$ 3, h $=$ 12 or 8, a $=$ 8, d $=$ 3, c $=$ 8

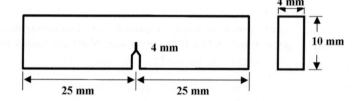

Fig. 5 Dimensions of the Charpy specimens

Typical dimensions of a Charpy Impact Specimens are given in Figure 5.

In the experiments reported here, a Hounsfield Charpy Plastics Impact Tester with an appropriate hammer was used. The instrument was calibrated and set up with minimal friction so that a zero value of friction was routinely obtained on a swinging hammer. These resins are quite brittle so that it was possible to ensure that the energy in the hammer was imparted to the specimen in fracture. Healing was achieved by aligning the two halves of the specimen in the healing bench as shown schematically in Figure 6. A gentle pressure of $\sim$0.2 MPa was applied to the specimens.

3.4.3 Single Edge Notch Tests

Single edge notch (SEN) specimens have also been examined for healing efficiency. The SEN test is a fracture mechanics test utilising simple 3-point loading and can be

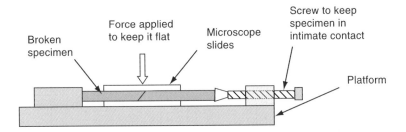

Fig. 6 Schematic of the healing platform for the fractured Charpy impact specimens

undertaken with smaller specimens. However, this methodology proved ineffective because of difficulties in realignment of the fracture surfaces. No data is presented here but future students should learn from our experience.

3.5   Assessment of Healing in Fibre Composites

Since the damage accumulation in most composite systems involves matrix cracks, the healability of resin is applicable to a fibre-reinforced material. However, these cracks usually hug the fibre–matrix interface where the interphase properties dominate so that the healing of composites needs to be studied. To study matrix or transverse cracking, it is best to use appropriate micromechanics tests where these form in a controllable manner. There are three laminate configurations in which matrix cracks form without the complete fracture of the composite:

(i) 0° unidirectional composite with brittle matrix so that multiple matrix cracks form on tensile loading. This is only applicable to high-failure strain fibres (such as glass) in a low strain-to-failure polymer matrix (such as a phenolic resin).

(ii) Angle ply laminates such as a 0°|90°|0° with high 90° ply thickness so that instantaneous transverse cracking occurs in the 90° ply on tensile loading. Multiple cracks form which are usually pinned at the 0|90° interface.

(iii) Angle ply laminates with thin plies. With 0°|90°|0° laminates in which the plies are highly dispersed transverse cracks are initiated at the coupon edge on tensile loading and propagate slowly across the width of the coupon. Smith and co-authors [26] have applied fracture mechanics to the growth of transverse cracks and therefore offer potential for healing studies which provide a more quantifiable approach.

For (i) and (ii) measurements of transverse crack density between healing and loading cycles provides a simple means of quantifying the healability of a composite. The major disadvantage is that with thermally induced healing, the change in thermal stresses within the composite need to be measured as well. In addition, for (iii) glass fibres in transparent resins are required so that the transverse crack growth can be monitored.

Multiple matrix (or transverse) cracking has been considered in detail in the seminal work of Kelly [27] and Bailey [28].

Using these microphenomena to study matrix healing in a composite has much potential.

However, most damage in composite structures is likely to involve impact loads where matrix or transverse cracks initiate delamination between plies. Delamination occurs because of the shear stresses involved in the stress transfer mechanism for reloading of the cracked 90° or 0° ply via the adjacent plies. However, the healing systems described here will have differing potential for the healing of transverse or matrix cracks as well as delaminated regions. Therefore, the healing of matrix cracks and delaminations in composites after impact, will also need study. For aerospace composites, the compression-after-impact test, has much potential but is highly demanding of a self healing system.

4   Solid-State Healing

4.1   Preliminary Compact Tension Test Data

Hayes et al. [23] used the solubility parameter concept to identify a linear-healing agent for a cross-linked epoxy resin, as described in Section 3.2. A bisphenol A based epoxy resin has been thermally healed by dissolving in it polybisphenol A-co-epichlorohydrin (Sigma Aldrich). The resin has the formulation shown in Table 4.

Figure 7 shows the relative recovery in fracture toughness, defined by the critical stress concentration factor (K_{1c}) obtained by compact tension testing of the cured resin. About 20% of the healing agent was included in the resin and each coupon was sequentially fractured three times and rehealed at each temperature for 1.5 h. The specimens at 140°C have therefore been repeatedly healed at least 15 times.

It is seen that the healing agent has reduced the fracture toughness K_{1c} value of the resin by $\simeq$ 20%. So that, at 20%, loading the matrix appears to have been plasticised. However, the resin containing the healing agent exhibits a recovery of 64% (K_{1c}) and 77% (G_{1c}) of the virgin properties after sequential healing.

Preliminary studies of the effect of concentration of healing agent are shown in Figure 8. Even with 5% of the healing agent, 37.5% of the initial properties were recovered.

4.2   Optimisation of Healing Agent Concentration

In this study, Charpy impact testing has been used to identify the optimum concentration of the healing agent. The resin used for this study was also an analogous DGEBA resin as shown in Table 4 except that the aliphatic epoxy resin GY298 was omitted (Resin B).

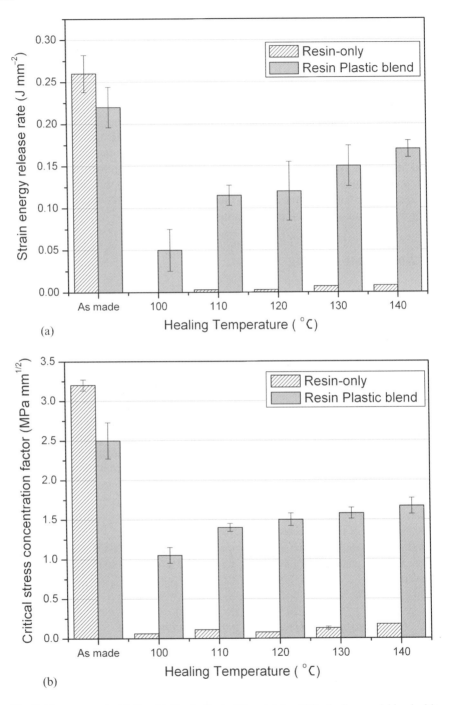

Fig. 7 The recovery in (**a**) G_{1c} (**b**) K_{1c} for Resin A2 containing 20% of a linear epichlorohydrin-co-bisphenol A polymer as a function healing temperature. (From Hayes et al. [11].)

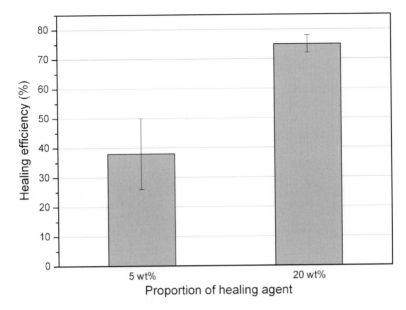

Fig. 8 Healing efficiencies of Resins A1 and A2 containing 5% and 20% of the linear polymer. Healing cycle was 1 h at 130°C. (From Hayes et al. [11])

Table 4 DGEBA-based epoxy resins used for healing studies. The resins were cured for 4 h at 80°C and post cured for 3 h at 130°C.

Component	A1 Resin A2 (phr)		Resin B (phr)
Epikote 828	–	–	100
Araldite LY1556	60	60	–
Araldite GY298	40	40	–
NMA	64	64	81.2
Capcure 3–800	24	24	30.4
Healing agent (%)	5	20	(2.5 – 15)
Cure temp (°C)	80	80	80
Cure time (h)	4	4	4
Post-cure temp (°C)	130	130	130
Post-cure time (h)	3	3	3

Figure 9 shows that the healing efficiency appears to be highest at a concentration of 7.5%. In this case the healing conditions were 6 h at 130°C.

Furthermore, it is observed that the resin without the healing agent also showed some residual healability. This could be attributed to post-curing since the initial cure schedule involved 3 h at 130°C and healing was given 6 h at 130°C. To assess the actual healing efficiency the following analysis has been used:

$$R_E = \frac{100 E_{heal}}{E_{init}} \qquad (17)$$

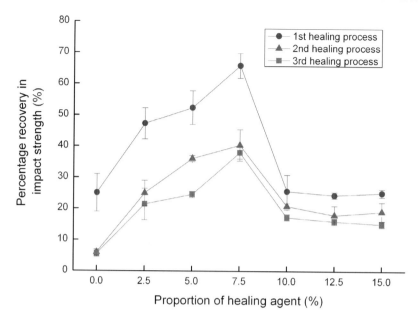

Fig. 9 The percentage recovery in Charpy impact strength of Resin B (R_E) as a function of the healing agent concentration from 0% to 15% by weight. Each sample has been subjected to three impact/heal cycles as shown. The healing conditions were 6 h at 130°C. (From Hayes et al. [23])

where R_E is the percentage recovery in impact strength, E_{init} and E_{heal} are the initial impact strength and post-healed impact strength.

To account for residual healing in the control resin, the healing efficiency (H_E) can be calculated from Equation (18).

$$H_E = R_E - R_E^o \tag{18}$$

where R_E^o is the residual healing of the unmodified resin. Table 5 indicates how the healing efficiency can be corrected for residual healing ability of the unmodified resin. Table 5 shows the data across a range of healing conditions.

Using this analysis, Table 5 shows that the Charpy impact energy measurements demonstrate that 7.5% of the healing agent provides the maximum healing and that this can be achieved using either 6 h at 130°C or 1 h at 160°C when 43% and 50% recovery of impact energy independent of residual healability. With all of the healing schedules, subsequent recovery of impact energy (R_E) is reduced. Figure 10 shows that even with the 1 h at 160°C, the recovery in Charpy impact strength is reduced after two further cycles of impact and heal. After the second healing H_E is still 40%.

4.2.1 Recovery of Fracture Toughness during Healing

There is some concern that rapid scanning of healability using Charpy testing may be lead to poor conclusions. Therefore, the experiments have been repeated using

Table 5 Recalculated healing efficiencies of resin B as a function time, temperature, and healing agent concentration

	R_E^o				R_E				H_E			
Healing temp (°C)	130	130	130	160	130	130	130	160	130	130	130	160
Healing time (h)	1	6	10	1	1	6	10	1	1	6	10	1
Healing agent concentration (%)												
0	15	25	28	16	–	–	–	–	–	–	–	–
2.5	–	–	–	–	41	49	40	23	26	24	12	7
5	–	–	–	–	47	54	44	60	32	29	16	44
7.5	–	–	–	–	41	68	61	66	26	43	33	50
10	–	–	–	–	27	25	24	36	12	0	0	20
12.5	–	–	–	–	26	25	24	29	11	0	0	13
15	–	–	–	–	21	25	24	25	6	0	0	9

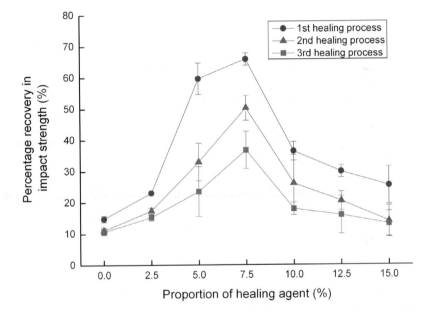

Fig. 10 The percentage of recovery in Charpy impact Strength (R_E) of Resin B as a function of healing agent concentration. Each sample has been subjected to three impact/heal cycles. Healing was accomplished after 1 h at 160°C. (From Hayes et al. [23])

the compact tensile test. Furthermore, the healing time at 130°C has been reduced to below that used for the original cure schedule.

The value of K_{1c} for the unmodified resin was 0.76 ± 0.01 MPa m$^{1/2}$. The addition of 5–15 w% of the healing agent did not influence the value significantly. However, the recovery in K_{1c} was found to be also optimised at a healing agent concentration of 7.5%. Using a similar approach to Equation (18) where

$$H_K = R_k - R_k^o \qquad (19)$$

H_K is the healing efficiency calculated from the recovery in K_{1c}.

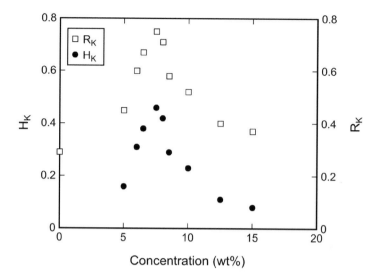

Fig. 11 Fractional healing efficiency (H_K) of Resin B containing a range of healing agent concentration from the recovery in compact tension fracture toughness (R_K). The healing period was 3 h at 130°C

Table 6 The recovery in critical stress concentration factor (K_{1c}) for resin B after healing for 3 h at 130°C. The effect of healing agent concentration

Concentration of Healing Agent (%)	R_K	H_K (%)
0	0.29	–
5	0.45	15
6	0.60	31
6.5	0.67	38
7.5	0.75	46
8	0.71	42
8.5	0.58	29
10	0.52	23
12.5	0.40	11
15	0.37	8

Figure 11 and Table 6 show the change in healing efficiency, H_K with healing agent concentration. It demonstrates that with a healing agent concentration of 7.5%, 46% recovery in K_{1c} was recorded independent of any residual healing ability. About 29% recovery in K_{1c} was recorded for the non-healable matrix despite a healing period of 3 h at 130°C even on top of post-curing of the coupons for 3 h at 130°C.

With a healing schedule of 2 h at 130°C the unmodified non-healable matrix also exhibited 22% recovery in K_{1c}. However, a maximum recovery of 61% was observed for the modified resin containing 8% of the healing agent. This gives a healing efficiency of 39%.

Interestingly, the lower healing efficiency appears to be associated with residual healing capability of the unmodified control resin. Further studies are required to understand of this phenomenon.

5   Healing of Damage in Fibre Composites

A glass fibre composite was prepared by accurate dry filament winding onto square frames to form a $0°|90°|0°$ configured preform for resin infusion by Resin B (Table 4). The panel was then vacuum bagged. After curing at 80° for 2 h and post-curing at 130°C for 3 h, the laminates were cut into seven $60 \times 60 \, mm^2$ coupons. Two laminates were prepared from the control resin and the self healing resin containing 7.5% of the healing agent. Each coupon was supported on a 50 mm ring and subjected to an impact event using a falling dart with hemispherical tip of 10 mm diameter. The impact energy was fixed at 2.6 J and care was taken to ensure there were no double impacts. For sequential impact events the coupons were carefully realigned for each of three impacts and two healing cycles. The area of the impact damage was recorded using a photomicrograph and recorded (Zeiss KS400 image analysis suite). The healing cycle involved heating of the plaque to 130°C for 2 h. The damaged area after impact was normalised to the initial damaged are of the first impact.

In this way, the relative change in impact damage are healing could be compared. Figure 12 shows how the healing stage reduced the damaged area by about 30% after each impact event.

It is also observed that the delamination area propagates during successive impacts but the normalised damaged area for the non-healable samples was always higher than that for the healable samples. It was also noticed that transverse cracks were readily healable as were the regions of the delaminations furthest away from the location of the impact event [23]. These observations are consistent with the ability of diffusional solid-state healing to mend resin cracks, within a fibre composite.

6   A Smart Composite System

For a truly smart composite system it is necessary to locate the damaged regions of the material and apply a suitable heating cycle locally to induce healing, Hayes and Hou [29] have reported that the carbon fibres which make up the load-bearing component of an aerospace composite can be used to detect impact damage. By designing the sensor so that adjacent plies of differing configuration were electrically isolated, measurement of electrical conductivity along the carbon fibres in a triangulation routine can identify not only fractured fibres but delaminated regions. Whereas the high-impact energy regions where fibres are fractured, increase in resistance, regions

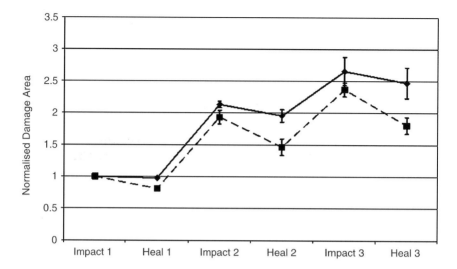

Fig. 12 The effect of repeated drop weight impact (2.6 J) and healing cycles on glass fibre composites with a heal (dashed) and non-healed (solid) matrices. The rate of growth of delamination area is lower for the system containing 7.5% of the healing agent in the matrix. (From Hayes et al. [23].)

Fig. 13 Schematic of the self-sensing arrangement for a carbon fibre laminate. Interleaves electrically isolate the sensing plies, facilitating damage location. The complete lay-up sequence used in this study was: $[0_2/R/90_2/R/0_3/90_6/0_3/90_2/0_2]$

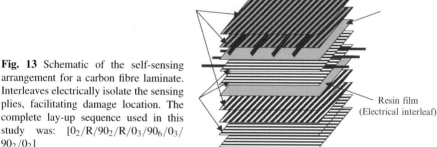

where the interfibre contacts are disrupted, the resistance decreases. Figure 13 illustrates a typical configuration for self sensing.

Figure 14 gives a typical pattern of impact damage sensed by the sensor arrangement shown in the previous Figure 13.

After impact damage, the self-sensing carbon fibres show both negative and positive changes in resistance. The former indicate delaminations while the latter the more characteristic fibre fracture. Full details of this self-sensing approach is given elsewhere by Hayes and Hou [29].

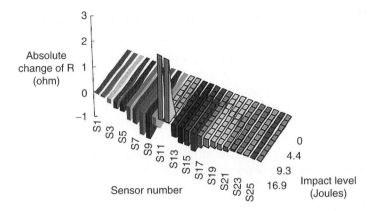

Fig. 14 Typical pattern of impact damage in a cross-ply laminates with a sensor density of 3 mm indicating damage size of 55.5 ± 1.5 mm for energy level of 9.3 J (Sensors: 3–21) 34.5 ± 1.5 mm for energy level of 4.4 J (Sensors: 6–17)

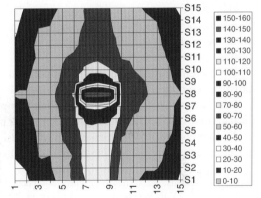

Fig. 15 The temperatures (in °C) achievable using the resistance of a damaged carbon fibre laminate of configuration given in Figure 13

6.1 Healing Activation

This paper describes epoxy resin matrices which can be healed by the addition of a linear polymeric healing agent. Healing is achieved by heating to a specific temperature, which mobilises the diffusion of the healing agent. Since impact damage is likely to be local and can be detected using self-sensing carbon fibres or other conductive reinforcing fibres, it follows that the increase in electrical resistance which arises at grossly damaged areas can be used to provide a localised heating suitable healing damage. To demonstrate this effect, a damaged panel was subjected to electrical current in an analogous way to the sensor in Figure 13.

Figure 15 shows that the high-resistance areas associated with impact damage can be used to provide a localised heating. Temperatures sufficient to activate healing were achieved.

6.2 The Concept of the Smart Self Healing Self-Sensing Composite

By combining the self-sensing ability of conductive reinforcing fibres with a heat-activated healable resin system, it is possible to design a smart material that can detect damage and apply a localised healing to the damaged region. To achieve this, the contacts used for measuring resistance need to be continuously monitored to provide damage detection. A protocol can be designed to monitor damage and activate healing.

All of the components of this smart material have been described in this article. Further work is in hand to develop this concept.

7 Conclusion

The mechanisms of self healing of polymeric materials have been described. In particular, the concept of a healable composite has been described in detail. The key to this is the development of a solid-state healing system for thermoset epoxy resins. It has been shown that the healing efficiency can be optimised at concentration of healing agent of 7.5% by weight.

The current healing agent is a linear polymer of similar solubility parameter to the cross-linked network, such as a poly bisphenol A-co-epichlorohydrin.

Both compact tension testing and Charpy impact testing can be used to assess the healability of the resins, although the former is much preferred.

A smart composite which combines self-sensing with self healing with an inherent means of activating the healing agent is described.

Acknowledgements We acknowledge financial support from the University of Sheffield, HEIF funds, Airbus UK, EPSRC through the Ceramics and Composites Laboratory. We also acknowledge the contributions of Liang Hou, Michael Branthwaite, Kamlesh Marshiya, Srikanth Hari, and Joel Foreman.

References

1. Dry C. (1996) Procedures developed for self repair of polymer matrix composites. Compos Struct 35:263
2. Bleay S.M., Loader C.B., Hawyes V.J., Humberstone L., Curtis P.T. (2001) A smart repair system for polymer matrix composites. Composites Pt A 32:1767
3. Pang J.W.C., Bond I.P. (2005) Bleeding composites damage detection and self-repair using baiomimetric approach. Composites Pt A 36:183
4. Pang J.W.C., Bond I.P. (2005) A hollow fibre reinforced polymer composite encompassing self-healing and enhanced damage visibility. Compos Sci Technol 65:1791
5. White S.R., Sottos N.R., Geubell P.H., Moore J.S., Kessler M.R., Sriram S.R., Brown E.N., Viswanathan S. (2001) Autonomic healing of polymer composites. Nature 409:794
6. Kessler M.R., Sottos N.R., White S.R. (2003) Self-healing structural composite materials. Composites Pt A 34:743

7. Brown E.N., White S.R., Sottos N.R. (2004) Microcapsule induced toughening in a self-healing polymer composite. J Mater Sci 29:1703
8. Chen X., Dam M.A., Ono K., Mal A., Shen H., Nutt S., Sheran K., Wudl F. (2002) A thermally remenable crosslinked polymeric material. Science 295:1698
9. Chen X., Wudl F., Mal A.K., Shen H., Nutt S.R. (2003) New thermally remendable highly crosslinked polymeric materials. Macromolecules 36:1803
10. Hayes S.A., Jones F.R. (2004) Self-healing composite materials. UK patent application GB0500242.3
11. Hayes S.A., Jones F.R., Marshiya K., Zhang W. (2007) A self-healing thermosetting composite material. Composites Pt A 38:1116
12. Szwarc M. (1968) Carbanions, living polymers and electron transfer processes. Interscience, New York
13. Dainton F.S., Ivin K.J. (1958) Quart Rev 12:61
14. Hoecker H., Jones F.R. (1972) Some aspects of the metathesis catalyst, Die Makromolekulare Chemie 161:251
15. Calderon N., Ofstead E.A., Ward J.P., Judy W.A., Scott K.W. (1968) J Am Chem Soc 90:4133
16. Grubbs R.H., Tumes W. (1989) Polymer synthesis and organotransition metal chemistry. Science 243:907
17. Flory P.J. (1953) Principles of polymer chemistry. Cornell University Press, Ithaca
18. Flory P.J. (1969) Statistical mechanics of chain molecules. Interscience, New York
19. Jacobson H., Stockmayer W.H. (1950) J Chem Phys 18:1600
20. Semlyen J.A. (1976) Ring-chain equilibria and the conformations of polymer chains. Adv Polym Sci 21:41
21. Eisenberg A., Kim J-S. (1998) Introduction to ionomers. Wiley, New York
22. Jones F.R. (ed) (1994) Handbook of polymer-fibre composites. Longman, Harlow, UK, pp 86–96
23. Hayes S.A., Zhang W., Branthwaite M., Jones F.R. (2007) Self-healing of damage in fibre reinforced polymer-matrix composites. JR Soc Interf 4:381
24. Hoy K. (1970) New values of solubility parameter from vapour pressure data. J Paint Technol 42:76
25. BS150 13586 (2000) Plastics – determination of fracture toughness. British Standards Institute
26. Smith P.A., Ogin S.L. (1987) ESA J 11:45
27. Aveston J., Kelly A. (1973) J Mater Sci 8:352
28. Bailey J.E., Curtis P.T., Parvizi A. (1979) Proc R Soc Land 366:599
29. Hayes S.A., Hou L. (2002) Smart Mater Struct 11:966

Ionomers as Self Healing Polymers

Russell Varley

Commonwealth Scientific Industrial Research Organisation (CSIRO), Manufacturing and Materials Technology
E-mail: Russell.Varley@csiro.au

1 Introduction

Ionomers are a class of polymers which have up to 20 mol% of ionic species incorporated into the structure of the organic polymer. These ionic species create interactions or aggregates [17, 28], not present in comparable non-ionic polymers that have a profound effect upon the mechanical and physical properties of the polymer. As a result there has been much research in both academia and industry over the last 30 years [4, 22] aimed at increasing the understanding of structure–property relationships in these polymers while also exploring new commercial applications. The self healing phenomenon exhibited by ionomers is a particularly interesting property arising from their unique chemical structure and is the subject of this chapter.

Despite the long-standing commercial availability of ionomers and the comprehensive structural characterisation of ionomers, there has been comparatively little research carried out (at least in the open literature) on their self healing characteristics [1, 8, 15]. This property in ionomeric systems occurs during high-energy impact or puncture where the cavity created by a projectile is immediately closed and sealed once the projectile has exited the polymer. This chapter, therefore, seeks to address the lack of information in relation to the self healing of ionomers by outlining the current understanding, research directions, and future applications for self healing ionomeric polymers. In addition to this, the chemical structure, morphology, and structure property relationships, which underpin the self healing process, will also be discussed.

2 Ionic Aggregation

The appearance of the "ionic peak" from SAXS, arising from the partial neutralisation of the acidic copolymers, has been used to determine the size of the ionic domains or "aggregates" within the polymer structure. Interpreting the meaning of this peak has provided significant challenges from the perspective of the chemical structure and morphology of the polymer. A range of models have been proposed to account for this peak [14, 16], but the most widely accepted is the Eisenberg–Hird–Moore (EHM) model [3, 5]. In this model the primary aggregates consist of several

S. van der Zwaag (ed.), *Self Healing Materials. An Alternative Approach to 20 Centuries of Materials Science*, 95–114.
© 2007 *Springer*.

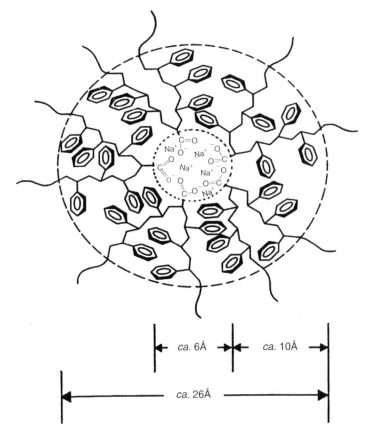

Fig. 1 Schematic representation of a multiplet and the region of restricted mobility in the surrounding polymer. (From Eisenberg et al. [3])

ion pairs known as *multiplets*. The size and number of ion pairs in any given multiplet will be governed by steric factors, such as the flexibility of the polymer chain as well as the size of the ion pairs, and the dielectric constant of the polymer backbone. An important factor arising from the multiplets is the formation of regions of restricted mobility. Each ion pair within the multiplet has two chains connected to it, so a multifunctional cross-linking site or "physical crosslink" is formed. A schematic representation of a multiplet and the region of restriction of mobility are shown in Figure 1.

The level of restriction in mobility will depend upon a number of factors, such as the anchoring of the polymer chain to the multiplet, molecular weight, crowding of the chains and chain extension in the vicinity of the multiplet.

A lone multiplet will tend not to have its own glass transition temperature as it will be too small to have separate phase behaviour. As can be seen from Figure 1, the total dimension and the region of restricted mobility will only be around ∼30 Å, well

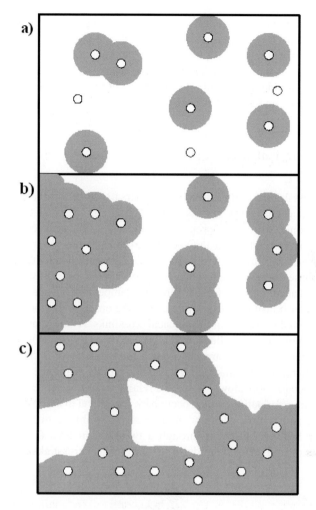

Fig. 2 Schematic representation of the morphologies of ionomers at increasing ion content (**a**) low ion content, (**b**) intermediate ion content, and (**c**) high content, showing the increasing influence of the restricted mobility region on morphology. (Image created by Varley [27])

below the approximate length of ~50 Å required to have separate phase behaviour. However, when the concentration of ionic species increases sufficiently such that the regions of restricted mobility begin to overlap, the threshold for separate phase behaviour is reached and a separate glass transition temperature can be observed. When this point is reached, the extended region of restricted mobility or "group of multiplets" is termed a *cluster* [2]. Figure 2 shows how the clusters (shaded areas) with increasing ion content increasingly begin to dominate and become the continuous phase. This combined effect of the multiplets and the variety of possible morphologies, is a key parameter controlling the enhancements in the mechanical and physical properties, such as the glass transition temperature, modulus, viscosity, and others. In this way ionomers can be viewed as being analogous to thermosetting polymers except in the case of ionomers the network is created from physical cross-linking that is thermo reversible in nature. The reversible nature of these physical crosslinks provides an insight into one aspect of the self healing mechanism that ionomeric systems undergo.

3 Morphology of Ionomers

Another process critical to the ability of ionomers to self heal is the understanding of the morphological changes which occur during heating. The order-to-disorder transition model of ionic clusters was proposed by Tadano et al. [24] to account for changes observed during thermal expansion and calorimetric studies. As shown in Figure 3, the model consists of three different phases that are controlled by temperature.

The first phase or the ordered state consists of ordered ionic clusters, polyethylene crystallites and an amorphous region. When the temperature is increased to above the order to disorder transition temperature (T_{dis}), the order within the ionic aggregates is lost along with the strength of the physical crosslinks. As the temperature is further increased, the polymer crystallites melt (T_m) even though the disordered aggregates persist in the molten state. Conversely, when the temperature is cooled from above the melting point, re-crystallisation takes place rapidly while the reordering of the ionic cluster occurs more slowly through a relaxational process.

Again, the transformations which occur here are critical to the self healing process. After ballistic penetration, when the thermal energy of impact dissipates, this model suggests that healing can occur through a two-step process: (1) solidification or re-crystallisation and then (2) reordering of the physical crosslinks. According to the

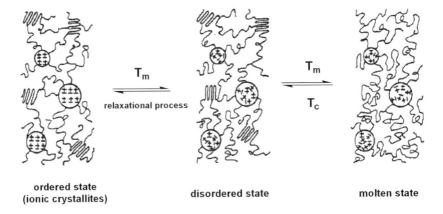

Fig. 3 Phase transition diagram highlighting the different transformations that the semi-crystalline polymer and ionic domains undergo during thermal activation. (From Tadano et al. [24])

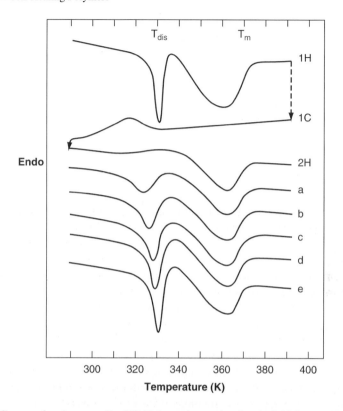

Fig. 4 DSC curves for zinc neutralised EMAA copolymer showing the initial trace (1H) followed by cooling and subsequent heating after 0 h (2H), 5 h (**a**), 1 day (**b**), 3 days (**c**), 9 days (**d**), and 38 days (**e**) at room temperature. (From Tadano et al. [27])

proposed model, the re-crystallisation will occur rapidly but the polymer will not return to full or near original strength until the reformation of the physical crosslinks which occur more slowly. Evidence for this two-stage transformational process is provided by differential scanning calorimetry (DSC) as shown in Figure 4.

Here, two distinct peaks, consisting of the above-mentioned transitions, T_{dis} and T_m are evident after heating. After cooling down and reheating again at increasing time intervals, T_m was shown to be unaffected, while T_{dis} was depressed initially but slowly returned over periods of hours and days at room temperature.

4 Dynamic Mechanical Properties of Clustered Ionomers

As discussed previously, the impact of ionic aggregation has a profound effect upon polymer properties. The dynamic mechanical properties of ionomers, were first studied by MacKnight et al. [13] who inferred structural phase information in relation to

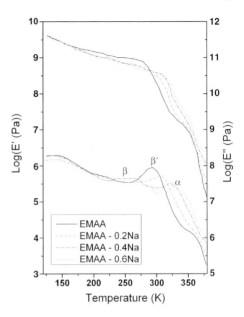

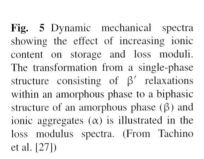

Fig. 5 Dynamic mechanical spectra showing the effect of increasing ionic content on storage and loss moduli. The transformation from a single-phase structure consisting of β′ relaxations within an amorphous phase to a biphasic structure of an amorphous phase (β) and ionic aggregates (α) is illustrated in the loss modulus spectra. (From Tachino et al. [27])

the network for varying levels of ionisation. Recent work performed by Tachino et al. [23] interpreted these transitions further and showed that increasing ionic aggregation transformed the network morphology from a single-phase polymer to a biphasic structure. Figure 5, shows the dynamic mechanical properties (loss and storage modulus) for an EMAA copolymer at increasing levels of neutralisation which illustrates these changes.

As shown, the loss modulus spectra for the non-ionic parent polymer EMAA, exhibits a single β′ peak (as noted on the plot) corresponding to the short-range motion of the segmental chains in the amorphous region. As the ionic content increases and ionic aggregates form, this single transition is split into a β transition related to the amorphous non-ionic regions and an α relaxation related to the glass transition of the ionic cluster, (i.e. the order-to-disorder transition temperature T_{dis}). The effect of increasing ionic strength is also evident with the increasing temperature T_{dis}, highlighting the increasing strength of the physical crosslinks and its effect upon thermal properties. The dynamic mechanical spectra therefore can be seen to support to the model proposed by Tadano et al. [24] providing evidence for the existence of the polymer having a non-ionic amorphous phase at low temperature, a melting point at higher temperature and an intermediate region containing an order-to-disorder transition. The analogous relationship to a cross-linked thermoset is also apparent from this description.

5   Ion Hopping

Another fundamental process that facilitates self healing in ionomer systems is the concept known as *ion hopping*. SAXS characterisation has shown [18, 29] that ionic aggregates persist in the melt (although disordered), up to 300°C and greatly increases the melt strength or viscosity compared to the parent non-ionic polymer. A balance of ionic associations however is required, so that ionomers can be processed conveniently, as too many aggregates render the polymer intractable, while also maintaining an elastomeric behaviour in the melt. To achieve this balance of flow and elastic behaviour ion hopping [25] is an important requirement. A schematic representation is shown in Figure 6 [26] where the ionic species within a given polymer are observed to "hop" from one aggregate to another.

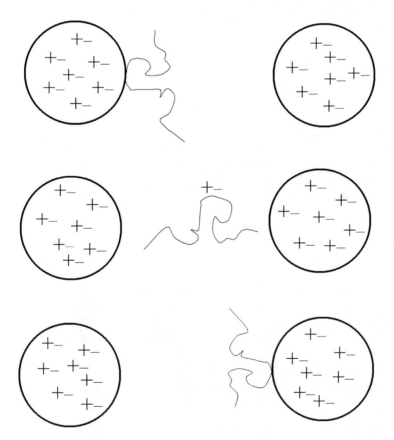

Fig. 6 Schematic representation of the "ion hopping" relaxation mechanism in ionomers. The circles represent the ionic clusters containing ionic groups. The curved line in the centre represents a charged group on the polymer chain "hopping" from one cluster to another. (From Vanhoorne and Register [26])

Ionic associations are dynamic, in that they have a finite lifetime τ, which is a measure of the average length of time an ionic group spends in a particular aggregate. The ions "hop" or diffuse from one aggregate to another thus allowing the relaxation of the chain segment of the macromolecule to which the ionic group is attached. With this mechanism occurring over the length of a macromolecule at random time intervals, the entire polymer can diffuse and flow without requiring all of the ionic associations to be released simultaneously. In relation to the self healing process, this provides the physical mechanism to allow the polymer structure to exhibit elastomeric behaviour in the melt while also maintaining an adequate level of processability.

6   The Self Healing Phenomena

As mentioned previously, self healing in ionomeric systems is activated during high-energy impact or puncture such as bullet penetration, as shown schematically in Figure 7.

It is therefore dissimilar to other self healing processes, which automatically sense a void, puncture, crack, or some other kind failure and heal themselves. However, because the process arises from the inherent chemical structure and morphology of the polymer, (i.e. there are no added "healing agents") the process can in principle be repeated many times and in that way is distinctive from other systems which can only repair themselves once (i.e. the systems where there is an added "healing agent"). The Dupont range of partially neutralised ethylene methacrylic copolymers, known under the trade name of Surlyn, possess all of the structural requirements discussed above,

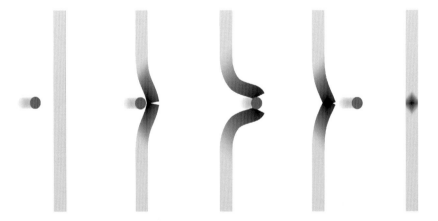

Fig. 7 Schematic representation of the self healing process which during impact and subsequent penetration by a high-energy projectile. The effects of impact, deformation and the elastic rebound and hole closure are illustrated. (From Freeling DCMat (2007))

with some specific grades having the appropriate balance of properties that facilitates the self healing process for a given set of experimental conditions.

Although ionomers have been extensively studies for many years, their application as self healing materials is much less widely investigated. The most significant experimental investigations of the self healing mechanism of ionomeric systems, available are those of Fall [6] and Kallista [10]. From their studies, the self healing process can be summarised as follows:

- Impact:
    - High-speed impact of bullet transfers energy to the polymer elastically (storage) and inelastically (dissipative).
    - The heat transferred to the polymer through frictional effects raises the localised temperature of the polymer from the ordered state into the melt.

- Deformation:
    - The elastomeric nature of the melt allows the polymer to be stretched or drawn to very high strain levels to before failure. Finally the polymer is broken and the bullet exits the polymer leaving a cavity.
    - Combined with the elastic (or stored) energy transferred during impact, and the relatively unaffected surrounding regions the ductile polymer is able to elastically rebound to its original position prior to impact.

- Healing:
    - The fragments of the still molten polymer are then able to reform, meld and interdiffuse together again to completely reseal or heal the cavity.
    - Over longer periods of time, the strength of the impact zone returns as the physical crosslinks reorder.
    - The structural integrity of the polymer provides sufficient strength in the melt to prevent other deleterious effects from polymer flow.

The importance of the above-mentioned phenomena in relation to self healing is now evident. Without each of them playing their role, self healing could not occur. The *ionic aggregate*, creates a physically cross-linked structure which enhances mechanical properties at both room temperature and in the melt, while also being thermoreversible. The different *phase morphologies* of ionomers show that self healing can occur rapidly through re-crystallisation but the increased strength takes somewhat longer to occur. The elastomeric nature of the molten polymer arising from the persistence (although disordered) of the ionic cluster, allows the enormous elongation of the polymer and also ensures that the polymer has sufficient molecular mobility in the melt while also retaining structural integrity. Ion hopping then facilitates the processing of the elastomeric polymer by creating the mechanism for the polymer chains to relax. Thus, successful healing of ionomeric polymer systems occurs when all of these processes function in balance, so that when the polymer strands rebound elastically, they will have sufficient molten characteristics to meld and interdiffuse together.

Some important experimental evidence for this mechanism is also discussed as follows. Fall [6] used a thermal IR camera to show that during bullet penetration the localised temperature of the polymer was raised to ∼98°C, while surrounding regions remain at around room temperature. As this was known to be about 3°C higher than the melting point, the impacted polymer is therefore thrust into the molten state briefly during impact. However, the ionic clusters, although disordered, persist and provide the molten polymer with sufficient elastomeric strength, to rebound when the bullet has passed. Another importance aspect of this rebounding process is that the surrounding polymer is not at an elevated temperature, and provides a stable framework or "anchor" for the elastomeric material to "pull against". However, the moment the bullet has passed through the polymer, the temperature reduces rapidly, shifting the equilibrium process back from a molten state (Tadano et al. [24]) to that of the *re-crystallised* state "instantaneously". In addition to this, the reforming of the physical crosslinks was shown to occur more slowly at the impact site, by Kallista [10] using DSC, further reinforcing the relationship to the Tadano model.

Despite the above discussion regarding the importance of the existence of the ionic domains within a polymer to the healing process, both Fall [6] and Kallista [10] were able to show that even weakly aggregated ionic clusters can facilitate self healing. Figure 8a shows typical examples of the healed impact sites for a range of ionomers at varying levels of neutralisation. The Surlyn 8920 and 8940 contain 60% and 30% acid neutralisation respectively, while the Nucrel polymers contain no neutralisation at all. Despite this, the Nucrel systems were shown to be weakly aggregated according to DSC measurements highlighting the powerful effect of aggregation upon polymer properties.

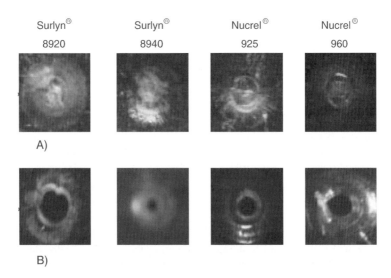

Fig. 8 Optical images of the healing scars from ballistic impact at (a) room temperature and (b) elevated temperature. The improved ability of the Surlyn 8940 to self heal compared to the other polymers s more clearly evident at elevated temperature. (From Kallista [10])

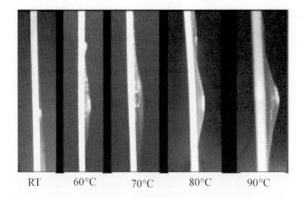

RT   60°C   70°C   80°C   90°C

Fig. 9 Viscoelastic response of the polymer to ballistic impact at varying temperature, showing the increasing plastic drawing or yielding at elevated temperature. (From Kallista [10])

The critical balance between polymer elasticity and flow was also demonstrated by Kallista [10]. Performing ballistic punctures at elevated temperatures was found to reduce the self healing capability of the EMAA ionomers as shown in Figure 8b. Artificially introducing thermal energy into the polymer means that when ballistic impact occurs, the entire polymer (not just the local impact area) is moved much further into the molten state than would typically occur. This in turn facilitates extra drawing and deformation of the entire polymer (Figure 9) compromising the balance between elastic rebound and the capacity of the molten polymer to interdiffuse. The results at elevated temperature do, however, exacerbate the differences between the polymers making comparison easier.

Clearly, for a given polymer there will be an optimum level of ionic aggregation required for self healing to occur as it can be seen here that the Surlyn 8920 with the highest ionic content performs as badly as the Nucrel systems which are only weakly aggregated. The Surlyn 8940 with 30% acid neutralisation appears to have the optimum level of ionic aggregation as the extent of healing is clearly superior. The morphology of the impact and exit points have been further investigated using scanning electron microscopy (SEM) after penetration of Surlyn 8940 by a 9 mm in bullet as shown in Figure 10 [27].

The overview of the crater zone (a) highlights possibly three distinct regions that are affected by ballistic impact such as: (1) a series of crazes radiating around and outwards from the centre of impact, (2) a smooth lip region surrounding the third and final distinct area a crater. Closer examination (i.e. higher magnification) of the outer region (b) shows a rough surface consisting of nodules of the order of around 1–2 μm in diameter. This highlights the level of ductile tearing occurring in these outer regions where the temperature of the polymer is not expected to be increased significantly above ambient compared to the polymer directly impacted. This region therefore retains the majority of its network properties and so the response is more akin to a ductile polymer network. The lip region of the crater (c) can be seen to exhibit very smooth surface indicative of polymer flow, thus confirming that the

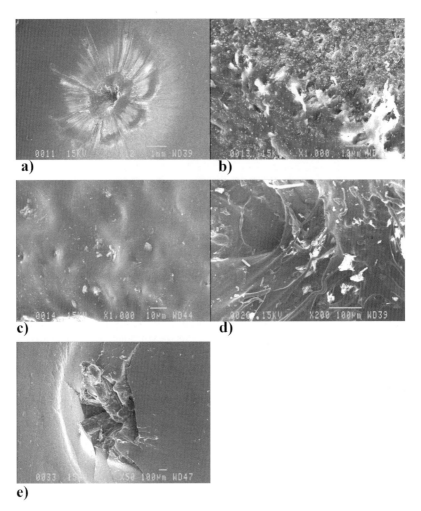

Fig. 10 Scanning electron micrographs of the impact zone after penetration of a 0.22″ bullet. (**a**) Shows an overview of the impact zone highlighting the three distinct regions affected. Higher magnifications of the affected regions are shown in (**b**) the radial crazes, (**c**) the smooth crater lip, and (**d**) crater caused by impact. The exit point of the bullet is also shown in (**e**). (From Varley and Van der Zwaag [27])

impacted region has been melted. Inside the crater (d) this polymer flow is also evident, but in addition to this there are substantial levels of fibrillation evident that show signs of stress relaxation and polymer failure. This shows that the during impact the polymer has undergone significant levels of molten behaviour while at the same time the polymer has been able to respond uniquely to high levels of elongation through fibrillation and stress relaxation. Importantly, while this has been occurring, the polymer that has not been subjected to elevated temperature has been able to hold the polymer together through its higher levels of mechanical strength at ambient conditions. The exit point of the bullet shown in (e) highlights the brittle puncturing of the polymer but the extent to which the polymer is able to elastically rebound after the impact. These micrographs provide microstructural support for the developing understanding of the self healing mechanism for ionomers.

A report by Huber and Hinkley [9] used low-strain rate impression-testing methods to investigate the self healing phenomenon. The strategy behind the investigation was to develop a simple method that could be used to investigate the phenomenon while developing a method to screen new candidate materials. The method essentially consisted of performing penetration tests at different impact rates, using a cylindrical probe that was removed after impact. The effect upon the cavity was then evaluated and monitored with time. It was found that high levels of hole closure (i.e. healing) were achieved at higher impact rates, while the cavity was also found to decrease (slightly) over the course of several days. This again supports the two-phase recovery process as defined by the fast- and slow-relaxational processes. In support of the work of Kallista [10], they also found that performing the indentation at elevated temperatures in fact reduced the level of healing and explained hole closure in terms of the an-elastic recovery of the large strain deformation. This work concluded, similarly to that of the Kallista [10] and Fall [6] studies, that sealing of a bullet hole is facilitated by the melt flow of the impacted polymer but that the recovery of inelastic strain is also important. They concluded that while this recovery can be accelerated by heating, the resultant loss of the stored elastic energy at higher temperature also inhibits healing and emphasises the importance of localised heating rather than overall heating.

Again in support of the findings of the Fall and Kallista work, Huber and Hinkley [9] showed that a polyurethane thermoplastic elastomer was able to demonstrate self healing capacity. This shows that ionic crosslinks or clusters are not the only morphology or structure that can facilitate self healing and that other types of reversible crosslinked morphologies such as the hydrogen bonding in urethane systems may facilitate the process.

Research by Varley and van der Zwaag [27] at the Technical University of Delft, the Netherlands; has recently investigated the self healing mechanism by separately studying both the elastic and viscous responses to high-speed deformation or impact. To do this, they developed methods which individually mimicked the response to impact and demonstrated that both healing mechanisms acting in concert together were required to achieve full healing. The elastic response of the ionomer system to impact was investigated by pulling a spindle with a disk of known diameter at the base of the spindle, through the polymer at high speed using a tensile-testing machine. A particular advantage of this method is that the spindle is pulled completely through and out of the polymer in a controlled manner, thereby mimicking penetration by a bullet in a manner that has not been studied previously. To mimic activation by thermal energy transfer, as typically occurs during ballistic penetration, the base of the spindle was heated in a controlled manner using a preheated block of metal. A schematic representation of this method is shown in Figure 11.

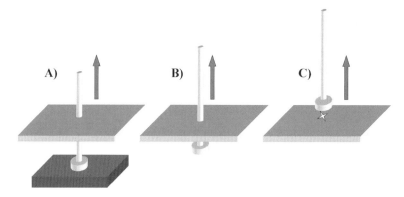

Fig. 11 Schematic representation of the method developed to investigate the elastic response of an ionomer to high-speed impact in a controlled manner: (**a**) represents the initial heating of the base of the spindle using a metal block as shown; (**b**) shows the spindle being pulled through the sample; and (**c**) shows the spindle completing the impact along with the corresponding impact zone

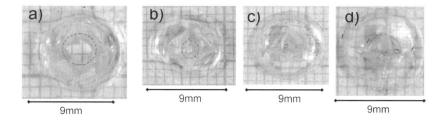

Fig. 12 Effect of increasing the local temperature from (**a**) 47°C, (**b**) 68°C, (**c**) 105°C, and (**d**) 140°C, upon the corresponding hole closure during impact

This method was used successfully to demonstrate the role of the impact temperature, the rate of impact, the effect of plasticiser modification and the physically cross-linked structure on the self healing process in ionomer systems. Examples of the self healing behaviour performed in a controlled manner are shown in Figure 12 for a spindle with a 9 mm disk attached to the base.

The effect of increasing the localised heating is clearly evident, as the diameter of the created cavity clearly decreases with increasing temperature. Complete hole closure can be seen to occur when the temperature is above the melting point of the polymer, at around 92°C and highlights the importance of the polymer being taken into the molten state. Here the unique structural morphology created by the ionic groups, are able to impart elastomeric properties to the polymer which promote hole closure through the subsequent rebound of the polymer. A simple fluid-based gravimetric approach was used to quantify the extent of elastic healing where the substantial improvement in hole closure is apparent, as shown in Figure 13, particularly once the temperature of the disk is above the melting point.

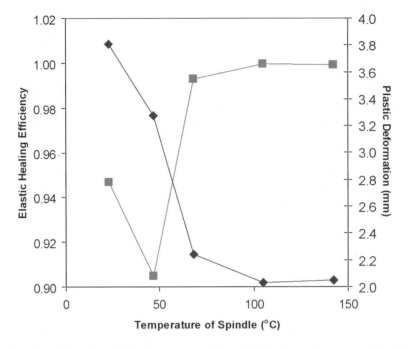

Fig. 13 Effect of increasing the local temperature of the spindle disk upon the corresponding healing efficiency and elastic rebound during impact

Also shown in Figure 13 is a measure of the plastic deformation behind the impact zone which also decreases with increasing temperature and further highlights the elastic behaviour of the polymer immediately after impact. Importantly it was also shown that non-ionic polymers with analogous chemical structures, such as polyethylene and polypropylene, did not show any elastic rebound or hole closure at similar temperatures above their melting point, compared to the ionomeric polymers. It is important to recognise that this experiment focuses upon hole closure, arising from the elastic response of the polymer and does not measure actual "healing" or sealing of the hole from polymer chain entanglements moving across the phase boundary. This was measured in a separate experiment using a tensile-testing machine and tensile test specimens which had been cut in two pieces. In this test, the tensile dog bone pieces were separated by 2 mm with the interfacial regions heated locally to above the melting point of the polymer. After a given time the samples were moved back together, rapidly cooled to room temperature and then pulled apart at a constant rate of displacement. A series of images in Figure 14 illustrate this method which was able to measure the "viscous healing", in the absence of any elastic response.

Figure 15 shows that the level of healing achieved for this method begins at around 56% (i.e. at time zero) while further heating for varying lengths of time and temperature, the viscous healing efficiency was increased substantially, in this case to over 90% of the original strength of the virgin polymer.

Fig. 14 Photographic images highlighting the different stages of the method developed to investigate the viscous response during self healing: (**a**) initial localised heating of the interfacial regions using a hot air gun; (**b**) moving the two sections back into place whereby the interface is cooled to room temperature using compressed air; and (**c**) pulling apart of the two sections under tension to determine the level of chain entanglement across the interface

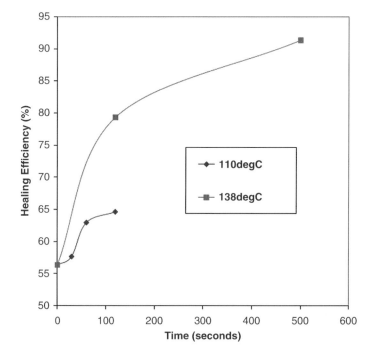

Fig. 15 Viscous healing efficiency of the tensile dog-bone samples as a function of heating time and temperature

Comparison with other types of polymers again highlights the unique properties that the combination of the chemical structure and morphology imparts to the polymer. A non-ionic thermoplastic, although exhibiting higher levels of viscous healing compared to the ionomer, has negligible elastic healing and therefore displays little capacity to self heal when the two mechanisms are considered together. Similarly,

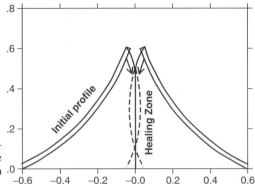

Fig. 16 Cross-sectional schematic diagram highlighting the capacity for hole closure after exit of the bullet. (From Rubinstein [19])

a covalently bonded low-crosslink density thermoset network, displayed no healing at all despite having been moved well into its elastomeric or rubbery phase above its glass transition temperature. Clearly, the covalent bonds prevent sufficient molecular mobility of the polymer chains across the interfacial region to enable healing to occur. Thus it can be seen that the ionomer systems can exhibit behaviour consistent with both thermoplastic and thermosetting behaviour which facilitate the self healing process. The thermoplastic behaviour promotes molecular mobility and chain entanglement, while the physical crosslinking through the ionic domains ensure that there is a high level of structural integrity of the polymer even at above the melting point and thus preventing wholesale melting of the polymer.

The most comprehensive working model of the ionomeric self healing process to date, has been presented by Rubinstein [19] where the macroscopic recovery process and the microscopic model of healing were discussed. The macroscopic recovery process largely confirmed the previously discussed conclusions where the healing process was considered to be a combination of the elastic response "closing the hole", followed by the localised viscous flow "sealing the hole". The actual recovery mechanism by which the polymer is able to close the hole, assuming satisfactory elastic rebound was illustrated. As shown in Figure 16, examination of the cross section of a theoretical puncture zone highlights that during impact, the length of the polymer is stretched and increased, so that when the bullet exits the other side and removes some polymer, there is sufficient quantity to cover the hole during the rebound or shape memory process.

While the elastic rebound ability of polymer to cover the cavity after impact is critical, polymers must have sufficient flow to melt and interdiffuse together so that the reformation of the physical crosslinks can occur over time. Rubinstein and Ermoshkin [19] have extended this concept to the molecular level using a molecular dynamics approach. Using the assumption that the interpenetration of polymer chains through the interface is the primary mechanism of healing, they showed that crack healing occurs by chain fluctuating back and forth across the interface and forming entanglements (or stitches) with other chains. The model was developed further to take into account temporary bonds between associating groups such as ionic species known as "stickers". Sticky dynamics (Liebler et al. [11, 12]) arises at longer timescale where the ionomer can flow by the dipoles (stickers) hopping from one multiplet to the next.

The self healing model was developed by introducing temporary bonds between associating groups, such as the ionic clusters in ionomers, to control the dynamics of reversible networks known as "stickers". When "sticky dynamics" are introduced into the model, it has been shown that it is able to reduce the required number of entanglements across the interface and significantly reduce the healing time. For further information on the dynamics of reversible networks by Leibler et al. [11, 12] and Rubinstein et al. [19–21] and the implications for self healing systems the reader is referred elsewhere.

The first applications of ionomers as self healing polymer systems has centred upon military applications in the areas of reducing the vulnerability of fuel tanks to combustion [15]. The application of ionomers as light weight and low cost alternatives to current methods of reducing fuel tank vulnerability to combustion has been investigated by Goldsmith [7]. It was found that the instantaneous healing of a fuel tank to impact by 12.7 mm armour-piercing bullets was a major improvement over the current self sealing systems, which were specified to heal within 2 min. This enormously reduced the level of fuel spray such that there was minimal fuel leakage, offering significant improvements over existing technologies.

7 Summary

To summarise therefore the self healing mechanism of ionomers is seen to be controlled by the inherent chemical structure and morphology of the polymer itself. Due to the fact that it is controlled by the chemical structure, it would be expected to be repeatable many times being dependent primarily upon the degradation of the polymer. It is not an autonomic process but is activated when thermal energy is transferred to the polymer. During impact, energy is absorbed which is elastically stored and dissipated as heat. This increases the local temperature of the impacted polymer to above the melting point (disrupting the physical crosslinks) while having little effect upon the temperature of the surrounding polymer. However, the ionic domains persist in the melt so that the polymer is able to be elongated to high levels of strain and rebound elastically when the stored energy is released at failure. The level of stretching enables the polymer to seal the cavity as it returns to its original position, but it is the viscoelastic properties and the capacity of physical crosslinks to reform that determines the final level and strength of healing.

Finally, it is critical for the success of healing that there is sufficient balance of elastic strength to rebound, yet at the same time, the polymer must be molten enough to be able to melt and chemically diffuse across interfacial boundaries.

Acknowledgements Support from the Delft Centre for Materials (DCMat), the Technical University of Delft and CSIRO to carry out this work is gratefully acknowledged. The ballistic firing conducted by Pieter van Nieuwkoop and technical assistance in the SEM operation by Frans van Oostrom is also gratefully acknowledged.

References

1. Anonymous (2004) Self-healing fuel tanks under fire. In Tech 51:14
2. Eisenberg A. (1970) Clustering of ions in organic polymers. A theoretical approach. Macromolecules 3:147–154
3. Eisenberg A., Hird B., Moore R.B. (1990) A new multiplet-cluster model for the morphology of random ionomers. Macromolecules 23:4098–4107
4. Eisenberg A., Kim J-S. (1998) Introduction to ionomers. Wiley, New York
5. Eisenberg A., Navartil M. (1973) Ion clustering and viscoelastic relaxation in styrene-based ionomers. II. Effect of ion concentration. Macromolecules 6:604–612
6. Fall R. (2001) Puncture reversal of ethylene ionomers – mechanistic studies. Virginia Polytechnic Institute and State University, VA
7. Goldsmith A. (2003) Ionomer-polymer self-healing material applications.
   Second AIAA "Unmanned Unlimited" Systems, Technologies, and Operations – Aerospace. American Institute of Aeronautics and Astronautics, San Diego, California.
8. Harris K.M., Rajagopalan M. (2003) Self healing polymers and composition for golf equipment. US 2002–229426
9. Huber A., Hinkley J.A. (2005) Impression testing of self healing polymers. NASA/TM-2005–213532
10. Kalista S.J. (2003) Self-healing of thermoplastic poly(ethylene-co-methacrylic acid) copolymers following Projectile puncture. Virginia Polytechnic Institute and State University, VA
11. Leibler L., Rubinstein M., Colby R.H. (1991) Dynamics of reversible networks. Macromolecules 24:4701–4707
12. Leibler L., Rubinstein M., Colby R.H. (1993) Dynamics of telechelic ionomers. Can polymers diffuse large distances without relaxing stress? J Phys II 3:1581–1590
13. Macknight W.J., McKenna L.W., Read B.E. (1967) Properties of ethylene-methacrylic acid copolymers and thier sodium salts: mechanical relaxations. J Appl Phys 38:4208–4212
14. Macknight W.J., Taggart W.P., Stein R.S. (1973) Model for the structure of ionomers. J Polym Sci Part C, Polym Symp 45:113–128
15. Manchor J.A., Bennett J.M., Weisenbach M.R. (2002) New concepts in passive fire protection. 43rd AIAA/ASME/ASCE/AHS SDM Conference. American Institute of Aeronautics and Astronautics, Denver, Colorado
16. Marx C.L., Caulfield D.F., Cooper S.L. (1973) Morphology of ionomers. Macromolecules 6:344–353
17. Rees R.W., Vaughan D.J. (1965) Physical structure of ionomers. Polymer preprints. American Chemical Society. Division of Polymer Chemistry 6:287
18. Register R.A., Yu X.H., Cooper S.L. (1989) Effects of matrix polarity and ambient ageing of the morphology of sulfonated polyurethane ionomers. Polym Bull 22:565–571
19. Rubinstein M., Ermoshkin A. (2005) Theoretical aspects of self-healing phenomenon. University of North Carolina at Chapel Hill, NC
20. Rubinstein M., Semenov A. (1998) Thermoreversible gelation in solutions of associating polymers. 2. Linear dynamics. Macromolecules 31:1386–1397
21. Rubinstein M., Semenov A.N. (2001) Dynamics of entangled solutions of associating polymers. Macromolecules 34:1058–1068

22. Shlick S. (ed.) (1995), Ionomers: characterisation. In: Theory and application. CRC Press, Boca Raton, p. 311
23. Tachino H., Hara H., Hirasawa E., Kutsumizu S., Tadano K., Yano S. (1993) Dynamic mechanical relaxations of ethylene ionomers. Macromolecules 26:752–757
24. Tadano K., Hirasawa E., Yamamoto H., Yano S. (1989) Order–disorder transition of ionic clusters in ionomers. Macromolecules 22:226–233
25. Tobolsky A.V., Lyons P.F., Hata N. (1968) Ionic clusters in high-strength carboxylic rubbers. Macromolecules 1:515–519
26. Vanhoorne P., Register R.A. (1996) Low-shear melt rheology of partially-neutralized ethylene-methacrylic acid ionomers. Macromolecules 29:598–604
27. Varley R.J., Van der Zwaag S. (2007) Evaluation of self healing capability of Surlyn 8940. (Unpublished results)
28. Wilson F.C., Longworth R., Vaughan D.J. (1968) Polymer preprints. American Chemical Society. Division of Polymer Chemistry 9:505
29. Yarusso D.J., Cooper S.L. (1985) Analysis of SAXS data from ionomer systems. Polymer 26:371–278

Self Healing Fibre-reinforced Polymer Composites: an Overview

Ian P. Bond, Richard S. Trask, Hugo R. Williams, and Gareth J. Williams

ACCIS – Advanced Composites Centre for Innovation and Science, Department of Aerospace Engineering, University of Bristol, Queen's Building, University Walk, Bristol, UK
E-mail: I.P.Bond@bristol.ac.uk

1 Introduction

Lightweight, high-strength, high-stiffness fibre-reinforced polymer composite materials are leading contenders as component materials to improve the efficiency and sustainability of many forms of transport. For example, their widespread use is critical to the success of advanced engineering applications, such as the Boeing 787 and Airbus A380. Such materials typically comprise complex architectures of fine fibrous reinforcement e.g. carbon or glass, dispersed within a bulk polymer matrix, e.g. epoxy. This can provide exceptionally strong, stiff, and lightweight materials which are inherently *anisotropic*, as the fibres are usually arranged at a multitude of predetermined angles within discrete stacked 2D layers. The direction orthogonal to the 2D layers is usually without reinforcement to avoid compromising *in-plane* performance, which results in a vulnerability to damage in the polymer matrix caused by *out-of-plane* loading, i.e. impact. Their inability to plastically deform leaves only energy absorption via damage creation. This damage often manifests itself internally within the material as intra-ply matrix cracks and inter-ply delaminations, and can thus be difficult to detect visually. Since relatively minor damage can lead to a significant reduction in strength, stiffness and stability, there has been some reticence by designers for their use in safety critical applications, and the adoption of a 'no growth' approach (i.e. damage propagation from a defect constitutes failure) is now the mindset of the composites industry. This has led to excessively heavy components, shackling of innovative design, and a need for frequent inspection during service (Richardson 1996; Abrate 1998).

However, the research community has made considerable steps in understanding damage modes and the development of robust failure models. A step change in composites technology could be achieved by adopting a philosophy in which some damage growth can be tolerated; this would provide considerable weight and cost savings and offer the designer greater freedom to formulate new designs. Furthermore, there are numerous applications in which a component is expected to tolerate significant damage growth yet still be fit for service; i.e. be "damage tolerant" or "fail-safe". If such an approach was coupled with an inherent "reparability" in composite materials, their more widespread use would be assured.

S. van der Zwaag (ed.), *Self Healing Materials. An Alternative Approach to 20 Centuries of Materials Science*, 115–138.

115

Traditionally, once damage has been detected within a polymer composite, the approach has been to design and undertake a cosmetic, temporary, or structural repair. These repair schemes range from simplistic external patches to complex intrusive tapered or stepped scarf repairs with the aim of restoring some or all of the material's stiffness and strength. However, composite repairs tend to be difficult and costly to apply, therefore, engineers have started to design polymer composites that have some form of damage tolerance, i.e. "an ability for the material to sustain weakening defects under loading and environment without suffering reduction in residual strength, for some stipulated period of service".

The concepts being considered for the development of improved damage tolerant composite materials are based on the suppression of either intra-ply matrix cracking or inter-ply delamination along weak interfaces. It is beyond the scope of this chapter to discuss these various concepts in detail and the reader is directed towards the numerous detailed works on these respective topics. However, they can be broadly classified into two categories of introducing either a material or structural approach.

Material	Structural
Tougher matrices	Woven/non-woven reinforcement
Toughened interlayers	3D fabrics
	Stitching, tufting, braiding,
	Z-pinning

An underlying theme for most of these techniques is to establish some degree of out-of-plane reinforcement to absorb energy and arrest crack growth. However, the introduction of such features, often has a negative influence on the innate mechanical performance of the host composite.

2   Self Healing Strategies in Engineered Structures

An alternative approach which can directly address a fibre-reinforced polymer composite's vulnerability to damage is to integrate some form of autonomic self healing which utilises a propagating crack to initiate a healing mechanism. Such an idea has gained considerable interest in the composites community over the recent years and it is the unique design freedom of a fibre-reinforced polymer composite material that offers the designer an ability to "tailor" material architecture, hybridise, and introduce innovative features.

A key focus of current scientific research is the development of bioinspired materials systems (Bar-Cohen 2006; Bruck et al. 2002). Naturally occurring *materials* in animals and plants have developed into highly sophisticated, integrated, hierarchical structures that commonly exhibit multifunctional behaviour (Curtis 1996). Mimicry of these integrated microstructures and micro-mechanisms offers considerable potential to engineers in the design and continual improvement of material performance

(Kassner et al. 2005) in the coming years. A great many natural *materials* are themselves self healing composite materials! The rest of this chapter will consider the ongoing developments in applying autonomic self healing to fibre-reinforced polymer composites.

2.1   Bioinspired Self Healing Approaches

The healing potential of living organisms and the repair strategies in natural materials is increasingly of interest to designers seeking lower mass structures with increased service life, who wish to progress from the more conventional conservative, damage tolerance philosophy. The conceptual inspiration from nature for self healing is not new, and many other engineering approaches could be considered to have been inspired by observing natural systems. These bioinspired approaches do not typically include mimicry of the biological processes involved because in many cases they are clearly too complex. The field of self healing in polymer composites has seen exciting developments in recent years using the inspiration of biological self healing applied with broadly traditional engineering approaches.

Nature's ability to heal has inspired many new ideas and mechanisms. Chemists and engineers have proposed different healing concepts that offer the ability to restore the mechanical performance of the material. One area of interest is the re-bonding of the failed surfaces. Polymeric materials possessing selective crosslinks between polymer chains that can be broken under load and then reformed by heat have been shown to offer healing efficiencies of 57% of the original fracture load (Chen et al. 2002). Another example is where a polymeric material hosts a second solid-state polymer phase that migrates to the damage site under the action of heat (Zako and Takano 1999; Hayes et al. 2005). Hayes and colleagues (2005) have developed a two-phase, solid-state repairable polymer by mixing a thermoplastic healing agent into a thermosetting epoxy matrix to produce a homogeneous matrix which contrasts with the discrete particles of uncured epoxy reported by Zako and Takano (1999). These systems offer the capacity for self healing, but are not autonomous; they require damage sensing, some form of higher decision-making via a feedback loop and heating requirements that would pose significant practical challenges in application.

Lee et al. (2004) have considered the possibility of using nanoparticles dispersed in polymer films for deposition at a damage site. In this work, the nanoparticles are dispersed in polymer films within a multilayer composite and are studied by integrated computer simulations. The model comprises a brittle layer containing a nanocrack sandwiched between two polymer films with analysis suggesting a self healing mechanism whereby nanoparticles congregate at the nanocrack. The numerical models also predict load transfer from the matrix to the stiff nanoparticles. This mechanism is considered applicable to optical communications, display technologies, and biomedical engineering. The authors report that mechanical properties of the composites repaired in this manner could potentially achieve 75–100% of the undamaged material strength. Later work by Gupta and colleagues (2006), using fluorescent nanoparticles, has shown that ligands on the nanoparticles can be selected to help drive nanoparticles into a crack in a microelectronic thin film layer. No restoration of mechanical properties were investigated. The applicability of this technology to structural composite materials is possibly limited as the target damage is on a very small scale.

A third area of study is based upon a biological "bleeding" approach to repair, i.e. microcapsules and hollow fibres. Microencapsulation self healing (White et al. 2001; Kessler and White 2001; Kessler et al. 2002) can involve the use of a monomer, dicyclopentadiene (DCPD), stored in urea-formaldehyde microcapsules dispersed within a polymer matrix. When the microcapsules are ruptured by a progressing crack, the monomer is drawn along the fissure where it comes into contact with a dispersed particulate catalyst (Ruthenium based "Grubbs" catalyst), initiating polymerisation and thus repair. The release of active components was clearly seen to restore a proportion of the loss in mechanical properties arising from microcracking within a polymer matrix and results confirmed that the dispersion of microcapsules within a bulk polymer matrix material was not detrimental to stiffness.

A key advantage of the microencapsulation self healing approach is the ease with which they can be incorporated within a bulk polymer material. The disadvantages are the need for microcapsule fracture and the need for the resin to encounter the catalyst prior to any repair occurring. In fibre-reinforced polymer composite materials, additional problems arise due to the size of microcapsules (typically 10–100 μm) disrupting the fibre architecture (i.e. fibre waviness and fibre volume fraction), the need for a good dispersion of the catalyst to provide uniform healing functionality, microcapsules having only limited resin volume, and the creation of a void in the wake of the crack after consumption of healing resin. Some results (Kessler and White 2001; Kessler et al. 2003) have indicated specific problems in terms of healing efficiency due to clumping of microcapsules into woven-roving wells whilst cracks propagate along woven-roving peaks. Recently, microcapsules have been applied to improve the fatigue life of an epoxy bulk polymer. Both manual infiltration of premixed DCPD monomer and catalyst (Brown et al. 2005a) and in situ healing using monomer filled microcapsules and dispersed catalyst (Brown et al. 2005b) have shown a performance advantage. With in situ healing, crack arrest and an improved fatigue life of up to 213% of the control specimens was demonstrated in high-cycle fatigue ($> 10^4$ cycles) (Brown et al. 2006). Sanada and colleagues 2006 used microencapsulated healing agent with a solid catalyst to repair interfacial debonding in fibre-reinforced composites with some success. Yin and colleagues (2007) have combined a microencapsulation self healing approach with a two-phase epoxy system, that is triggered by the application of heat. When used to impart self healing to an E-glass/epoxy composite, a significant proportion of healing under DCB loading was observed.

Self healing using liquid resin-filled hollow fibres embedded within an engineering structure, similar to the arteries in a natural system, has been investigated at different length scales in different engineering materials by various authors, for example, in bulk concrete (Dry 1994; Dry and McMillan 1996; Li et al. 1989; Dry 2000;

Dry et al. 2003), in bulk polymers (Dry 1996), and in polymeric composites at a millimetre length scale (Motuku et al. 1999) and at a micrometer length scale – see Figure 2a (Bleay et al. 2001; Pang and Bond 2005a,b).

The use of hollow glass fibres (see Figure 1a) embedded in a composite laminate was pioneered by Bleay et al. (2001). In their work, commercial hollow fibres were consolidated in to lamina and then manufactured into composite laminates, i.e. the self healing material acts as the structural fibres. The key advantages of the hollow fibre self healing concept are that the fibres can be located to match the orientation of the surrounding reinforcing fibres, thereby minimising Poisson ratio effects. The fibres can be placed at any location within the stacking sequence to address specific failure threats (see Figure 1b), different healing resins can be used depending upon the operational requirements of the structure, different activation methods can be used to cure the resin and crucially, a significant volume of healing agent can be made be available. The disadvantages are the relatively large diameter of the fibres compared to the reinforcement, the need for fibre fracture, the need for low-viscosity resin systems to facilitate fibre and damage infusion and the need for an extra processing stage for fibre infusion.

The ability to "see" and become aware of internal damage in composite materials is as critical as in the human body. The ability to form a "bruise" within a hollow fibre self healing composite material was investigated by Pang and Bond (2005a,b). In this work they designed a damage visual-enhancement method, by the bleeding action of a fluorescent dye from discrete self healing lamina housed alongside the structural fibre lamina (see Figure 1c). This approach permitted a "bruising" of the laminate to assist in the identification of regions for subsequent non-destructive evaluation.

2.2 Biomimetic Self Healing Approaches

The variety of different healing concepts discussed above have been considered and assessed from an engineering perspective. It is only recently that studies are beginning into the underlying biological methods, mechanisms and processes in order to deliver a truly biomimetic self healing solution. The challenge for the future is the evolution of "engineering self-healing" towards a biomimetic solution. This work is still in its infancy but mimicry of blood clotting, tissue bruising and tailoring healing networks to address damage formation are all being actively considered.

To date, autonomous healing materials in engineering structures have been distributed randomly throughout the structure (i.e. solid-state polymers of microcapsules) or spaced evenly through the composite laminate structure. In nature the network is tailored for a specific function with the healing medium often being multifunctional. The first reported instance of tailoring the location of self healing functionality in engineering to match the damage threat is by Trask and Bond (2006). In this work the key failure interfaces were identified and then the hollow fibre self healing network was designed for a specific composite component and application, in this

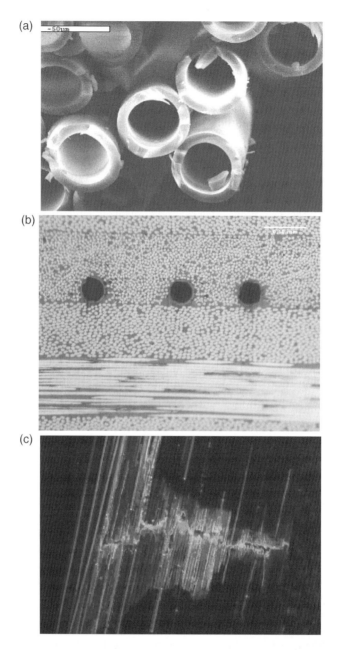

Fig. 1 (**a**) Hollow glass fibres; (**b**) hollow glass fibres embedded in carbon fibre-reinforced composite laminate; (**c**) damage visual enhancement in composite laminate by the bleeding action of a fluorescent dye from hollow glass fibres

case a space environment. The need for self healing in the space environment was found to put significant demands on the repair agent in terms of mechanical properties, processability, and environmental compatibility. The self healing mechanism was found to restore 100% of the strength when compared to undamaged laminates containing healing plies.

Verberg and colleagues (2006) have computationally studied a biomimetic "leukocyte" consisting of microencapsulated nanoparticles that are released by diffusion, while the microcapsule is driven along microvascular channels. The surface chemistry of the capsule and nanoparticle could be selected to enable the microcapsule to "roll" along the inner wall of the microchannel, but damage to the surface in its path would cause the movement to pause until the diffusing nanoparticles collect at, and ideally repair, the defect thus allowing movement to recommence.

3   Self Healing Using Resin-filled Hollow Fibres

Let us consider in detail one such self healing approach for fibre-reinforced composites, based on the use of healing resin-filled hollow fibres. A comprehensive study (Trask et al. 2007a) has been undertaken to establish a methodology for manufacturing such a system followed by mechanical characterisation under impact loading (Zhou 1998) to ascertain self healing effectiveness and the critical aspects of applying such an approach.

3.1   Introduction

Hollow glass fibres (Hucker et al. 1999, 2002, 2003a, b] are used in preference to embedded microcapsules (Brown et al. 2003; Rule et al. 2005) because they offer the advantage of being able to store functional agents for self repair, as well as integrating easily with and acting as a reinforcement. A typical hollow fibre self healing approach used within composite laminates could take the form of fibres containing a one-part resin system, a two-part resin and hardener system, or a resin system with a catalyst or hardener contained within the matrix material (Bleay et al. 2001).

A bespoke hollow glass fibre (HGF)-making facility (Hucker et al. 1999, 2002) was used to produce HGF between 30–100 μm diameter with a hollowness of around 50% (Figure 2). These were then embedded within either glass fibre-reinforced plastic (GFRP) or CFRP and infused with uncured resin to impart a self healing functionality to the laminate. During a damage event some of these hollow fibres fracture, initiating recovery of properties by "healing" whereby a repair agent infiltrates the damage zone from within broken hollow fibres and acts to ameliorate any critical effects of matrix cracking and delamination between plies and, most importantly, prevent further damage propagation. This release of repair agent is analogous to the bleeding mechanism in biological organisms (e.g. Human thrombosis).

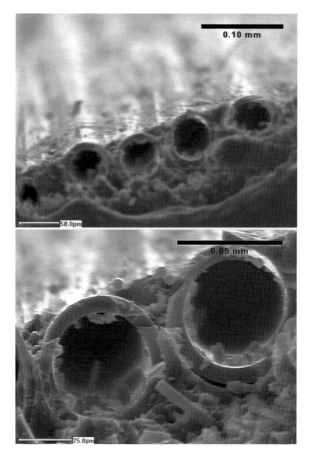

Fig. 2 Typical hollow glass fibres (60 μm external diameter with 50% hollowness fraction) embedded within CFRP

The exact nature of the self healing method will depend upon: (1) the nature and location of the damage; (2) the choice of repair resin; and (3) the influence of the operational environment. The self healing fibres can be introduced within a laminate as additional plies at each interface, at damage critical interfaces or as individual filaments spaced at predetermined distances within each ply. In order to more fully understand and optimise the healing process, two parallel studies in glass-reinforced and carbon-reinforced epoxy systems have been undertaken. A translucent glass/epoxy laminate provides good visualisation of damage occurrence and the healing process when viewed with transmission microscopy, however, a carbon/epoxy laminate is opaque and to enhance visualisation an UV fluorescent dye (Ardrox 985) was added to the healing resin in both studies.

3.2   Specimen Manufacture

3.2.1   Self Healing Glass Fibre-Reinforced Plastic

The HGF chosen for this study had an external diameter of $60 \pm 3\,\mu m$ and an internal diameter of $\sim 40\,\mu m$ yielding a hollowness fraction (ratio of internal to external area) of $\sim 55\%$. This larger fibre diameter (compared to Figure 3) gives a significant volume for healing agent storage. Once manufactured, the individual fibres were consolidated within an epoxy resin film (913 Hexcel Composites), which was selected to match the host laminate material. The healing resin is then infused into individual filaments using a vacuum-assisted capillary action. Once the ends have been sealed (Bostik BondFlex 100HMA high-modulus silicone sealant) the infused hollow fibre layers (which can now be considered as standard "prepreg" plies) are incorporated into a laminate-stacking sequence as required.

A 16 ply composite laminate with a $[0°/+45°/90°/-45°]_{2s}$ stacking sequence manufactured from pre-impregnated E-glass/913 epoxy resin (Hexcel Composites) was selected for the first evaluation of the HGF self healing approach. Self healing filaments were introduced at four $0°/45°$ damage critical ply interfaces that were identified and reported previously (Trask and Bond 78), as shown in Figure 3. An epoxy resin system (CYTEC Cycom 823) was selected as the healing resin because of the need to match the chemistry of the host laminate, its availability as a two-part system permitting inclusion in separate storage filaments, its low-viscosity profile, and its time to gelation of 30 min after mixing. Furthermore, it was observed experimentally that the individual components of the two-part Cycom 823 were sufficiently robust to survive the host laminate curing process ($120°C$ for 1 h) after infiltration into the hollow filaments within the laminate stack.

Panels of $200 \times 200 \times 2.5\,mm$ were prepared according to the pre-preg manufacturer's instructions. Each panel was then sectioned into coupons $20 \times 50\,mm$ for

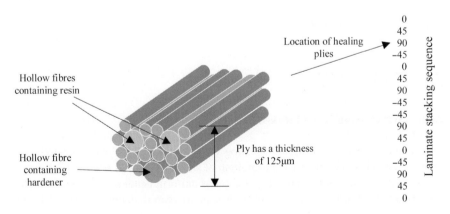

Fig. 3 Location of resin and hardener self healing filaments intermingled within an E-glass ply in the 16 ply stacking sequence of the composite laminate

testing. After cutting, the edges of the samples are sealed with a two-part rapid curing epoxy system (Araldite Rapid) to prevent any healing resin loss through the exposed ends of the hollow fibres.

3.2.2 Self Healing Carbon Fibre-reinforced Plastic

It is imperative that any embedded HGF does not detrimentally affect the innate mechanical performance of a FRP, but provides a sufficient volume of healing resin to address any damage. Thus, the incorporation of HGF as discrete plies was deemed unsuitable for CFRP laminates as it would effectively produce a hybrid glass/carbon laminate and result in a significant reduction in mechanical performance. A less intrusive approach was devised whereby a small number of individual HGFs were distributed within individual CFRP plies to act as dispersed storage vessels for the healing agent. Therefore, the distribution of HGF within a CFRP laminate poses a problem of optimising disruption of the host laminate architecture against delivery of adequate healing resin volume.

Preimpregnated T300 carbon fibre/914 epoxy resin (Hexcel Composites) was selected as the host material as it is widely used in aerospace applications. A quasi-isotropic stacking sequence of 16 plies was prepared as a $230 \times 160 \times 2.5$ mm plate. Two different HGF distributions (fibre spacing of 70 µm and 200 µm respectively) were wound directly onto uncured CFRP plies prior to lamination to investigate the effect of HGF on the host laminate properties and the healing efficacy of different healing agent volumes. HGF was located at two $0°/-45°$ interfaces within the lay-up as follows:

$(-45°/90°/45°/0°/\text{HGF}/-45°/90°/45°/0°/0°/45°/90°/-45°/\text{HGF}/0°/45°/90°/-45°)$

The inclusion of HGF within the CFRP stack was such that short lengths (10–20 mm) of exposed HGF protruded from the panel edges. This facilitated vacuum assisted infiltration of the HGF after laminate cure with a two-part epoxy resin healing agent (CYTEC Cycom823) immediately prior to testing.

3.3 Mechanical Testing

3.3.1 Self Healing Glass Fibre-reinforced Plastic

Four-point bend flexural strength testing (ASTM D6272–02) was selected as an appropriate way to characterise the strength of standard and self healing GFRP specimens as this configuration ensures a region of uniform bending stress in the area of the damage between the loading noses. A support span to depth ratio of 32:1 and a support span to load span ratio of 3:1 were selected. Six repeat tests were conducted using a loading rate of 5 mm/min. A linear potentiometric displacement transducer (LPDT) was used to record mid-span deflection.

Table 1   Summary of flexural strength and healing efficacy for GFRP

Specimen Description	Flexural Strength (MPa)	% Retained Strength
Control laminate – no damage	668	100
Self healing laminate – no damage	559	84
Control laminate – damaged (2500N peak load)	479	72
Self healing laminate – damaged, no repair	494	74
Self healing laminate – damaged + 2 h at 100°C	578	87

In the case of the damaged specimens, a three-point bend indentation test using a hardened steel hemisphere (4.63 mm diameter) with the specimen back face supported by a steel ring (OD = 34 mm, ID = 19 mm) was used to create repeatable damage. A peak load of 2500N was used to initiate matrix shear cracks and delamination within the laminate. In the case of the laminate containing the self healing specimens this was permitted to heal by heating to 100°C (from ambient) in an air circulation oven and held at this temperature for 2 h. The application of heat to augment healing was deemed acceptable as it allowed some degree of control during the study and matched the typical working environment of the intended aerospace application.

The results in Table 1 indicate that the inclusion of HGF imparts an initial strength reduction of 16%. The control laminate and self healing laminate with no healing had comparable damage tolerance both in terms of damage size and residual failure strength (typically 72–74%). After healing it was found that a laminate had a residual strength of 87% compared to the undamaged control and 100% compared to an undamaged self healing laminate albeit with greater standard deviation.

3.3.2   Self Healing Carbon Fibre-Reinforced Plastic

As described earlier, the configuration and manufacture of self healing CFRP was somewhat different to the GFRP. Ten specimens (100 × 20 × 2.5 mm) were cut from a plate with the use of a water-cooled diamond grit saw. The sample edges were polished with SiC paper to avoid any unwanted edge effects. Samples were then dried, sealed in bags, and stored in a temperature and humidity controlled environment prior to testing. Immediately prior to testing, the HGF within each specimen were infiltrated, using a vacuum assist technique, with premixed, two-part epoxy healing resin (CYTEC – Cycom 823).

Quasi-static impact damage was imparted to each specimen using a hardened steel (5 mm) spherical indenter supported by a steel ring of 27 mm outer diameter and 14 mm inner diameter mounted on a test machine. The indentations were stopped at a peak load of 2000N to simulate barely visible impact damage (BVID) in the composite laminate. At this point the damage is contained within the laminate and can be likened to BVID, as the impact surface suffers a minor indent and the back face experiences minimal distortion due to back face delamination.

After indentation, the specimens were subjected to 70°C for 45 min to reduce healing resin viscosity (25 cps) and facilitate infiltration into damage sites, followed by cure at 125°C for 75 min. Whilst this process diverges from the original aim of achieving autonomous healing, the use of a premixed resin and elevated temperature is an attempt to mitigate some of the shortcomings of the Cycom 823 and attempt to demonstrate the greatest healing efficiency possible with this system. No resin system exhibiting all desirable attributes (i.e. low viscosity, insensitivity to mix ratio, rapid cure under ambient conditions, and unlimited shelf life) is currently available. However, from a practical perspective, temperature activation provides excellent control of cure initiation, eliminating time constraints on the testing/manufacturing process.

Four-point bend flexural testing (ASTM-D6272-02) was again used to assess the self healing efficiency of the resulting CFRP. A support span to depth ratio of 32:1 and a support span to load span ratio of 3:1 were selected. Results were obtained from ten undamaged, five damaged, and five healed specimens.

The results of the four point bend flexural testing are shown in Table 2. This compares the performance of undamaged, damaged and healed specimens for the two HGF pitch spacings alongside a control CFRP laminate with no HGF.

Analysis of Table 2 shows that the 70 μm fibre spacing resulted in the largest reduction in undamaged strength (8%). This is attributed to a significant disruption in fibre architecture (see Figure 7 below). However, after being damaged this configuration exhibited a significant amount of damage tolerance compared to the 200 μm fibre spacing and control laminates. The large volume fraction of HGF also provides a considerable reservoir of healing agent as shown by a 97% recovery of undamaged strength (equivalent to 89% of the undamaged strength of the control laminate).

Figure 5 shows typical load-displacement curves from flexural testing. A damaged specimen containing HGF under load experiences a number of intermittent decreases in load as failure is approached. This is probably attributable to the propagation of matrix cracks and delaminations up to a critical level at which they affect the load bearing performance of the laminate. Conversely, the healed specimen appears to suppress these damage sites, inhibiting crack propagation, and thereby delaying the failure of the laminate to higher load levels. Figure 4 also highlights the effectiveness of self healing as the load-displacement curve for the healed laminate lies close to the undamaged curve.

The 200 μm HGF-spacing specimens exhibit little reduction in undamaged strength (2%) attributable to the reduced disruption to the host laminate. These

Table 2 Summary of flexural strength and healing efficacy for CFRP

Specimen Type		Undamaged	Damaged at 2000N	Healed at 2000N
Control CFRP	Strength (MPa)	583	405	–
	% of baseline	100	69	–
HGF spaced at 70 μm	Strength (MPa)	535	444	520
	% of baseline	92	76	89
HGF spaced at 200 m	Strength (MPa)	569	401	467
	% of baseline	98	69	80

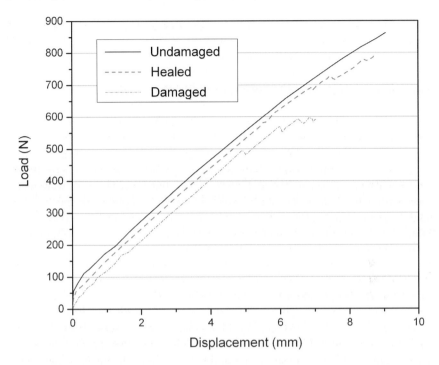

Fig. 4 Comparative load-displacement curves under 4-point bend for undamaged, damaged, and healed for 70 μm spaced HGF

specimens behaved similarly to the control when damaged, presumably due to the limited amount of HGF available for crushing. However, healed samples achieved 82% of their undamaged strength (equivalent to 80% of the undamaged strength of the control laminate), despite the significantly lower volume of available healing resin. Both HGF spacings investigated (70 μm and 200 μm) showed similar trends, experiencing an initial reduction in flexural strength in the undamaged state compared to an unmodified control. This can be attributed to three effects:

1. Distortion of the reinforcing fibre architecture
2. Generation of resin-rich regions (crack nucleation/propagation sites)
3. Displacement of carbon fibres by non-structural HGF (reduction in fibre volume fraction)

3.4   Laminate Microstructure and Damage Analysis

Specimens from different test panels were sectioned and polished for microstructural examination. This provided a qualitative assessment of the disruption caused by the different HGF configurations on the host GFRP and CFRP laminates and identified any inconsistencies. Furthermore, it provided images to examine the damage and repair of the material.

3.4.1 Self Healing Glass Fibre-Reinforced Plastic

A damaged and healed GFRP specimen was examined to determine whether healing resin had infiltrated the damage zone. A cross section of the healed damage illuminated under UV light is shown in Figure 5 which clearly illustrates the extent of the infiltration by the healing resin into the damage zone when viewed along the 0° fibre direction. This result suggests that the four self healing layer locations were ideally placed within the complex damage network to fully infuse the damage site. To understand the self healing mechanism involved in more detail further microscopic examinations were undertaken; crushed hollow fibres under the impact zone (Figure 6a, b) and healing resin bridging fracture surfaces (Figure 6c, d).

3.4.2 Self Healing Carbon Fibre-Reinforced Plastic

A fibre spacing of 70 μm (Figure 7) highlighted several issues about the consequences of embedding HGF within a laminate. This small pitch spacing was selected to ensure the HGF were in close proximity and thereby facilitate a high degree of healing efficiency. However, HGF clumping did occur resulting in disruption to the host ply (Figure 7b). A fibre spacing of 200 μm resulted in a much reduced laminate disruption (Figure 8). The resin rich regions surrounding the HGF are minimised and there was no evidence of fibre clumping. The large spacing between HGF permitted excellent consolidation during cure further improving overall HGF embedment.

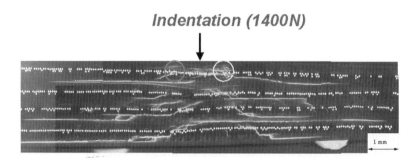

Fig. 5 Impact damaged cross section of $[0°/+45°/90°/-45°]_{2s}$ GFRP laminate containing healing filaments at the $+45°/90°$ and $-45°/90°$ interfaces

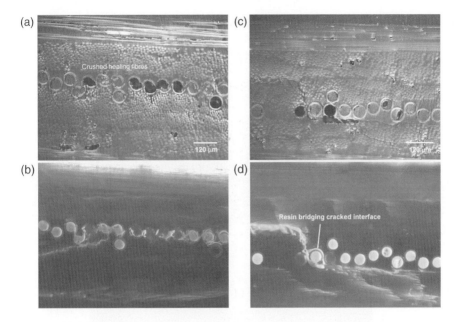

Fig. 6 Crushed-healing fibres located under the impact site viewed under normal (**a**) and UV (**b**) illumination. Healing resin bridging cracked interface viewed under normal (**c**) and UV (**d**) illumination

The process of self healing is reliant upon the occurrence of two phenomena:

1. HGF fibre fracture initiated by a damage event
2. Connectivity between HGF and damage network within material

The presence of intra-ply shear cracks linking delaminations provide connectivity between different damage sites in the laminate and facilitate healing at multiple interfaces. Shear cracks can be seen to initiate HGF fracture by similar mechanisms to delaminations (Figure 9). However, the width of shear cracks is generally smaller than delaminations ($\sim$10 μm compared to $\sim$30 μm) and so encourage a stronger capillary action to transport healing resin.

The number of fractured HGF depends very much on the severity of the impact and their location within the laminate stack. This provides an ability to tailor for a specific impact energy threshold. In the specimen shown in Figure 9, approximately 40% of the overall HGF were damaged but with a very uneven distribution throughout the laminate thickness. With regard to the proportion of healing resin which emerges from individual HGF upon fracture, our studies have shown a range from entire evacuation at gross delamination sites to localised evacuation after matrix cracking. The rate at which a healing agent fills a damage site is very much a function of the healing resin properties. Thus it is critical that any candidate system is optimised to undertake a healing function. Post-evacuation, the HGF are still capable of sustaining some load and there was no evidence that empty HGF act to initiate new damage.

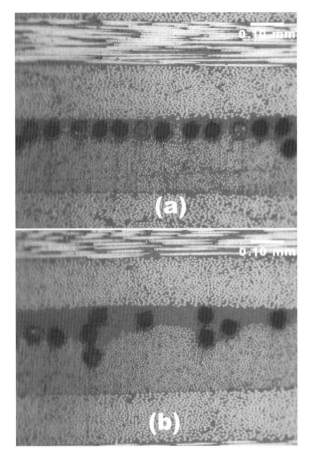

Fig. 7 HGF spaced at 70 μm showing (**a**) good embedment and (**b**) fibre clumping within host laminate

4 Biomimetic Self Healing in Composites: The Way Forward

The preceding sections have surveyed the current level of self healing research in composite engineering structures and discussed in detail an approach based on the use of resin filled hollow fibres. Most of the self healing work to date has been bioinspired and not biomimetic, although this influence is beginning to change. The definition of biomimetic self healing applied in this work is that some systematic study of biological healing approaches must be used to influence the approach adopted. Clearly, the timescale for the realisation of self healing within engineered structures will be considerably reduced by a comprehensive exploration and study of the many examples of how the natural world undertakes the process (Trask et al. 2007b). Biomimicry of the complex integrated microstructures and micromechanisms found in

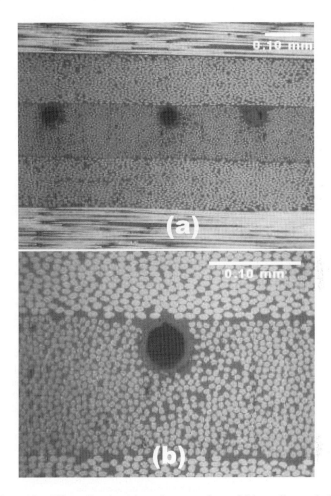

Fig. 8 HGF spaced at 200 μm showing (**a**) consistent spacing and (**b**) excellent embedment within host laminate

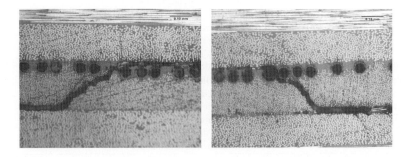

Fig. 9 Intra-ply shear cracks intercepting HGF within a self healing CFRP laminate

biological organisms offers considerable scope for the improvement in the design of future multifunctional materials.

Table 3 provides an overview of healing methods used in nature and cross-referenced them against the engineering approaches currently employed by the different research groups. The table is by no means exhaustive but gives a general overview of the characteristic similarities of biological and engineering systems.

Let us conclude this chapter by considering examples of biological self healing and repair strategies that may offer some potential for initiating new engineering approaches (Vincent and Mann 2002) and examine what level of research activity, if

Table 3 Biomimetic self healing inspiration in advanced composite structures (Trask et al. 2007b)

Biological attribute	Composite/polymer Engineering	Biomimetic self healing or repair strategy	Reference
Concept of self healing	Remendable polymers	Bioinspired healing requiring external intervention to initiate repair	Chen et al. (2002, 2003), Hayes et al. (2005)
Bleeding	Capsules	Action of bleeding from a storage medium housed within the structure; two-phase polymeric cure process rather than enzyme "waterfall" reaction	White et al. (2001), Kessler and White (2001), Kessler et al. (2002)
	Hollow fibres	Action of bleeding from a storage medium housed within the structure; two-phase polymeric cure process rather than enzyme "waterfall" reaction	Bleay and colleagues (2001), Pang and Bond (2005a, b), Trask and Bond (2006)
Blood cells	Nanoparticles	Artificial cells that deposit nanoparticles into regions of damage	Lee et al. (2004), Verberg et al. (2006)
Blood flow vascular network	Hollow fibres	2D or 3D network would permit the healing agent to be replenished and renewed during the life of the structure	Toohey et al. (2006), Williams et al. (2006)
Blood clotting	Healing resin	Synthetic self healing resin systems designed to clot locally to the damage site; remote from the damage site clotting is inhibited and the network remains flowing	–
Skeleton/bone healing	Reinforcing fibres	Deposition, resorption, and remodelling of fractured reinforcing fibres	–
Tree bark healing – compartmentalisation	–	Formation of internal impervious boundary walls to protect the damaged structure from environmental attack	–

any, is currently exploiting biological features and characteristics to improve engineering technologies. The references provided are by no means exhaustive, but serve as a starting point for more detailed studies.

4.1 Vascular Networks

Biological organisms have a highly developed, multifunctional vascular network to distribute fuel, control internal temperature and effect self healing, among many other roles. A key feature of these systems is that they supply fluid to an area from a point reservoir, giving a branching network. Studies show that the branching and size of these vessels have evolved to minimise the power required to distribute and maintain the supporting fluid within many other constraints (Murray 1926; Sherman 1981; McCulloh et al. 2004). The system is also reconfigurable in response to circumstances by adjusting the radius of individual vessels by vasoconstriction and dilation in mature tissue, or by growth in embryonic blood vessels (Taber et al. 2001).

The future of the self healing concept for composite materials relies on the development of a continuous healing network embedded within a composite laminate that delivers healing agent from a reservoir to regions of damage to permit the repair of all types of composite failure modes. New and novel approaches must be found to ensure that the healing agent can be replenished and renewed during the life of the structure. It must restore the matrix material properties and restore the structural efficiency of fractured fibres.

4.2 Healing Agent

Mammalian blood clotting has evolved around the chemical reactions of a series of active enzymes and their inactive precursors known as clotting factors. The intrinsic system takes the form of an enzyme "cascade" or "waterfall" of reactions (Hougie 2004) involving clotting factors. It was first proposed by Macfarlane 1964. It is initiated by damage that breaches the endothelial cells that line the blood vessels and culminates in the production of fibrin, a fibrous polymer. One of the most notable features of the haemostatic system is that despite the rapid response to injury, system malfunction is extremely rare (Davies and McNicol 1983). This is achieved by the rapid removal of the activated enzymes upon the production of fibrin, and the action of endothelial cells (Macfarlane 1964).

Synthetic self healing resin systems need to be developed to duplicate the blood clotting approach found in the human body (Runyon et al. 2004). At present, self healing in man-made structures requires the intimate contact of two or more components for initiation. In blood clotting the once inactive blood cells are triggered when the endothelial cells are breached. This only occurs locally to the damage site, remote from the damage the clotting sequence is stable and inhibited. It is desirable to develop a resin system that can mimic this in order to allow multiple, localised repair events. A key inspiration from haemostasis is to have the reaction initiated by the absence of an inhibitor (e.g. endothelial cells). Anaerobic resin systems could be considered to function in this way.

4.3 Compartmentalisation

While a cut to mammalian flesh triggers the blood-clotting process, in other natural systems, e.g. trees, it is internal scabbing or the formation of internal impervious boundary walls that develop over time to protect the tree from further damage (Biggs 1985). This defence mechanism is termed compartmentalisation and is the main healing mechanism that protects plants and trees from pathogen infection through wounds (Bostock and Stermer 1989), ensuring that fluid diffusion and infectious microorganisms never extend beyond the wound site (Biggs 1992). There is debate whether this approach can be really be termed self healing (Shigo 1984; Bostock and Stermer 1989). The xylem vascular conduits are themselves compartmentalised and there is no provision for sealing-breached vessels; redundancy allows the complete loss of injured vessels (McCulloh et al. 2003). Nevertheless, it has shown to be an evolutionarily successful way of managing injury.

There is a parallel concern in the use of composite materials; the effect of environmental attack on damaged structures. Moisture ingress can reduce the strength of composite structures significantly over time. The approach that has evolved in trees could be considered a form of damage-tolerant design by engineers. A biomimetic system for producing an impervious internal boundary in a damaged structure would allow damage-tolerant design to consider only the direct loss of structural performance caused by damage, rather than having to also consider the secondary performance loss associated with moisture ingress through damage. This could, for example, be especially valuable for sandwich structures operated in the marine environment, providing safety, inspection, and maintenance benefits.

4.4 Reinforcement Repair

To date, self healing in fibre-reinforced composite materials has been primarily focussed on the potential offered by polymer matrices because typical impact damage is primarily in the matrix. However, the reinforcing phase provides the majority of the strength and stiffness within any fibre-reinforced composite material, and it is this component that would benefit significantly from a self healing capability for other failure modes.

The healing potential of fractured bone in the human body is influenced by a variety of biochemical, biomechanical, cellular hormonal, and pathological mechanisms (Kalfas 2001). The healing process is a continuous state of bone deposition, resorption, and remodelling. To date the repair of fractured synthetic-reinforcing fibres has yet to be undertaken in an engineering context. Furthermore, the in situ growth of additional fibres to support additional design loads, similar to branch growth in trees, has not been considered. The problem with these approaches is likely to be the timescales required for growth, which are unrealistic for engineering applications.

5   Concluding Remarks

It is the possibility of self healing a damaged structure that is increasingly of interest to composite designers seeking lower mass structures with increased service life, who wish to progress from the more conventional conservative, damage tolerance philosophy. Self healing approaches applied in composite materials to date have primarily been bioinspired.

For example, a resin-filled hollow fibre approach has been shown to be a very effective method of imparting self healing to a high-performance glass or carbon fibre-reinforced polymer. The specific placement of self healing plies or fibres to match a critical damage threat and repair internal matrix cracking and delaminations is crucial to the approaches effectiveness. Such a system does offer significant potential in restoring structural integrity to a composite component during service and prolonging residual life after a damage event, however, their are limitations.

Future studies will need to undertake detailed analyses of healing in nature to allow true biomimetic self healing composite systems to be developed. Tailored placement of healing components and the adoption of biomimetic vascular networks for self healing are becoming increasingly active research topics. Novel biomimetic healing agents are a requirement closely linked to the adoption of delivery systems. Compartmentalisation is an alternative strategy that bridges the gap between self healing and the more traditional engineering damage-tolerance design philosophy and could be particularly applicable to the problem of moisture ingress into damaged composite structures. Self healing reinforcement could extend the concept of self healing beyond the repair of matrix dominated failure modes.

Such concepts outline the future of self healing in composites. Their exploitation will require extensive multidisciplinary research because many natural systems have evolved to be enormously complex but robust systems. The key engineering challenge is to understand and extract the key functional aspects of natural systems in order to produce systems that can feasibly and cost-effectively be applied to composite structures.

Acknowledgements The authors wish to acknowledge the financial support of the European Space Agency, the UK Engineering and Physical Sciences Research Council, and University of Bristol for funding various aspects of this research under ESTEC Contract No.: 18131/04/NL/PA, Grant Nos GR/TO3390 and GR/T17984 and a Convocation Postgraduate Scholarship.

References

Abrate S. (1998) Impact on composite structures. Cambridge University Press, Cambridge, UK, pp 1–5

Bar-Cohen Y. (2006) Biomimetics – using nature to inspire human innovation. Bioinspiration and Biomimetics 1:1–12

Biggs A.R. (1985) Suberized boundary zones and the chronology of wound responses in tree bark. Phytopathology 75(11):1191–1195

Biggs A.R. (1992) Anatomical and physiological responses of bark tissues to mechanical injury. In: Blanchette RA, Biggs AR (eds) Defense mechanisms of woody plants against fungi. Springer, Berlin, pp 13–40

Bleay S.M., Loader C.B., Hawyes V.J., Humberstone L., Curtis P.T. (2001) A smart repair system for polymer matrix composites. Composites A. 32:1767–1776

Bostock R.M., Stermer B.A. (1989) Perspectives on wound healing in resistance to pathogens. Annu Rev Phytopathol 27:343–371

Brown E.N., Kessler M.R., Sottos N.R., White S.R. (2003) In situ poly(urea-formaldehyde) microencapsulation of dicyclopentadiene. J. Microencapsul 20(6):719–730

Brown E.N., White S.R., Sottos N.R. (2005a) Retardation and repair of fatigue cracks in a microcapsule toughened epoxy composite – Part I: manual infiltration. Compos Sci Technol 65:2466–2473

Brown E.N., White S.R., Sottos N.R. (2005b) Retardation and repair of fatigue cracks in a microcapsule toughened epoxy composite – Part II: in-situ self-healing. Compos Sci Technol 65:2474–2480

Brown E.N., White S.R., Sottos N.R. (2006) Fatigue crack propagation in microcapsule toughened epoxy. J Mater Sci 41(19):6266–6273

Bruck H.A., Evan J.J., Peterson M.L. (2002) The role of mechanics in biological and bioinspired materials. Experim Mech 42:361–371

Chen X., Dam M.A., Ono K., Mal A.K., Shen H., Nutt S.R., Sheran K., Wudl F. (2002) A thermally re-mendable cross-linked polymeric material. Science 295(5560):1698–1702

Chen X., Wudl F., Mal A.K., Shen H., Nutt S.R. (2003) New Thermally remendable highly cross-linked polymeric materials. Macromolecules 36(6):1802–1807

Curtis P.T. (1996) Multifunctional polymer composites. Adv Perform Mater 3:279–293

Davies K.A., McNicol G.P. (1983) Haemostasis and thrombosis. In: Weatherall DJ, Ledingham JGG, Warrel DA (eds) Oxford Textbook of Medicine, Chapter 19. Oxford University Press, Oxford

Dry C. (1994) Matrix cracking repair and filling using active and passive modes for smart timed release of chemicals from fibres into cement matrices. Smart Matls Structs 3(2):118–123

Dry C. (1996) Procedure developed for self-repair of polymer matrix composite materials. Comp.Structs 35(3):263–269

Dry C. (2000) Three designs for the internal release of sealants, adhesives and waterproofing chemicals into concrete. Cement Concrete Res 30:1969–1977

Dry C., Corsaw M. (2003) A comparison of the bending strength between adhesive and steel reinforced concrete with steel only reinforced concrete. Cement Concrete Res 33:1723–1727

Dry C., McMillan W. (1996) Three-part methylmethacrylate adhesive system as an internal delivery system for smart responsive concrete. Smart Matls Structs 5(3):297–300

Gupta S., Zhang Q., Emrick T., Balazs A.C., Russell T.P. (2006) Entropy-driven segregation of nanoparticles to cracks in multilayered composite polymer structures. Nat Mater 5:229–233

Hayes S.A., Jones F.R., Marhiya K., Zhang W. (2007) Self-healing composite materials. Composites A 38(4):1116–1120

Hucker M., Bond I., Foreman A., Hudd J. (1999) Optimisation of hollow glass fibres and their composites. Adv Comp Lett 8(4):181–189

Hucker M.J., Bond I.P., Haq S., Bleay S., Foreman A. (2002) Influence of manufacturing parameters on the tensile strengths of hollow and solid glass fibres. J Mater Sci 37(2):309–315

Hucker M.J., Bond I., Bleay S., Haq S. (2003a) Experimental evaluation of unidirectional hollow glass fibre/epoxy composites under compressive loading. Composites A 34(10):927–932

Hucker M.J., Bond I., Bleay S., Haq X.S. (2003b) Investigation into the behaviour of hollow glass fibre bundles under compressive loading. Composites A 34(11):1045–1052

Hougie C. (2004) The waterfall-cascade and autoprothombin hypotheses of blood coagulation: personal reflections from an observer. J Thromb Haemost 2:1225–1233

Kalfas I.H. (2001) Principles of bone healing. Neurosurg Focus 10:1–4

Kassner M.E., Nemat-Nasser S., Suo Z., Bao G., Barbour J.C., Brinson L.C., Espinosa H., Gao H., Granick S., Gumbsch P., Kim K-S., Knauss W., Kubin L., Langer J., Larson B.C., Mahadevan L., Majumdar A., Torquato S., van Swol F. (2005) New directions in mechanics. Mech Mater 37:231–259

Kessler M.R., White S.R. (2001) Self-activated healing of delamination damage in woven composites. Composites A. 32(5):683–699

Kessler M.R., White S.R., Sottos N.R. (2002) Self healing of composites using embedded microcapsules:repair of delamination damage in woven composites. 10th European conference on composite materials (ECCM-10), Bruges, Belgium, 3–7 June 2002

Kessler M.R., White S.R., Sottos N.R. 2003 Self-healing structural composite materials Composites A 34:743–753

Lee J.Y., Buxton G.A., Balazs A.C. (2004) Using nanoparticles to create self-healing composites. J Chem Phys 121(11):5531–5540

Li V.C., Lim Y.M., Chan Y.W. (1989) Feasibility of a passive smart self-healing cementitious composite. Composites B 29B:819–827

Macfarlane R.G. (1964) An enzyme cascade in the blood clotting mechanism, and its function as a biochemical amplifier. Nature 202:498–499

McCulloh K.A., Sperry J.S., Adler F.R. (2003) Water transport in plants obeys Murray's law. Nature 421:939–942

McCulloh K.A., Sperry J.S., Adler F.R. (2004) Murray's law and the hydraulic vs mechanical functioning of wood. Funct Ecol 18:931–938 (doi:10.1111/j.0269–8463.2004.00913.x)

Motuku M., Vaidya U.K., Janowski G.M. (1999) Parametric studies on self-repairing approaches for resin infused composites subjected to low velocity impact. Smart Matls Structs 8:623–638

Murray, C.D. (1926) The physiological principle of minimum work. I. The vascular system and the cost of blood volume. Proc Natl Acad Sci USA 12:207–214 (doi:10.1073/pnas.12.3.207)

Pang J.W.C., Bond I.P. (2005a) Bleeding composites—damage detection and self-repair using a biomimetic approach. Composites Part A 36:183–188

Pang J.W.C., Bond I.P. (2005b) A hollow fibre reinforced polymer composite encompassing self-healing and enhanced damage visibility. Compos Sci Technol 65:1791–1799

Richardson M.O.W., Wisheart M.J. (1996) Review of low-velocity impact properties of composite materials. Composite Part A 27A:1123–1131

Rule J.D., Brown E.N., Sottos N.R., White S.R., Moore J.S. (2005) Wax-protected catalyst microspheres for efficient self-healing materials. Adv Mater 17(2):205–208

Runyon M.K., Johnson-Kerner B.L., Ismagilov R.F. (2004) Minimal functional model of hemostasis in a biomimetic microfluidic system. Angewandte Chemie International Edition 43:1531–1536

Sanada K., Yasuda I., Shindo Y. (2006) Transverse tensile strength of unidirectional fibre-reinforced polymers and self-healing of interfacial debonding. Plast, Rubb Compos 35(2):67–72

Sherman T.F. (1981) The meaning of Murray's Law. J Gen Physiol 78:431–453

Shigo A.L. (1984) Compartmentalization: a conceptual framework for understanding how trees grow and defend themselves. Ann Rev Phytopathol 22:189–214

Taber L.A., Ng S., Quesnel A.M., Whatman J., Carmen C.J. (2001) Investigating Murray's Law in the chick embryo. J Biomech 34:121–124

Trask R.S., Bond I.P. (2006) Biomimetic self-healing of advanced composite structures using hollow glass fibres. Smart Mater Struct 15:704–710

Trask R.S., Williams G.J., Bond I.P. (2007a) Bioinspired self-healing of advanced composite structures using hollow glass fibres. J Roy Soc Interf 4(13):363–371 (doi: 10.1098/rsif.2006.0189)

Trask R.S., Williams H.R., Bond I.P. (2007b) Self-healing polymer composites: mimicking nature to enhance performance. Bioinspiration Biomimetics 2(1):1–9.

Toohey K.S., White S.R., Sottos N.R. (2006) Microvascular networks for self-healing polymer coatings. Fifteenth United States National Congress of Theoretical and Applied Mechanics. Boulder, CO, USA, 25th–31st June 2006

Verberg R., Alexeev A., Balazs A.C. (2006) Modelling the interaction between particle-filled capsules and substrates: potential for healing damaged surfaces. Fifteenth United States National Congress of Theoretical and Applied Mechanics. Boulder, CO, USA, 25th–31st June 2006

Vincent J.F.V., Mann D.L. (2002) Systematic technology transfer from biology to engineering. Philos Trans R Soc Lond A 360:159–173

White S.R., Sottos N.R., Geubelle P.H., Moore J.S., Kessler M.R., Sriram S.R., Brown E.N., Viswanatham S. (2001) Autonomic healing of polymer composites. Nature 409:794

Williams H.R., Trask R.S., Bond I.P. (2006) Vascular self-healing composite sandwich structures. 15th US National Congress of Theoretical and Applied Mechanics. Boulder, CO, USA, 25–31 June

Williams G.J., Trask R.S., Bond I.P. (2007) A self-healing carbon fibre reinforced polymer for aerospace applications. Composites A. 38(6):1525–1532 (doi: 10.1016/j.compositesa.2007.01.013)

Yin T., Rong M.Z., Zhang M.Q., Yang G.C. (2007) Self-healing epoxy composites – preparation and effect of the healant consisting of microencapsulated epoxy and latent curing agent. Compos Sci Technol 67:201–212

Zako M., Takano N. (1999) Intelligent material systems using epoxy particles to repair microcracks and delamination damage in GFRP. J Int Mat Sys Struct 10(10):836–841

Zhou G. (1998) The use of experimentally-determined impact force as a damage measure in impact damage resistance and tolerance of composite structures. Compos Struct 42:375–382

Self Healing Polymer Coatings

Rolf A.T.M van Benthem, Weihua (Marshall) Ming, and Gijsbertus (Bert) de With

Laboratory of Materials and Interface Chemistry, Eindhoven University of Technology, Den Dolech 2, 5600 MB Eindhoven, The Netherlands
E-mail: r.a.t.m.v.benthem@tue.nl

1   Abstract

This chapter deals with aspects to be considered with respect to the self healing of polymer coatings. After discussing the scope and limitations of self healing concepts in polymer coatings, an overview of present approaches and technologies is given. The differences between reversible and irreversible networks are discussed. Surface self-replenishing as well as active species self-replenishing are dealt with. A currently applied industrial example is elaborated: scratch-healing automotive coatings. Finally, possible future scenarios are discussed.

2   Introduction to Polymer Coatings

There are not many materials in our modern society that are both as ubiquitously visible and as usually unnoticed as coatings. Some are used to give a good appearance to, for example, buildings, vehicles, and furniture without attracting the eye's conscious attention to their own existence; others are highly functional but invisibly embedded in devices, or just transparently covering other materials. Whether they are perceived as relatively "simple" materials such as for example paint layers, or are a functional part of "high-tech" devices, they are in many aspects playing quite vital roles in the quality of our life. As such, the attention to the potential of self healing effects in these material layers is more than justified.

This chapter deals with self healing phenomena and considerations in coatings, polymer coatings to be more precise. Where the term "coating" generally implies any thin layer covering another material, the prefix "polymer" excludes inorganic coatings, such as metallic and ceramic layers. This is a deliberate choice in the outline of this chapter. For self healing phenomena in such coating materials, reference is made to earlier chapters (2–6) dealing with these material classes. In general, self healing phenomena relate to a material class as such, irrespective of the specific scale of use, i.e. in thin layers/films or as structural materials. This statement is valid for these inorganic materials because the material chemistry is highly similar in bulk and coatings and, thus, the considerations of self healing effects follow similar lines of thought. The need for a separate chapter on polymer coatings in this book, in parallel with the chapters on polymer materials, is explained by the invalidity of a similar statement on the consideration of self healing effects in polymers and polymer coatings.

First of all, the material chemistry is distinctively different between "polymers", i.e. polymeric materials as generally used in practice (see Chapters 2–6) and polymer coatings, although by name they sound similar. The term "polymer coating" suggests that these coatings comprise predominantly, if not solely, organic polymers, but there are also many examples in which inorganic materials actually dominate by weight. What these coatings do have in common is an organic polymer matrix or binder, a continuous polymer phase holding all constituents together. Also bulk polymer materials can have quite high inorganic filler fractions, but in this case their use is more related to structural properties (hardness/modulus, flexural strength, flame retardation) than to surface properties (color, hiding power, gloss, and reflectance). Moreover, bulk polymer materials are not as highly filled as coatings usually are. The performance of a polymer coating is strongly dominated by three main interfaces: coating-substrate (bottom), coating-air (top), and binder-filler (internal). Compared to structural polymer materials, in coatings the ratio of the values of the interface area to polymer matrix volume is much larger.

More importantly, the material chemistry and topology of the polymer phase is highly different in coatings as compared to bulk polymers. Whereas the vast majority of polymeric materials is thermoplastic in nature, most polymer coatings are thermosets. Thermosets are infinite three-dimensional networks of covalently interconnected polymer segments. The reason for most coatings to be thermosets, is that the network architecture offers the highest resistance of these thin polymer layers to solvents, chemicals and mechanical stresses [1]. An intriguing feature of most polymer coatings is that they are applied as liquids, for ease of handling and spreading, but that they need to be transformed from this obviously thermoplastic phase into the final thermoset state immediately after being applied to the underlying material (in a process called "cure" or "curing"). This need for curing is inevitably reflected in the choices of material chemistry of such materials. Another aspect is material transport and obviously this plays a central role in self healing materials. Material transport in a thermoplastic polymer material can easily be envisaged by molecular motions of individual macromolecules which are mostly entangled but still largely independent. Material transport in a thermoset material is highly hindered by the network structure, whether one considers the motions of the polymeric network segments themselves, or the motions of unconnected small molecules diffusing through the network, as exemplified in section 2.2.

Finally, the functional requirements for a polymer coating are generally quite different from those for a structural polymer material. The two primary functions of a surface coating are decoration and protection. The decorative function relates to the appearance of the coated substrate, largely dominated by the appearance of the coating layer itself, and covers many aspects such as specular gloss and diffuse reflection, surface undulations, color perception, hiding power, transparency, cleanability, etc. The protective function relates to the underlying substrate, which has to be sealed off from external influences such as light, humidity, air, fungi, bacteria, dirt, chemicals, mechanical abrasions, and the like. Often this protective function is in close interplay with the underlying substrate. Corrosion protection is an example. In this chapter we will not deal comprehensively with all these functions but it should be clear that most coatings have to provide for a delicate mix of functional properties. This is an essential point when one considers the design of a self healing effect for one of those properties: the envisaged effect and the measures to provide for it should not compromise any of the other essential functional properties of the polymer coating. Although once stated, this remark may seem obvious, design does not always adhere to this rule. In fact this rule applies to almost all materials to be applied technologically.

2.1 Self Healing: Scope and Limitations

This section will elaborate further on that last remark: whereas the scope of self healing is as wide as the reach of any property of a polymer coating, the total required set of properties determines the limitations of self healing. Let us, for example, consider a coating that has the ability to reflow after serious scratches or cracks. Ideally we would want to design the coating in such a way that it simply reflows into the damaged area, spontaneously and within a couple of minutes. We all know the archetype of such a coating: a wet paint layer. We also know the disadvantage of spontaneous flow: on vertical surfaces the layer tends to drip down. When touched, it feels extremely sticky and can be damaged easily. In other words, this coating cannot simultaneously have the property of withstanding flow in order to feel hard and the property of easy flow in order to repair scratching. This brings us to the first type of limitations: contradictions. But there is more to this example than just the contradiction between viscous (re)flow and resistance to flow. How should the coating distinguish between a damaged area, into which material transport is desired, and a neighboring uncoated area, into which material transport is not desired? The coating, no matter how "smart" it may be, cannot make that decision: desired versus undesired material transport is a contradiction by itself. It is up to us, while designing the self healing coating, to keep those decisions for ourselves and use specific intrinsic driving forces and natural barriers to let the coating behave to our desire.

This contradiction can be resolved, however, when time is taken into account. We can, for example, use an external trigger to temporarily change the properties of a coating, to allow self healing, and let the coating return to its "ground state" after a specific time interval. It goes without saying that during that time interval the coating should not be exposed to the influences it normally is supposed to withstand; the contradiction remains active during that stage. Examples of triggered damage recovery will be given in sections 3.1 and 3.3.

A second type of limitations is based on the intrinsic features of the material chemistry used, organic polymers in this case. Foremost, organic polymers are sensitive to weathering and ageing: under the influence of moisture, UV radiation, and oxygen from the air, progressing degrading processes such as photolysis, photooxidation, and hydrolysis are inevitable. Although the thought of designing a coating that continuously or repetitively self heals such damages is very tempting, an ideal everlasting coating is simply not possible on the basis of organic polymers. It is a more appropriate mindset to think of self healing mitigation strategies to slow these degrading processes down. Similar intrinsic limitations of polymer coatings are damages caused by thermo-oxidative degradation (burning) and reactions with aggressive chemicals.

2.2   Damage Recovery on Different Size Scales: Pre-emptive Healing

Before we discuss different healing strategies in polymer coatings, let us first consider the meaning of damage and damage recovery. Damage can be seen as an immediate or imminent loss of function of a coating, and can be viewed from different size scale angles.

A superficial scratch that distorts the surface, and hence, the reflection of the light, degrades the decorative function. A crack that penetrates right through the coating from the surface to the interface with the substrate disables the protective function of that coating by allowing direct access of environmental influences to the substrate. These are examples of immediate damage on or close to the level and scale of the film thickness itself, which we shall denote as macroscopic from now on. It is important to realize that there is microscopic damage indispensably associated with the macroscopic damage: the crack through the coating layer could not arise if not the coating constituents, i.e. macromolecular network segments or possibly even filler particles, have been broken (Figure 1). It is even possible to discern a mesoscopic equivalent process in crack formation: the local deformation behavior of the polymeric network segments, comprising yielding and crazing that lead to the initiation and propagation of the crack.

An imminent loss of function must also be considered as damage. The crack discussed above would not have formed if there had not been any buildup of stress

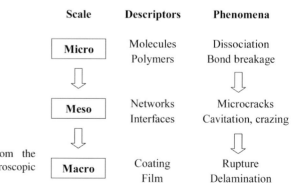

Fig. 1 Damage escalation, from the microscopic level to the macroscopic level

in the coating. Besides the obvious stresses directly related to immediate damage, there can be several other kinds of stress that influence the eventual crack formation. They are built up prior to the moment of macroscopic failure. Firstly, there can be internal stresses in the polymer coating as a result of the chemistry of network formation, mostly a combination of diminished reaction volume and loss of hydrodynamic volume: shrinking. Secondly, there can be internal stresses in the polymer coating, as well as interfacial stresses at the substrate interface, as a result of the difference in thermal expansion coefficients between polymer coating and substrate, in combination with the thermal history of the coated material. Thirdly, previous external stresses may have left their traces on the microscopic and mesoscopic level, whereas they have not been revealed on the macroscopic scale yet. Such prior external stresses can be of the same kind as the final stress that leads to the macroscopic failure, for example repetitive mechanical strains eventually leading to cracking, but also of a different kind. For the latter we distinguish between reversible and irreversible effects. An example of a reversible effect is hygroscopic strain due to the changing relative humidity of the environment. As an example of an irreversible effect, it is well known, for example, that weathering exposure of a coating enhances the probability of cracking when mechanical strain is imposed. This is due to progressive embrittlement of the coating leading to easier crack propagation.

In damage recovery, we have to consider the same micro-meso-macro hierarchy. Damage recovery always involves material transport in the organic polymer coating, either the individual motion of small molecules or the collective motion in viscoelastic flow of the polymer binder (Figure 2). The material transport needed for recovery purposes can take place within the bulk of the coatings, or from the bulk to each of the three interfaces (substrate, air, filler). As stated before, material transport is rather limited in organic polymer networks. It is therefore imperative to try and limit the need for material transport in the self healing process as much as possible. Appropriate timing is therefore of the essence. If one waits for damage to reach the macroscopic scale (e.g. the crack discussed above), it is obvious that the amount of material transport to fill the crack again is very likely to be too much to ask from a networked polymer binder. If we could stop the crack from growing through crazing

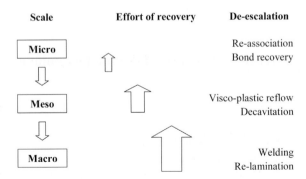

Fig. 2 Damage recovery and pre-emptive de-escalation

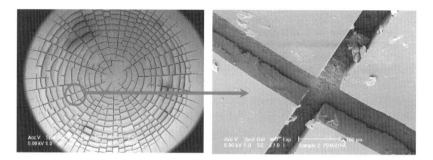

Fig. 3 SEM photographs of a coated steel panel with a reverse impact test damage area. (Courtesy DSM Research, Dr. M. Bulters)

on the mesoscopic scale, or even prevent the initiation of the crack on the microscopic scale, progressively less material transport would be needed to keep the coating in the pristine state on the macroscopic scale, that is, in relation to its macroscopic function.

This, then, is the principle of "pre-emptive healing". If we see damage as a process that starts on the microscale and escalates via the mesoscale ultimately to a loss of function on the macroscale, we can see an obvious strategy for self healing in preventing the damage process from escalating to the next higher level. Although the term pre-emptive could be interpreted such that a healing action starts before there is any damage, it actually implies that the healing action starts before the damage process reaches the level in which the function of the coating is affected. In other words, the imminent damage is averted by an invisible action, at least on the level of the macroscopic observer.

Figure 3 is an electron microscope photograph of a ruptured coating on steel panel after a falling-bullet reverse impact testing. It is clearly visible that after the film ruptured in both radial and normal direction, the elastic energy present in the film as a result of internal and external stresses, once released by the rupture, not only caused widening of the cracks but also delamination from the steel substrate. Such catastrophic damage is beyond repair, let alone self healing. Pre-emptive relaxation of the stress could have prevented this damage.

3   Current Overview of Approaches and Technologies

3.1   Self Healing Binders

3.1.1   Encapsulated Liquid Binders

In Chapter 2, self healing polymer composites using microcapsules containing a healing agent have been described in detail. The concept is on the basis of an ingenious autonomic healing strategy adopted by White et al. [2], in which a living polymerization (ring-opening metathesis polymerization, ROMP) catalyst and urea-formaldehyde-microencapsulated dicyclopentadiene (DCPD) monomer were embedded in an epoxy composite. Once mechanically induced microcracks ruptured the capsules, DCPD monomer was released, filled the crack, and polymerized upon the contact with the catalyst. The crack was healed, with a reported 75% recovery of the original strength of the epoxy composite.

A similar encapsulation approach has been recently used in self healing coatings [3]. Microcapsules containing liquid film formers are incorporated in self healing coatings; the microcapsules used are typically of about 60–150 μm in diameter and the shell is made of urea formaldehyde or gelatin. After the coating has been cured, any physical compromise of the coating results in microcapsules bursting to release liquid that fills and seals the compromised volume. The coating was developed for the purpose of suppressing lead dusts.

One potential limitation for the microencapsulation approach is that the coating lacks the capability of multiple self healing. Moreover, long-term stability of the catalyst remains an issue of concern. The size of the microcapsules (50 μm and above) will only allow possible applications in relatively thick coatings able to accommodate the microcapsules. In addition, the voids (in the order of μm), left behind because of the diffusion of reactive binder, would become new loci of stress concentration.

Recently, Toohey et al. [4] reported that they obtained continuous self healing polymer coatings by using a replenishable supply of healing materials in a microvascular network. The reaction system used is based on ROMP of DCPD. Instead of being stored in microcapsules, DCPD is kept in interconnected microchannels. Similar to the microcapsule-based system, propagating cracks can rupture the microchannels, allowing DCPD to flow into the crack plane and to react with the ROMP catalyst embedded in the coatings. In the event of the crack reopening, more healing agent can flow through the microchannel and heal the damage again. However, the fabrication of the microvascular network appears to be challenging before widespread applications are envisaged.

3.1.2 Deformation Recovery in Networks

Polymer coatings, even networks, show visco-elasto-plastic behavior. This implies that any deformation is composed of a viscous response, plastic flow, and elastic deformation [5]. The elastic, time-independent component of the response is related to stored energy, which can be used to recover from the deformation, at least partly. The plastic, time-independent component does not contribute to healing, while the viscous, time-dependent component in principle can also contribute to self healing via its memory effect. While in thermoplastic materials viscoelastic behavior is dominant, in thermosets, as used in coatings, both plasticity and viscoelasticity play a role. A more complete overview of both deformation phenomena can be found in [6].

Let us consider a polymer coating which is not cross-linked, a thermoplastic polymer composition, being subject to a surface deformation: in particular a scratch,

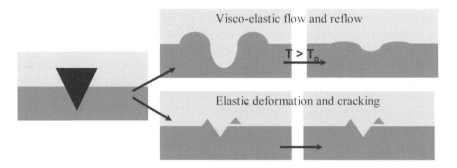

Fig. 4 Scratch formation and recovery in thermoplastic (top) and thermoset (bottom) films

Figure 4 (top). The damage, a mesoscopic ditch visible as the scratch, is actually the result of material transport. Under the mechanical stress of the moving indenter, some polymer flows from the indentation area, leaving the ditch, to the sides, where it is piled up. The energy put into the system to create the scratch is lost in the process of viscous flow unless residual stress due to viscoelastic (or plastic) deformation is present. After this damage, ideal recovery would imply reflow of the material from the sides into the ditch, completely leveling to the original state.

This is possible, but the driving forces for this desired material transport are quite small. Gravity, for example, is utterly insufficient to drive this surface leveling. The most important driving force is the surface tension, which will try to keep the surface area as small as possible. Resistance against damage recovery, however, is the same as the resistance against the damaging process itself: the viscoelasticity of the thermoplastic polymer. This means that if we want to design a coating that easily recovers from scratching by the action of the surface tension, i.e. a polymer composition with a low glass transition temperature, this coating will also easily scratch to begin with. Vice versa, a coating which has a high glass transition temperature will have a high resistance towards scratching but will also show slow damage recovery. In the unfortunate event, however, that material transport during damaging is not a gradual flow process but accompanied by microcracks in the bottom of the ditch, also those cracks will eventually have to have the ability to reflow. This is governed by the same resistance as described for the scratch.

In Figure 4 (bottom), the response of an ideal thermoset polymer coating is depicted. Also in this case, the scratch is formed as a result from material transport from the indented area to the sides. In this case, however, the resistance is stronger and proportional to the extent of the deformation: the elastic and/or plastic response. As a matter of fact, all or part of the energy put into the system to create the scratch, dependent on whether the yield strength is exceeded or not, is stored in the polymer network segments in the vicinity of the scratch. As soon as the external, mechanical stress is taken away, the stored elastic energy is relieved, and like a spring bouncing back after being compressed, the scratch bounces back into a flat level surface. The timescale on which this happens is related to the mobility of the polymer network chains, i.e. the glass transition temperature, as in the case of the thermoplastic coating, but the recovery will in principle be complete (provided that the yield strength is not exceeded) even if the surface tension forces are neglected.

This seemingly ideal behavior, however, can be drastically spoiled if at some point the mechanical stress results not only in elastic deformation, but in crack formation. During these fractures, the stored elastic energy is released untimely and this energy is no longer useful for a directed bounce-back movement. Material on either side of the fracture will try to bounce back by itself, no longer behaving like a unity and some of the cracks will be left. Unlike in the case of a thermoplastic polymer coating, these fractures cannot reflow by a viscous process anymore, driven by surface tension, because now the elastic forces are opposing the desired material flow. It consequently follows that a scratch in an elastic thermoset which is accompanied by fractures is a permanent damage.

In practice, the behavior of a polymer coating will be a superposition of the viscoelastic response (top) and the pure elastic/cracking response (bottom). The search for the optimal mix of these responses can lead to a coating with self healing properties to some extent, especially when use can be made of a thermal trigger for healing. In the ideal case the coating has a sufficiently high glass transition temperature, at least higher than the temperature of its exposure to the scratching, in order to minimize the indentation driven material transport during the damaging process. A relatively high proportion of the response is elastic, so that most minor scratches, those without crack formation, will disappear immediately and entirely. In case fracture does occur in the damage process, plastic residual strain allows for some reflow of the cracks but healing should be mainly driven by surface tension. A temperature trigger, i.e. a short heating cycle to bring the polymer above its glass transition temperature will enhance mobility and allow the surface tension to drive the healing of the surface as much as possible.

It must be clear though that such scratch self healing is restricted to quite shallow scratches in the order of maximal micrometer range, visible and annoying to the eye nevertheless, but insignificant in relation to the coating layer thickness. The mechanism is, therefore, mainly acting to restore the aesthetical (decorative) function of the coating.

An example of this strategy is given in section 3.3, the industrial use of scratch-healing in automotive clear coats.

3.1.3   Stress Relaxation in Reversible Networks

Chapters 3 and 5 are dealing with self healing phenomena based on polymers which are constituted, at least partially, by non-covalent bonds, e.g. multiple-hydrogen bonds [7] and ionomers. In contrast to covalent bonds, which are permanent and not dissociable, dissociation and association of monomers and/or polymeric segments in reversible polymers are part of a dynamic equilibrium ruled by thermodynamics. The essence of the self healing phenomena in these reversible polymeric materials is that damage, in the form of a broken bond, can be easily undone by reassociation, whereas broken covalent bonds cannot be reassociated anymore (except perhaps reversible covalent bonds, as will be discussed in section 3.1.4).

The same principle can be considered for polymeric networks. One can imagine a three-dimensional network composed of monomers (having at least two or three reversible bonds with the other monomers), as well as one composed of ordinary covalent polymer segments cross-linked through reversible bonds. A reversible network can have all typical characteristics of a network, even though there is a dissociation/association equilibrium. If there is a sufficiently high cross-link density, a momentary inspection will reveal that there still is always a three-dimensional network, although a part of the cross-links is temporarily in dissociated state. An inspection on a different moment will reveal that another fraction of the cross-links is in the temporary dissociated state, but that the same average amount of cross-links persists to uphold the three-dimensionally bonded architecture.

Notwithstanding the true network architecture, there are two main differences with a covalent polymer network. The first relates to the equilibrium state of the cross-links: if the equilibrium is shifted because of external influences this will result in a changing cross-link density, in some cases even the loss of the three-dimensional architecture. One example of such an external influence is the addition of a solvent. A covalent network will, as a result of penetrating solvent molecules, swell to a certain degree determined by the cross-link density, stretching the polymer segments between the cross-links to their maximal extent, but it will not completely dissolve. In contrast, a reversible network will not have a maximal swelling degree if the equilibrium of the reversible cross-links is shifted to the dissociated state by the dissolution in the solvent. The dissolution rate is strongly depending on the rate of solvent penetration and bond dissociation, and the network structure could eventually be lost entirely. To a coating's disadvantage, such behavior would imply a high vulnerability to chemical solvents, e.g. gasoline on an automotive lacquer, or stains, like coffee or wine markings on a tabletop lacquer. To its advantage, however, similar to the reversible polymers of the previous chapters, such a coating could self heal a structural damage of broken bonds by reassociation of the cross-links.

The second main difference relates to the time scale of the equilibrium cross-links: a covalent cross-link is able to store elastic energy upon deformation of the network for an indefinite time, but a reversible cross-link may dissociate at one point in time and space and later reassociate at another locus, effectively releasing energy. When the timescale of dissociation/reassociation is larger than the timescale of the deformation stress, the reversible network will behave in the same way as a covalent network, but when it is much shorter it will not have the structural integrity of a normal covalently cross-linked material as we know it. For example, a rubber band based on a reversible polymer network could be stretched around a rolled-up newspaper, but unlike normal rubber bands it would gradually loosen its grip around the newspaper and widen up [8]. It should be stressed, however, that this creep behavior may be considered a disadvantage for structural materials (bands, tires, bars, moldings) but not necessarily so for a coating. Coatings generally do not have a structural function; they are applied on other materials that do have one. Their main function is to decorate and protect the underlying material. In this way, the ability to show creep can also be regarded as an advantage: it allows relaxation of stresses.

As outlined in the introduction 2.2, relaxation of both internal stresses and externally imposed stresses is an important feature for pre-emptive self healing of coatings: it helps to prevent escalation of damage from the microscopic level to the macroscopic level, i.e. film cracking.

The relation between physical ageing and internal stresses in coating performance is already well known for covalently cross-linked coatings, as are methods to measure such stresses [9]. The usual way to make an internal coating stress visible and measurable, is to apply the coating on a metal strip of known thickness and stiffness. A developing tensile stress, e.g. while cooling the coated strip after curing at elevated temperature, will reveal itself in a concave bending of the strip (coated face inside), while a compressive stress will reveal itself in a convex bending (coated face outside). The curvature of the metal strip is proportional to the stress in the coating and can be used for quantification.

As indicated before, stress relaxation in polymers, and therefore polymer coatings, is related to plastic and viscoelastic behavior. An overview of both phenomena can be found in [6]. The residual stress as a result of plasticity can be relieved via either a thermal treatment or via viscoelasticity. Viscoelastic stress relaxation of covalently cross-linked coatings is possible to a limited extent too. Close to or above the glass transition temperature the polymer backbone segments have sufficient mobility to yield to the stress. The crosslink density determines the maximal degree of yielding relaxation, stretching the polymer segments between the cross-links to their maximal extent, in close analogy to the solvent swelling discussed earlier. Reversible networks will not be limited by such a maximal extent of relaxation, if the timescale of dissociation/reassociation is much shorter than the time available for stress relaxation. Such a "creeping network" should be able to completely relax internal stresses such as cure shrinkage and thermal stresses, but also large imposed stresses such as convex bending of the underlying substrate material, given sufficient time for relaxation in relation to the kinetics of the dissociation/reassociation equilibrium. The influence of the polymer backbone segments will remain important however: there is sufficient molecular mobility to address the cross-links for yielding relaxation only close to or above the glass transition temperature.

There have been attempts to photo-induced stress relaxation in coatings [10], but to the best of our knowledge, there are no examples known yet in literature in which reversible networks are reviewed for their ability to autonomously and pre-emptively heal stresses in coatings. At the publishing date of this book, a project program will have started in our group on this subject [11].

3.1.4   Reversible Covalent Networks

As detailed in Chapters 3, self healing polymers have been achieved by reversible covalent bonds. There have been a few reports on different types of reversible covalent bonds, which can be potentially used as self healing coatings, even though there have been, to the best of our knowledge, no reports so far in literature.

Wagener et al. [12] reported in 1991 a thermally reversible polymer on the basis of azlactone rings, as shown in Scheme 1a. But a temperature increase to above 200°C is needed for reversible reactions. Reversible cross-linked polymers based on urea linkage were also reported [13], in which a gel-sol transition was observed in a DMF solution of two polymers, poly(4-vinylimidazole-*co*-methyl methacrylate) and poly(3-isopropenyl-α, α-dimethylbenzyl isocyanate-*co*-methyl methacrylate), as the temperature varies between room temperature and 110°C (Scheme 1b). However, the reversibility of the urea linkage in bulk was not discussed. Nonetheless, the

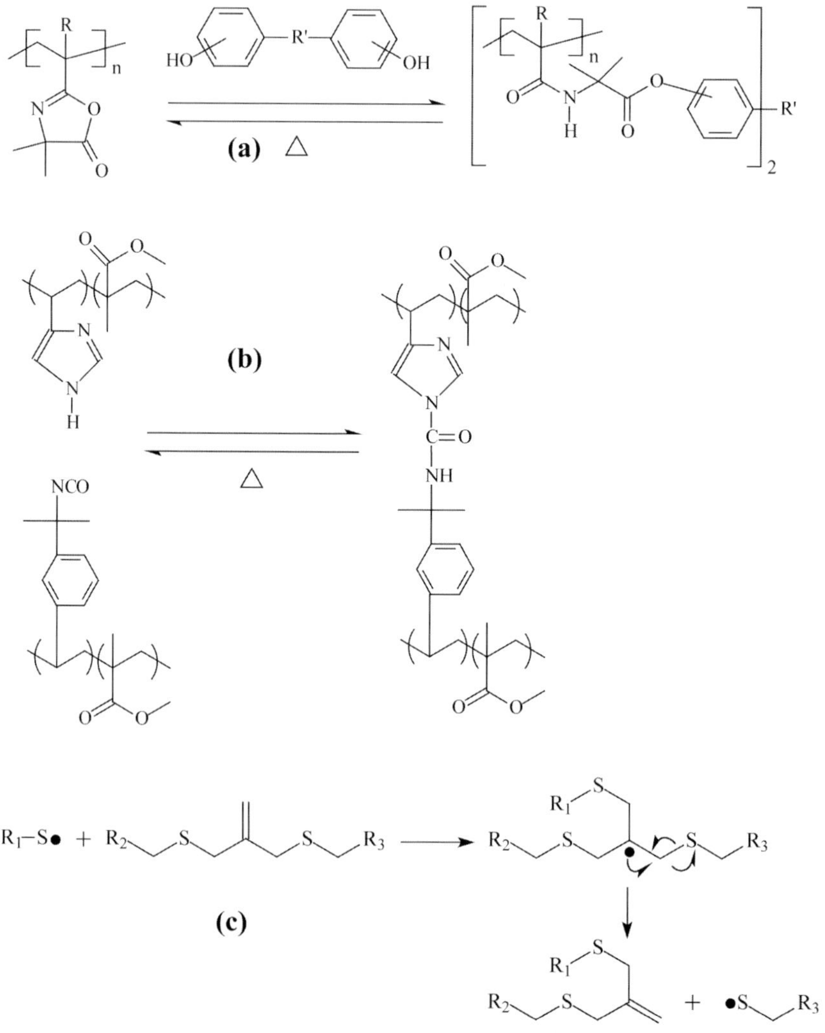

Scheme 1: Reversible covalent bonds on the basis of **(a)** azlactone and phenol [12], **(b)** imidazole and isocyanate [13], and **(c)** thiol-ene combination [14]

popularity of urethane in coating industry may enable this type of chemistry to be applied as self healing polymer coatings.

Diels–Alder reactions have been known to be reversible for decades. An outstanding example is a thermally remendable polymer material, developed by Chen et al. [14], on the basis of a thermally reversible Diels–Alder reaction. The material can, when heated to ~120°C, heal multiple times and retain more than 60% of its original strength. More details can be found in Chapter 3.

Bowman et al. [10] recently reported photo-induced plasticity in thiol-ene-based cross-linked polymer materials, which exhibit stress/strain relaxation without material property changes upon UV exposure. The result was achieved by introducing radicals via photocleavage of residual photoinitiator in a polymer matrix, which then diffuse via addition-fragmentation chain transfer of mid-chain functional groups (Scheme 1c). But applications may be limited to rubbery networks in which sufficient diffusion of segments can be assured. Similar reversible polymers on the basis of radical crossover reaction between alkoxyamine-based polymers have been reported by Otsuka et al. [15]. The reversible reaction is triggered by temperature increase.

Although reversible covalent polymers can in principle be used as self healing coatings that are capable of healing multiple times, there are some disadvantages. First, an *external* trigger such as heat, UV, or in other form, is normally necessary to induce the reversible reaction. Therefore, the healing process is not truly autonomous. Second, when the external trigger is applied, the *whole* material, including the parts where no healing is necessary, will be affected. For bulk materials whose shape is of importance, this may pose as a major drawback. For polymers used as coatings, this disadvantage does not necessarily apply since the shape of a thin coating depends primarily on the substrate it covers. Moreover, local heating can be much more easily applied. Therefore, we believe that reversible covalent polymers hold promise in obtaining self healing polymeric coatings.

3.2   Self-Replenishing Coatings

As mentioned earlier in this chapter, apart from the formation of a crack, a damage of a coating can be exemplified by the loss of a certain function in time, such as deterioration of the adhesion between the coating and the substrate, loss of UV protection, undesired change of surface wettability, loss of anticorrosion function for metallic substrates, and so on. In these cases, a coating does not suffer from loss or reduction of mechanical strength, therefore the re-bonding of broken bonds, as discussed in the previous section, may not be necessary. Instead, an intelligent way is needed to heal the coating damage by recovering the coating function. Certain coating functions are due to the presence of a specific species, which can be a low molecular-weight free molecule or a dangling chain attached to a cross-linked network. For instance, low surface wettability of a coating comes from low-surface-energy fluorinated or silicone-based chains; for anticorrosion purpose, an active corrosion inhibitor can be added. In order for the coating function to recover, the specific species need to be transported, or they can move by themselves in a "self-replenishing" fashion, to the loci where the coating function diminishes. The driving force of these specific species may include difference in surface energy, environmental changes (such as pH and temperature), and so on; in addition, they should enjoy a certain degree of mobility in a coating system for a relatively rapid transportation.

One example of a self-replenishing coating is an anticorrosion coating in which corrosion inhibitor self-replenishes at the coating/metal interface [16]. Others [17] used a layer-by-layer assembly technique to entrap a corrosion inhibitor (benzotri-azole) to silica nanoparticles, and then incorporated the nanoreservoirs in epoxy-functionalized ZrO_2/SiO_2 sol-gel coatings deposited onto an aluminum alloy. The release of benzotriazole is initiated by pH changes during corrosion of the aluminum alloy.

Another example is self-replenishing low-adherence coatings. The surface of a coating is very important, which governs wettability and friction of the coating, among others. It is well known that the surface composition of a coating may differ significantly from the bulk composition. Low-adherence coatings are widely used today, since their water/oil repellency makes them easily cleanable (a well-known example is PTFE). The low-surface tension is provided by fluorine- or silicon-containing species that are present at the film surface. Low-adherence coatings have already been developed via surface segregation of fluorinated species [18]. However, it has been shown that the fluorine-enriched layer is very thin [18e], and the coating may not sustain low adherence upon mechanical abrasion.

We recently developed a self-replenishing approach to maintain the low-adherence character of fluorinated polyurethane coatings [19]. The approach includes the following two key features. (1) The miscibility of the fluorinated tail with non-fluorinated components is enhanced by introducing a spacer between the perfluo-roalkyl group and a reactive group to allow more fluorinated species to be present in the film bulk; the spacer has the same structure with the polyester polyol used to prepare the polyurethane. (2) The spacer is based on poly(ε-caprolactone) (PCL), which renders the fluorinated tail enough mobility when incorporated into cross-linked networks. In case of surface damage that leads to the loss of the top layers of the coating, fluorinated tails from sublayers will be able to reorient themselves to minimize the air/film interfacial energy, owing to the flexibility of the PCL spacer. Model polyester precursors with controlled functionality were synthesized via controlled ring-opening polymerization of ε-caprolactone using perfluoroalkyl alcohol or polyol as initiators [20]. The as-prepared precursors were cured with a polyiso-cyanate cross-linker to obtain films with low surface energy. The self-replenishing behavior of fluorinated species has been demonstrated by angle-resolved XPS in combination with microtoming [19]. As shown in Figure 5, the F/C molar ratio appears to be constant throughout the coating, indicating a relatively homogeneous distribution of fluorinated species in the bulk of the film. What is significant is that, for each slice, the F/C ratio in the top 4 nm is about 60% greater than that in the top 8 nm of the slice. The greater F/C ratio in the top 4 nm clearly suggests that the replenishing of fluorinated tails already takes place after the microtoming and before the XPS measurements; otherwise the F/C ratios in the top 4 and 8 nm would

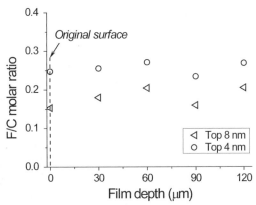

Fig. 5 F/C ratios at two different probe depths (4 and 8 nm) for thin slices cut from a polyurethane coating containing perfluorooctyl-PCL (R_f8-PCL). The coating examined was prepared from a reaction mixture (R_f8-PCL, a PCL-based polyol, and a polyisocyanate cross-linker). The coating was about 200 μm thick and contained 1wt% of fluorine. Thin slices of about 30 μm thick were cut by microtoming and the top surfaces of the slices were then examined by XPS

be the same for the slices from the bulk of the film. Since the glass transition temperature of the coating is about −20°C, it is not surprising that the self-replenishing of the fluorinated species takes place at room temperature after microtoming.

The replenishing of fluorinated species depends on the mobility of polymer spacer and network, as well as temperature. When the polymer network has a glass transition temperature close to room temperature, the replenishing of fluorinated species can take place simultaneously at the time a new surface is formed (shown above). In this respect, the replenishing behavior can be regarded as autonomous. This self-replenishing strategy may be extended to other functional coatings, such as antimicrobial coatings, biocidal coatings, and so on.

3.3 Industrial Practice

What can be more annoying in our modern life than an ugly scratch on our mobile status symbol, our new and otherwise shiny car? It is not surprising that the first examples of self healing coatings can be found in the automotive finishes industry. Well, the claims of the paint and car manufacturers relate to relatively small "car wash" scratches only (we are still vulnerable to the big scratch coming from our envious neighbor's key) and there is a considerable dose of sunshine necessary to trigger the healing process, but all is fair: scratches do disappear.

Automotive coatings usually consist of four layers, the upper two being the colored base coat and the transparent topcoat. The latter, also referred to as clear coat, protects the underlying layers from environmental influences (moisture, sunlight-UV, bird droppings, scratches, etc.) and determines for a large part the "optical appearance" of the car, most importantly the gloss. Low- and middle-priced cars generally have a topcoat of acrylic polymers cross-linked with melamine-resins; higher price level cars can have topcoats of acrylic polymer cross-linked with polyurethane resins, or all-polyurethane systems.

Automotive paint manufacturers have designed and optimized their clear coat formulations for the best preservation of the car's glossiness during its lifetime and use, determined by the consumer's geographical preferences and perceptions [21]. Typically, a car in the USA or Australia has a clear coat with high cross-linking density and high glass transition temperature, withstanding solar irradiation, sand blasting and stone chipping. A European car on the other hand sees much less sunlight, but sees the inside of the car wash much more frequently than its US counterpart. European automotive topcoats therefore, have lower glass transition temperature and lower crosslink densities. It is a well-known phenomenon for some 20 years, self healing avant la lettre, that annoying car wash bristle marks on such clear coats easily disappear or at least become less visible when the car is exposed to heat for some time. The surface of a car can easily reach 70°C when exposed to the summer sun for an hour. The principles of this phenomenon are described in section 3.1.2.

The customer's perception of the added value of a "self-healing car", however, has been an inspiration to some paint material suppliers and car manufacturers to elaborate further on this scratch-healing phenomenon and use it in branding. Nissan, in cooperation with Nippon Paint, introduced the X-Trail model in Europe in 2006, proclaiming it to be the first car with a self healing topcoat that makes car wash bristle marks disappear completely within weeks, and with a guarantee on that property for 2 years.

Bayer Material Sciences, supplier of polyurethane resins for acrylic-polyurethane and all-polyurethane clearcoats (Desmodur, Desmophen), started research in 2003 into reconciliation of the requirements for etch and weathering resistance with the ability to self heal small scratches. The classic difference between acrylic-melamine clearcoat performance (high cross-link density, etch, and weather resistant, but brittle) and polyurethane clear coat performance (low cross-link density, weatherable and flexible, but less etch resistant) formed the basis of their approach. In 2006, they announced a new all-polyurethane concept that managed to combine a relatively low glass transition ("flexible network bows" [22]) with a considerably increased cross-link density, as the solution to this paradox. This new concept is now being used by major automotive paint suppliers [23]. Figure 6, taken from a Bayer Materials Science brochure [24] illustrates that the mechanism of scratch reflow is indeed not different from earlier scratch-healing coatings (section 2.2), but optimized for the effect: while the high crosslink density increases the elastic response of the coating to the imposed visco-plastic deformation, the low glass transition temperature (and the low-associated yield strength) prevents crack formation in the scratch and assists in the reflow process triggered by increased temperature.

4 Future Scenarios

One thing is for sure: we will see more examples of commercial self healing coatings in the coming years than the automotive clear coats now breaching their way to the market. It remains doubtful whether our car will ever be able to self heal

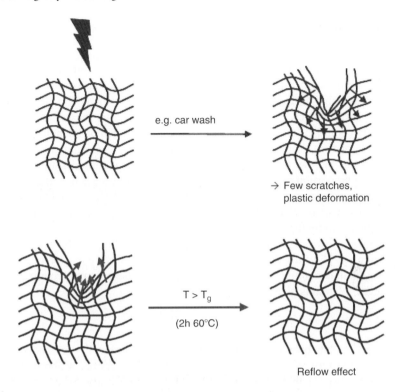

e.g. car wash

→ Few scratches,
plastic deformation

$T > T_g$

(2h 60°C)

Reflow effect

Fig. 6 Bayer Material Science's representation [24] of scratch-healing polyurethane coatings

deep scratches, but then what can we expect? Will there also be new and more sophisticated self healing mechanisms? The examples of self healing mechanisms as discussed in the previous section definitely do not cover the entire potential of conceivable self healing mechanism for polymer coatings and thin films. What more options will be found useful in the (near) future? We cannot give a comprehensive answer to that question, but we can point to a number of mechanisms that we think could be of value but have not been tried to an appreciable extent yet.

It is clear that surface tension is the most important driving force for healing of polymer surfaces and interfaces between polymers. As a first general mechanism, the surface or interfacial tension could be used for more than just material transport of surface active groups: would it not be nice to have some reactive groups left in the network that enrich at damaged interfaces and are able to form new bonds across the interface? Delamination could be undone, cracks could disappear.

Another general mechanism is decoupling of functions. In many respects the demands on a material are of an opposing nature. Separation of the contradicting functions in two (or more) different layers and/or materials could provide a solution for self healing abilities that do not withstand the normally expected coating properties of decoration and protection.

We will deal with these options in the sequel.

4.1   Residual Network Reactivity

A full conversion of all reactive groups during network formation is never accomplished in traditional thermoset polymer materials, simply because of reasons of non-exact stoichiometry (if at least two different, but complementary functional groups come together), kinetics, or hindered molecular mobility after the gel-point or even a vitrification point. Leftover reactive groups could be located on "dangling ends" of the network, or even located on unreacted polymer segments (oligomers). But could we not use this fact deliberately and design the thermoset formulation such that a considerable proportion of the reactive oligomers or chain ends is not attached to the network during the curing but left untouched for later use? They could then be used to form new bonds in a damaged area, as a reservoir for self healing capability, their segregation driven by the surface tension.

It has been shown experimentally [25] that there is difference between the surface tension γ_∞ of a fully cross-linked polymer and the surface tension of the same polymer γ_N still containing oligomers. If N denotes the degree of polymerization, an approximate relation is given by

$$\gamma_\infty - \gamma_N \propto N^{-x}$$

where $x = 2/3$ has been observed experimentally. Theoretically an exponent $x = 1/2$ is expected [26] for the so-called normal attraction regime, which is due to a decrease in entropy of the chain ends when they approach a surface. The entropy decrease leads to a certain enrichment of chain ends to the surface and may be either enhanced or counteracted by enthalpy effects. The increase in surface tension with increasing degree of polymerization indicates that, after a nearly full cure of a certain material system, segregation of oligomers to the interface is limited. For options at later stage to cure a bit further, a certain minimal amount of oligomers is thus required.

It will be clear, though, that are also limitations to this approach. What if the extra reactive groups are consumed by side reactions? Are the groups that did react during cure sufficient for obtaining basic properties including weatherability? Moreover, the theoretical expectation of chain end segregation has not been confirmed unequivocally by experiments. For a uniform distribution of chain ends the value for x is 1. The experimentally observed value $x = 2/3$ may indicate a transition between these two behaviors or be due to segregation of small molecule impurities.

4.2   Segregation of Interactive Chain Ends

In section 3.2, the principles of self-replenishing have been demonstrated. The self-replenishing species were either network-bound low-surface energy groups or unbound low molecular weight species. Could the same mechanism be also used to segregate chain ends to a damaged interface that are able to interact with similar chain ends at the opposite side of the damaged interface, and hence, to reestablish molecular interaction across the interface? Such a mechanism could result in readhesion of a delaminated coating or film, or even firmly reattach the opposite faces of a crack, whereas such phenomena are not possible with conventional polymer network materials.

Since the entropy of the chain ends results in a limited increase of chain ends present in the surface, in order to have sufficient segregation to be able to entangle with chains from the opposite, homo-interface surface, and enthalpy must help the segregation. This can be done by modifying the chain ends with strongly surface favoring groups, such as fluorine or silicone-based species. However, the segregated groups must also be able to react or at least interact with the opposite surface in order to be able to induce (self) healing. Since reactive groups are usually less prone to surface segregation than the aforementioned groups, sufficient flexibility in the end chains should be ensured.

A similar effect can be realized for hetero-interfaces. The conventional solution for hetero-interfaces is to use block copolymers. It seems possible to predict the increase in the (thermodynamic) work of adhesion for these interfaces from relatively simple molecular simulations. A recent example is provided by Kisin et al. [27] by the adhesion between polystyrene-co-acrylonitril (SAN) and copper metal by using polystyrene-co-maleic anhydride (SMA)-SAN block copolymers. Experimentally this modification resulted in an increase of adherence force of 100%, as measured with the pull-off test [28, 29]. Similar effects can be envisaged for the interface between filler particles and matrix in polymer coatings as well as between coating matrix and substrate if a release of block copolymer can be realized. A small residual amount of these block copolymers in the matrix might do the trick provided sufficient mobility is present.

4.3 Multilayer and Graded Coatings

As elaborated in the introduction, the demands on coatings are multiple and often of opposite nature. The solution for this is, of course, decoupling of the effects involved. To realize this decoupling the coating must be multilayered or, less easy, graded. In the case of coatings on a stiff substrate, a well cured, and therefore, hard coating may be applied to the substrate first and subsequently a soft, and therefore, self healing top layer may be added. Although the hard layer obviously cannot self heal, the top layer can. For flexible substrates this approach will not work since then fracture of the hard coating will occur upon bending. A related but slightly different approach is to use graded coatings with a gradient from hard to soft from the substrate to the top. Obviously, these approaches are at the expense of a more elaborate application procedure as compared to a single-layer coating, probably resulting in a more expensive solution.

References

1. Hill L. (1992) J Coat Tech 64 (nr808):29–41.
2a. White S.R., Sottos N.R., Geubelle P.H., Moore J.S., Kessler M.R., Sriram S.R., Brown E.N., Viswanathan S. (2001) Nature 409:794
2b. Rule J.D., Brown E.N., Sottos N.R., White S.R., Moore J.S. (2005) Adv Mater 17:205
2c. Cho S.H., Andersson H.M., White S.R., Sottos N.R., Braun P.V. (2006) Adv Mater 18:99
3. Kumar A., Stephenson L.D. Self-healing coatings using microcapsules containing film form-ers and lead dust-suppression compounds. US Pat. Appl. Publ. 2006, 2006042504
4. Toohey K.S., White S.R., Sottos N.R. (2005) Self-healing polymer coatings. Proceedings of the 2005 SEM annual conference and exposition on experimental and applied mechanics. pp 241–244
5. Jardret V., Morel P. (2003) Prog Org Coat 48:322–331
6. De With G. (2006) Structure, deformation, and integrity of materials. Wiley-VCH, Weinheim
7a. Bosman A.W., Sijbesma R.P., Meijer E.W. (2004) Mater Today 7:34–39
7b. Folmer B.J.B., Sijbesma R.P., Versteegen R.M., van der Rijt J.A.J., Meijer E.W. (2000) Adv Mater 12: 874–878
7c. Sijbesma R.P., Beijer F.H., Brunsveld L., Folmer B.J.B., Hirschberg J.H.K.K., Lange R.F.M., Lowe J.K.L., Meijer E.W. (1997) Science 278:1601–1604
8. Kaufmann G.B., Seymour R.B. (1990) In fact, the natural rubber as used by the native Indians to make a ball to play their basketball-like game, as observed by the first discoverers of America, suffered from this defect, just because of the absence of crosslinks. J Chem Educ 67:422
9a. Perrera D.Y., VandenEynde D. (1981) J Coat Tech 53 (nr 677):39 and ibidem 59 (nr 748):55
9b. Perrera D.Y., Oosterbroek M. (1994) J Coat Tech 66 (nr 833):55
9c. Van der Linde R., Belder E.G., Perrera D.Y. (2002) Prog Org Coat 40:215
9d. Piens M., De Deurwaerder H. (2001) Prog Org Coat 43:18
10. Scott T.F., Schneider A.D., Cook W.D., Bowman C.N. (2005) Science 308:1615
11. Funded by the Dutch IOP (Innovation Oriented Research) Self-healing Materials, we intend to use the bending curvature of coated metal strips to quantify the relaxation behavior of quadruple-hydrogen bond cross-linked polyurethane coatings, as function of (glass transition) temperature and time. The cross-links are based on ureido-pyrimidine (UPy, see ref. [7]) units also in combination with covalent urethane cross-links. The ultimate goal is to demonstrate that these coatings do not show macroscopic failure (film cracking) under conditions where conventional coatings would fail, as a result of the pre-emptive stress relaxation mechanism. At the same time it is necessary to preserve as much as possible of the typical covalent network characteristics, such as resistance to solvent swelling.
12. Wagener K.B., Engle L.P., Woodard M.H. (1991) Macromolecules 24:1225
13. Chang J.Y., Do S.K., Han M.J. (2001) Polymer 42:7589
14a. Chen X., Dam M.A., Ono K., Mal A., Shen H., Nutt S.R., Sheran K., Wudl F. (2002) Science 295:1698
14b. Chen X., Wudl F., Mal A.K., Shen H., Nutt S.R. (2003) Macromolecules 36:1802
15a. Otuska H., Aotani K., Higaki Y., Takahara A. (2003) J Am Chem Soc 125:4604
15b. Higaki Y., Otsuka H., Takahara A. (2006) Macromolecules 39:2121
16. Palaivel V., Huang Y., Van Ooij W.J. (2005) Prog Org Coat 53:153
17. Shchukin D.G., Zheludkevich M., Yasakau K., Lamaka S., Ferreira M.G.S., Möhwald H. (2006) Adv Mater 18:1672
18a. Ming W., Laven J., Van der Linde R. (2000) Macromolecules 33:6886
18b. Ming W., Tian M., van de Grampel R.D., Melis F., Jia X., Loos J., Van der Linde R. (2002) Macromolecules 35:6920
18c. Ming W., Melis F., Van de Grampel, R.D., Van Ravenstein L., Tian M., Van der Linde, R. (2003) Prog Org Coat 48:316

18d. Van Ravenstein L., Ming W., Van de Grampel R.D., Van der Linde R., De With G., Loontjens T., Thuene P.C., Niemantsverdriet J.W. (2004) Macromolecules 37:408

18e. Van de Grampel R.D., Ming W., Van Gennip W.J.H., Van der Velden F., Laven J., Niemantsverdriet J.W., Van der Linde R. (2005) Polymer 46:10531

18f. Dou Q., Wang C., Cheng C., Han W., Thüne P.C., Ming W. (2006) Macromol Chem Phys 207:2170

19. Dikic T., Ming W., Van Benthem R.A.T.M., De With G. (2006) PCT Int Appl WO 2007046687

20. Dikic T., Ming W., Van Benthem R.A.T.M., De With G. (2005) Synthesis of well-defined perfluoroalkyl-end-capped polymers by ring-opening polymerization. International conference on fluorine and silicone coatings. Manchester, UK, Paper 3.

21a. Gregorovich B.V., Adamsons K., Lin L. (2001) Prog Org Coat 43:175–187

21b. Hara Y., Mori T., Fujitani T. (2000) Prog Org Coat 40:39–40

22. Meier-Westhues U., Mechtel M., Klimmasch T., Tillack J. On the road to improved scratch resistance Bayer Materials Science, www.bayercoatings.com, accessed on 14–11–2006

23. Dössel K.F., (2006) Proceedings of car body painting. Berlin, July 2006

24. Mechtel M., Shining prospects for two-component polyurethane clearcoats Bayer Materials Science, www.bayercoatings.com, accessed on 14–11–2006

25. Legrand D., Gaines G. (1969) J Coll Interf Sci 31:162 and (1973) 42:181

26. De Gennes P.G., (1992) in: Sanchez IC (ed) Physics of polymer surfaces and interfaces. Butterworth-Heinemann, Boston, p 55

27. Kisin S., Bozovic J.,. Van der Varst P.G.T., Koning C.E., De With G (2007) Estimating the polymer–metal work of adhesion from molecular dynamics simulations. Chem Mater 19:903–907

28. Bozovic J.S., Höppener S., Kozodaev D., Kisin S., Klumperman B., Schubert U., De With G., Koning C.E. (2006) Chem Phys Chem 7:1912

29. Kisin S., Bozovic J.S., Klumperman B., De With G., Koning C.E. J Mater Chem submitted

Self Healing in Concrete Materials

Victor C. Li and En-Hua Yang

Department of Civil and Environmental Engineering, University of Michigan, Ann Arbor, USA
E-mail: veli@umich.edu

1   Introduction

1.1   Background

The phenomenon of self healing in concrete has been known for many years. It has been observed that some cracks in old concrete structures are lined with white crystalline material suggesting the ability of concrete to self-seal the cracks with chemical products by itself, perhaps with the aid of rainwater and carbon dioxide in air. Later, a number of researchers [1, 2] in the study of water flow through cracked concrete under a hydraulic gradient, noted a gradual reduction of permeability over time, again suggesting the ability of the cracked concrete to self-seal itself and slow the rate of water flow. The main cause of self-sealing was attributed to the formation of calcium carbonate, a result of reaction between unhydrated cement and carbon dioxide dissolved in water [1]. Thus, under limited conditions, the phenomenon of self-sealing in concrete is well established. Self-sealing is important to watertight structures and to prolonging service life of infrastructure.

In recent years, there is increasing interest in the phenomenon of mechanical property recovery in self healed concrete materials. For example, the resonance frequency of an ultra high-performance concrete damaged by freeze–thaw actions [3], and the stiffness of pre-cracked specimens [4] were demonstrated to recover after water immersion. In another investigation, the recovery of flexural strength was observed in pre-cracked concrete beams subjected to compressive loading at early age [5]. In these studies, self healing was associated with continued hydration of cement within the cracks. As in previous permeability studies, the width of the concrete cracks, found to be critical for self healing to take place, was artificially limited using feedback-controlled equipment and/or by the application of a compressive load to close the preformed crack. These experiments confirm that self healing in the mechanical sense can be attained in concrete materials.

Deliberate engineering of self healing in concrete was stimulated by the pioneering research of White and coworkers [6, 7] who investigated self healing of polymeric material using encapsulated chemicals. A number of experiments were conducted on methods of encapsulation, sensing, and actuation to release the encapsulated chemicals [8–12] into concrete cracks. For example, Li et al. [10] demonstrated that air-curing polymers released into a crack could lead to a recovery of the composite elastic modulus. The chemical release was actuated by the very action of crack formation

S. van der Zwaag (ed.), *Self Healing Materials. An Alternative Approach to 20 Centuries of Materials Science*, 161–193.
© 2007 *Springer*.

in the concrete, which results in breaking of the embedded brittle hollow glass fibers containing the polymer. Thus, the healing action took place where it was needed. Another approach, taken by Nishiwaki et al. [13], utilized a repair agent encapsulated in a film pipe that melts under heating. A heating device was also embedded to provide heat to the film pipe at the cracked location when an electric current is externally supplied. Yet another approach, suggested by the experiments of Bang et al. [14] and Rodriguez-Navarro et al. [15], used injected microorganisms to induce calcite precipitation in a concrete crack. These novel concepts represent creative pathways to artificially inducing the highly desirable self healing in concrete materials.

From a practical implementation viewpoint, autogenous self healing is most attractive. Compared to other engineering materials, concrete is unique in that it intrinsically contains micro-reservoirs of unhydrated cement particles widely dispersed and available for self healing. In most concrete and particularly in those with a low water/cement ratio, the amount of unhydrated cement is expected to be as much as 25% or higher. These unhydrated cement particles are known to be long lasting in time. Autogenous self healing is also economical, when compared with chemical encapsulation or other approaches that have been suggested. As indicated above, the phenomenon of autogenous self healing has been demonstrated to be effective in transport and mechanical properties recovery. Unfortunately, the reliability and repeatability of autogenous self healing is unknown. The quality of self healing is also rarely studied, and could be a concern, especially if weak calcite is dependent upon for mechanical strength recovery. Perhaps the most serious challenge to autogenous healing is its known dependence on tight crack width (CW), likely less than 50 μm, which is very difficult to achieve in a consistent manner for concrete in the field. In practice, concrete CW is dependent on steel reinforcement. However, the reliability of CW control using steel reinforcement has been called into question in recent years. The latest version of the ACI-318 code has all together eliminated the specification of allowable CW. Thus, a number of serious material engineering challenges await autogenous healing before this phenomenon can be relied upon in concrete structures exposed to the natural environment.

To create practical concrete material with effective autogenous self healing functionality, the following six attributes are considered particularly important. For convenience, we label a material with all these six attributes of autogenous self healing as "robust":

- *Pervasiveness*: Ready for activation when and where needed (i.e. at the crack when cracking occurs)
- *Stability*: Remain active over the service life of a structure that may span decades
- *Economics*: Economically feasible for the highly cost-sensitive construction industry in which large volumes of materials are used daily
- *Reliability*: Consistent self healing in a broad range of typical concrete structure environments
- *Quality*: Recovered transport and mechanical properties as good as pre-damage level
- *Repeatability*: Ability to self-repair for multiple damage events

In this chapter, we review recent experimental findings of the autogenous recovery of mechanical properties and transport properties in a specially designed cementitious material known as Engineered Cementitious Composite, or ECC. ECC has been deliberately engineered to possess self-controlled CW that does not depend on steel reinforcement or structural dimensions [16]. Instead, the fibers used in ECC are tailored [17] to work with a mortar matrix in order to suppress localized brittle fracture in favor of distributed microcrack damage, even when the composite is tensioned to several percent strain. ECC with CW as low as 30 μm has been made. The ability of ECC to maintain extremely tight CW in the field has been confirmed in a bridge deck patch repair [18] and in an earth-retaining wall overlay [19].

The deliberately pre-cracked ECC specimens were exposed to various commonly encountered environments, including water permeation and submersion, wetting and drying cycles, and chloride ponding. The mechanical properties studied include dynamic modulus, tensile stiffness, strength, and ductility. The transport properties studied include water permeability and chloride diffusivity.

The rest of this introduction is devoted to a concise review of the conditions required for reliable autogenous self healing, based on published literature (section 1.2). Further background information on ECC, with emphasis on tight CW control, is provided in section 1.3.

1.2 Review of Conditions for Reliable Autogenous Self Healing

Previous researchers have engaged in limited studies in the phenomenon of concrete self healing, the formation of self healing products, and the necessary conditions to experience self healing in concrete materials. These studies have resulted in identifying three general criteria which are critical to exhibit reliable autogenous self healing – presence of specific chemical species, exposure to various environmental conditions, and small CW. These are summarized below. In some instances these findings are contradictory, as in the case of maximum allowable CW in which some specify maximum CWs of 10 μm, while others specify 300 μm to exhibit self healing in various environmental conditions:

- Essential environmental exposure – water (submerged) [20, 21], environmental pH [22, 23], wet–dry cycles (capillary suction) [24], temperature above 80° C [2], temperature above 300° C [25]
- Essential chemical species – bicarbonate ions (HCO_3^-) [1, 23, 26], carbonate ions (CO_3^{2-}) [1, 23, 26], free calcium ions (Ca^{2+}) [1, 23, 26], unhydrated cement (C_3A) [25, 27], free chloride ions (Cl^-) [28–30]
- Maximum CW −5–10 μm [31], 53 μm [32], 100 μm [2], 200 μm [1], 205 μm [33], 300 μm [20]

While the individual findings may differ, trends within these studies are clear. First, self healing can occur in a variety of environmental conditions ranging from underwater to cyclic wet–dry exposures. These conditions are readily available for many infrastructure types. Second, adequate concentrations of certain critical chemical species are essential to exhibit self healing mechanisms. This too, is readily available due to the chemical make up of cementitious materials and incomplete hydration, as well as the presence of CO_2 in air and NaCl in seawater and deicing salt. Finally, and potentially most important, is the requirement for tight CWs below roughly $200\,\mu m$, possibly $50\,\mu m$. This condition is difficult to achieve consistently, and explains why reliable formation of self healing products in most concrete structures has not been realized. This set of material physical and chemical properties, and exposure conditions, may serve as a reference base towards systematic design of robust self healing concrete.

Very little work has been carried out in determining the quality of self healing and its repeatability.

1.3 ECC Engineered for Tight Crack Width Control

In recent years, fiber-reinforced cement-based composites optimized for ultra high-tensile ductility while minimizing the amount of fibers have been designed based on micromechanics design tools [34]. Such materials, defined as ECC, attain desirable inelastic deformation mechanisms through ingredients tailored to interact synergistically under load, rather than relying on high-fiber content. Specifically, the type, size, and amount of ingredients of ECC, including cement, fiber, sand, fly ash, water, and other chemical additives commonly used in fiber-reinforced cementitious composites are determined from micromechanics such that the composite "plastically yields" under excessive loading through controlled microcracking while suppressing brittle fracture localization. Table 1 gives an example of composition of ECC. The volume fraction of fiber is 2%. ASTM Type I portland cement and low-calcium ASTM class F fly ash are used. Large aggregates are excluded in ECC mix design, and only fine sand is incorporated. The silica sand has a maximum grain size of $250\,\mu m$ and an average size of $110\,\mu m$. The PVA fiber has a diameter of $39\,\mu m$, a length of $12\,mm$, and overall Young's modulus of $25.8\,GPa$. The apparent fiber strength when embedded in cementitious matrix is $900\,MPa$. The fiber surface is treated with a proprietary oil coating of 1.2% by weight.

Figure 1 shows an example of the uniaxial tensile stress–strain curve of ECC that exhibits a "yielding" behavior similar to that of ductile metals. The tensile strain

Table 1 Mix proportions of a PVA-ECC (kg/m^3)

Cement	Sand	Class F Fly Ash	Water	Superplasticizer	PVA Fiber
583	467	700	298	19	26

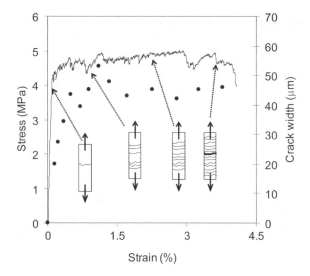

Fig. 1 Typical tensile stress–strain curve of ECC

capacity of ECC is 400–500 times that of normal concrete and the fracture toughness of ECC is similar to that of aluminum alloys [35]. Furthermore, the material remains ductile even when subjected to high shear stress [36]. The compressive strength of ECC ranges from 40–80 MPa depending on mix composition, the high end similar to that of high-strength concrete.

During tensile deformation up to about 1% strain, microcracks are formed in the specimen with CW increasing from zero to a steady-state value (about 50 μm shown in Figure 1). When the material is subjected to additional straining, more microcracks are formed, but the CW remains more or less constant. This steady-state CW is an intrinsic property of ECC and depends only on the fiber and fiber/matrix interface properties [37] which are deliberately tailored. It does not depend on the structural size or the amount of steel reinforcement. The ability of the material to exhibit tight CW control in the field, beyond laboratory conditions, is important in realizing robust self healing in actual structures. The relation between reliable self healing and tight CW is emphasized in this chapter.

2   Self healing in ECC

The chemical makeup and physical characteristics, inherent tight CW control in particular, of ECC makes the self healing behavior prevail under proper environmental conditions. This section summarizes recent laboratory test results on the self healing behavior in ECC material.

2.1   Self Healing Examination Methods

In the published literature, a number of techniques have been used to detect self healing and/or to quantify the quality of self healing in cementitious materials. Permeability test and acoustic emission technology are two common such techniques [1, 2, 28–31,33, 38–41]. In many studies, a falling head or constant head test is used to examine the extent of self healing by monitoring the flow rate or quantity of water passing through the cracked specimens. The change in the coefficient of permeability of concrete with respect to time is used to measure the amount of self healing which has occurred. Typically, a gradual reduction in this coefficient is used to infer self healing taking place in the specimen. This approach is restricted to describing self healing of transport property through permeation, and also requires that the promotion of self healing derive from fluid (typically water which may carry dissolved CO_2) flow through the crack. Formation of chemical products such as calcite in the crack has been cited as a reason for increase of sealing function over time [1, 2, 28, 29, 38]. The permeability test is typically used to examine self healing of a single crack although it can be applied to water permeability through multiple cracks.

Acoustic emission technology based on ultrasonic pulse velocity (UPV) measurements has also been used to assess crack healing. Although UPV measurement can detect the occurrence of crack healing, it has been shown that this method cannot accurately determine the extent of crack healing [33]. Resonant frequency or dynamic modulus measurements [30, 31, 42, 43] and pulse echo technique [25, 33] have also been used by a number of researchers to quantify the self healing process. Recently, one-sided stress wave transmission measurements were used to characterize the process of self healing. Based on the experimental observation; however, this transmission measurement is unable to clearly distinguish among CWs above 100 μm [33]. The advantage of these techniques is that they can be conducted relatively fast. However, these methods cannot explicitly differentiate between the nature of self healing – recovery of mechanical and/or transport properties – taken place. Material bulk properties are inferred from the test data.

In order to develop a more comprehensive understanding of self healing in ECC, four methods have been used in examining its self healing behavior. The dynamic modulus measurements provide a quick means to assess the presence of self healing. The uniaxial tension test is used to determine self healing of mechanical properties. Water permeability is used to examine the recovery of transport property through permeation. Surface chemical analysis (XEDS) and environmental scanning electron microscopy (ESEM) are used to analyze the chemical composition and morphology of the self healing product. Each of these test methods is described in more detail below. Together, they show unequivocally the presence of self healing in ECC in both the transport sense and in the mechanical sense.

2.1.1 Dynamic Modulus Measurements

The material dynamic modulus measurement based on ASTM C215 (Standard Test Method for Fundamental Transverse, Longitudinal, and Torsional Resonant Frequency of Concrete Specimens) appears to be a particularly promising technique to monitor the extent and rate of autogenous healing. This test method (ASTM C215), which relies on changes in resonant frequency, has proven a good gauge of material degradation due to freeze thaw damage and is specifically referenced within ASTM C666 for freeze thaw evaluation. Rather than quantifying damage; however, it has been adapted to measure the extent and rate of self healing in cracked concrete [43], when healing is seen as a reduction in material damage.

Prior to using resonant frequency as an accurate measure of "healing" within a cracked ECC specimen, it is essential to verify it as a valid measurement of internal damage/healing. Therefore, a series of ECC specimens measuring $230 \times 76 \times 13$ mm were prepared and subjected to varying levels of strain deformation ranging from 0% to 4% (i.e. different damage level) under uniaxial tension (see section 2.1.2) at 28 days. After unloading, the resonant frequency of each cracked specimen was determined.

From this series of tests, a relationship between tensile strain (i.e. damage) and change in resonant frequency was determined. Further, this relation extends to the number of cracks within a specimen versus resonant frequency. These relations are shown in Figure 2 and b, respectively. The resonant frequency has been normalized by that at zero strain, i.e. the resonant frequency measured with the virgin ECC without preloading. These figures show a distinct bilinear relationship between the resonant frequency and the tensile strain deformation or number of cracks. Below approximately 1% strain, a sharp drop in resonant frequency with strain/crack number can be seen, while above 1% strain this trend softens. The bilinear relationship may be attributed to the increase in number and CW (from 0 μm to 50 μm) of the multiple-microcracks at low-strain level, while only increase in crack number at steady state CW (50 μm) has been observed after about 1% strain. These results indicate that a change of resonant frequency can be used to quantify the degree of damage (i.e. tensile strain beyond the first crack) to which an ECC specimen has been subjected. Therefore, this technique should prove useful in quantifying both the rate (with respect to cycles of exposures, see section 2.2) and extent of self healing, or "negative damage", within cracked ECC specimens.

2.1.2 Uniaxial Tensile Test

Unlike conventional concrete material, tensile strain-hardening behavior represents one of the most important features of ECC material. To assess the quality of self healing in such materials, the magnitude of recovered mechanical properties were measured under uniaxial tensile loading. First, deliberate damage was introduced by tensioning a coupon specimen to predetermined strain levels followed by unloading.

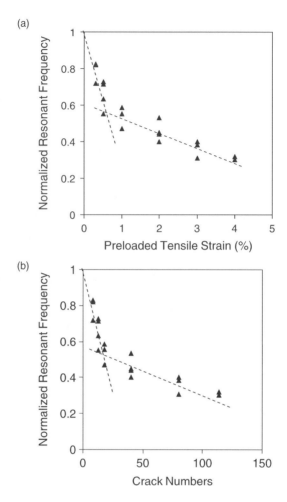

Fig. 2 Resonant frequency as an indicator for internal damage: (**a**) Resonant frequency as a function of preloaded tensile strain; (**b**) Resonant frequency as a function of crack numbers

After exposure to a healing environment, the specimen is then reloaded in direct tension to analyze the recovery magnitude of tensile strength, stiffness and strain capacity in ECC as shown in Figure 3. These properties were then compared with those measured before damage (in the case of elastic stiffness) and with those after damage but before self healing. A servohydraulic testing system was used in displacement control mode to conduct the tensile test. The loading rate used was 0.0025 mm/s to simulate a quasi-static loading condition. Aluminum plates were glued both sides at the ends of coupon specimens to facilitate gripping. Two external linear variable displacement transducers were attached to the specimen to measure the specimen deformation.

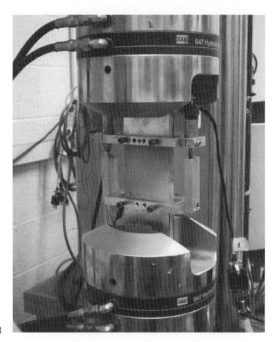

Fig. 3 Setup of the uniaxial tensile test

2.1.3 Water Permeability Test

Water permeability test was carried out to measure the transport property, permeability coefficient, of material either virgin (uncracked), preloaded (cracked/damaged), or rehealed specimen. To conduct permeability test, two experimental setups were used. A falling head test was used for specimens with a low permeability, while a constant head test was used for specimens (such as those with large CW) with a permeability too high to practically use the falling head test. These two setups are shown schematically in Figure 4a and b, respectively. The falling head and constant head permeability test setups have been adapted from Wang et al. [40] and Cernica [44].

The permeability of specimens in the falling head test can be determined using Equation 1 [44], while the permeability of specimens in the constant head test can be determined using Equation 2 [44].

$$k = \frac{a \cdot L}{A \cdot t_f} \left(\frac{h_0}{h_f} \right) \tag{1}$$

$$k = \frac{V \cdot L}{A \cdot h_0 \cdot t_f} \tag{2}$$

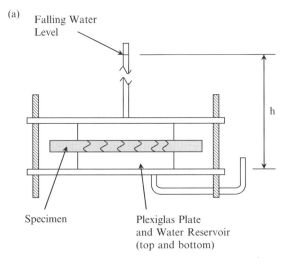

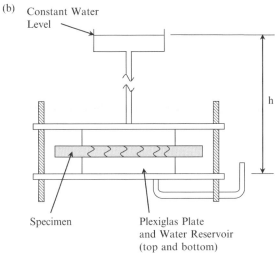

Fig. 4 Permeability test setups: (**a**) Falling head permeability test setup; (**b**) Constant head permeability test setup

where k is the coefficient of permeability, a is the cross-sectional area of the stand-pipe, L is the specimen thickness in the direction of flow, A is the cross-sectional area subject to flow, t_f is the test duration, h_0 is the initial hydraulic head, h_f is the final hydraulic head, and V is the volume of liquid passed through the specimen during the test.

2.1.4   Microscopic Observation and Analysis

The quality of healing is likely influenced by the type of self healing products formed inside the crack. Analyses of these products were conducted using ESEM and x-ray energy dispersive spectroscopy (XEDS) techniques. The crystalline and chemical properties of self healing products were determined. These techniques are particularly useful in verifying the chemical makeup of self healing compounds, essential in identifying the chemical precursors to self healing and ensuring their presence within the composite.

2.2   Environmental Conditioning

As described in the previous section (section 1.2), the robustness of self healing should be examined under a variety of environmental exposures typically experienced by concrete infrastructure systems. In the investigation of self healing of ECC, various environmental conditioning regimes have been adopted. These include cyclic wetting and drying, conditioning temperature, and immersion in water or chloride solution. Specifics of the conditioning regimes are summarized below:

- CR1 (water/air cycle) subjected pre-cracked ECC specimens to submersion in water at 20°C for 24 h and drying in laboratory air at 21°C ± 1°C, 50% ± 5% RH for 24 h, during which no temperature effects are considered. This regime is used to simulate cyclic outdoor environments such as rainy days and unclouded days.
- CR2 (water/hot air cycle) consisted of submersion of pre-cracked ECC specimens in water at 20°C for 24 h, oven drying at 55°C for 22 h, and cooling in laboratory air at 21°C ± 1°C, 50% ± 5% RH for 2 h. This regime is used to simulate cyclic outdoor environments such as rainy days followed by sunshine and high temperatures in summer.
- CR3 (water permeation) consisted of continuous permeation through cracked ECC specimen in water at 20°C till the predetermined testing ages. This regime is used to simulate environmental conditions of infrastructure in continuous contact with water with a hydraulic gradient, such as water tank, pipelines, and irrigation channels.
- CR4 (chloride solution submersion) considered direct exposure of pre-cracked ECC specimens to a solution with high chloride content. This regime is used to simulate the exposure to deicing salt in transportation infrastructure or parking structures, or in concrete containers of solutions with high salt content.
- CR5 (water submersion) consisted of submersion in water at 20°C till the predetermined testing ages. This regime is used to simulate ECC in some underwater structures.

2.3   *Effect of Crack Width on Self Healing*

To examine the effect of CW on self healing, mortar coupon specimens measuring $230 \times 76 \times 13$ mm reinforced with a small amount (0.5vol.%) polyvinyl alcohol (PVA) fiber were prepared. These specimens were deliberately made to exhibit tension-softening response typical of normal fiber reinforced concrete so that a single crack of controlled CW can be introduced. Each specimen was first preloaded under uniaxial tension to produce a single crack with a CW between 0 μm and 300 μm. Negligible closing of the crack was detected upon unloading. After unloading, the specimen was exposed to ten wetting and drying cycles (CR1). Resonant frequency was measured after preloading and 10 wetting and drying cycles (CR1) to monitor the extent of self healing of specimen with different CWs.

Figure 5 shows the resonant frequency of specimens before and after wet–dry cycles as a function of CW. The y-axis gives the resonant frequency of preloaded specimens before and after the prescribed wet–dry conditioning, normalized to the resonant frequency of uncracked (virgin) material. Therefore, 100% represents a total recovery of the resonant frequency. It is expected that further hydration and moisture content changes during the specimen conditioning regimes may contribute to some fraction of the resonant frequency recovery. To account for this, the averaged resonant frequency of virgin uncracked specimens under the same 10 cyclic conditioning regimes (10 cycles of CR1) was used in the normalization. Each data point represents the average of three test results. As seen in Figure 5, the resonant frequency of specimens after 10 cyclic wet–dry exposures can recover up to 100% of the uncracked value provided that CWs are kept below 50 μm. With an increase of CW; however,

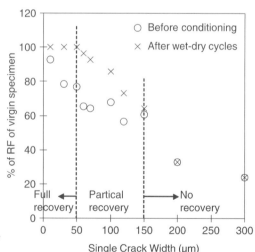

Fig. 5 Recovery of resonant frequency (RF) as a function of crack width

the degree of material damage indicated by the drop in resonant frequency increases and the extent of self healing diminishes. When the CW exceeds 150 μm, the specimen resonant frequency remains unchanged after undergoing the wet–dry cycle conditioning, signifying the difficulty of repairing microstructural damage within these cracked materials. Therefore, maintaining a CW below 150 μm, and preferably below 50 μm, is critical to enable the process of self healing.

Along with the resonant frequency monitoring, permeability tests were conducted on those specimens after 10 wet–dry cycles. Figure 6 shows the permeability coefficient of preloaded specimen after 10 cyclic conditioning exposures as a function of CW. Although it is known [45] that self healing can occur during the very act of performing a permeability test, the data shown here were initial values so that permeability changes during the test were deliberately excluded. Thus any self healing detected here were due to the wet–dry cycle exposures. From Figure 6, it can be seen that after conditioning, the permeability of specimens with CWs below 50 μm is essentially identical to that of virgin uncracked specimens, which represents a full recovery of transport property, permeability. With increasing CW, the permeability increases exponentially.

The resonant frequency measurements and the permeability measurements together suggest that complete and autogenous self healing within cement-based materials in both mechanical and transport properties can be achieved, provided that damage must be restricted to very tight CWs, below 50 μm. This extremely tight CW is difficult to attain reliably in most concrete materials, even when steel reinforcement is used. However, tailorable ECC materials with inherent tight CW control have been intentionally designed to meet this rigorous requirement (see section 1.3).

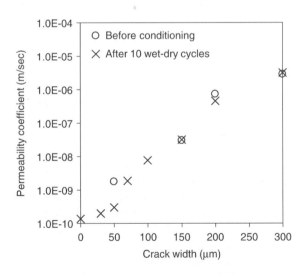

Fig. 6 Permeability coefficient as a function crack width before and after conditioning

2.4 Quality of Self Healing in ECC

In this section, the quality of autogenous self healing in ECC is summarized. The ECC material utilized for these studies has a tensile strain capacity of 3% and steady-state CW of 50 μm. The recovery of resonant frequency, mechanical tensile properties, and transport properties (water permeability and chloride diffusivity) were examined under selected environmental conditioning regimes (section 2.2).

2.4.1 Recovery in Dynamic Modulus

Dynamic modulus measurement was used to monitor the rate and extent of self healing. ECC coupon specimens measuring $230 \times 76 \times 13$ mm were prepared and preloaded to different predetermined uniaxial tensile strain levels from 0.3% to 3% at 6 months. On unloading, a small amount of crack closing, 15–20%, was observed depending on the preloading strain magnitude. These specimens were subsequently exposed to wet–dry cycles. Figure 7a and b shows the resonant frequency of ECC specimens with various pre-damage levels (0.3% – 3%) under cyclic wetting and drying CR1 and CR2, respectively. The shaded area indicates the range of resonant frequencies of virgin ECC specimens which had undergone the same cyclic wetting and drying conditioning regime. From these two figures, it can be seen that the resonant frequencies of all preloaded ECC specimens gradually recovers under both conditioning regimes. Ultimately, the resonant frequencies stabilize after 4–5 cycles. These results demonstrate that roughly 4–5 wetting and drying cycles are adequate to engage noticeable self healing of cracked ECC material. Specimens subjected to higher pre-tensioning strains exhibit a lower initial frequency after cracking, due to a larger number of cracks (i.e. damage), and ultimately lower recovery values after wet–dry cycles.

The extent of self healing within preloaded ECC specimens can be evaluated by calculating the ratio of the final resonant frequency after wet–dry cycles to the initial uncracked resonant frequency as depicted in Figure 8. From Figure 8 it can be seen that the resonant frequencies of most specimens approached or even exceeded 100% of their initial resonant frequency when increases due to additional hydration were not considered, the highest of which reaches 110% of the initial frequency. To account for this, the resonant frequencies of virgin uncracked ECC specimens under the same conditioning regimes (10 cycles of CR1 or CR2) were measured. After normalizing the results by taking account of the hydration increase, the resonant frequencies for CR1 tests after preloading were 40–82% of initial, while after wet-dry cycles had regained dynamic modulus 87–100% of initial values. For CR2 conditioning, the resonant frequencies after pre-loading were 31–83% of the initial value and after self healing had stabilized at 77–90% of initial.

Of particular interest is the relation between the extent of self healing and level of strain in the preloaded ECC specimens under CR1. Preloaded testing series with tensile strain of 0.5% exhibited a reduction in resonant frequency of only 18%, while

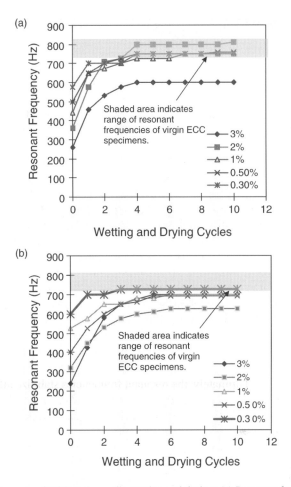

Fig. 7 Self healing rate of ECC under cyclic wetting and drying: (**a**) Resonant frequency recovery under CR1 (water/air cycle); (**b**) Resonant frequency recovery under CR2 (water/hot air cycle)

those preloaded to 3% strain showed an initial reduction of 60%. Self healing in 0.5% strained specimens showed rebounded resonant frequencies back to 100% of initial values, while specimens preloaded to 3% strain returned to only 87% of initial frequencies. This phenomenon is captured in Figure 9, which highlights the rebound in resonant frequency versus number of cracks within the ECC specimen. As the number of cracks grows, the rebound in resonant frequency due to self healing grows. However, the ultimate self healed condition may not be as complete as in specimens strained to a lower deformation. This is likely due to the presence of a greater number of cracks within the highly strained specimens. Within self healed ECC specimens, the material which heals the cracks is typically much weaker than the surrounding mortar matrix. With an increasing number of cracks, while the opportunity for a greater amount of healing exists, the likelihood of healing all these cracks to a level

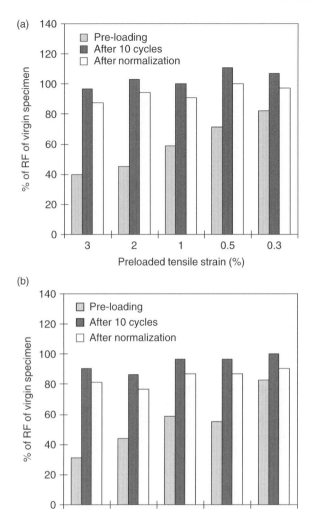

Fig. 8 Extent of self healing in ECC under cyclic wetting and drying: (**a**) Extent of self healing in ECC under CR1 (water/air cycle); (**b**) Extent of self healing in ECC under CR2 (water/hot air cycle)

similar to the uncracked state drops. Therefore, the accompanying reduction in ultimate self healing state (i.e. final resonant frequency) with an increase in strain capacity is not altogether surprising.

In addition to this, a noticeable difference exists in the extent of self healing within specimens subjected to CR1 and CR2. This is most evident in Figure 8. While most specimens subjected to CR1 recovered their full initial dynamic modulus, those subjected to CR2 did not. This may be due to the temperature effects associated with

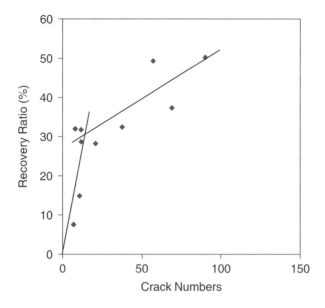

Fig. 9 Rebound in resonant frequency versus number of cracks within the ECC specimen

the CR2 conditioning regime. After submersion in water at 20°C, these specimens are then oven dried at 55°C. During this process, moisture escapes from the specimens through evaporation. As the water evaporates, steam pressure builds up within the pores, resulting in internal damage and potential microcracking. This additional damage which takes place during the self healing process, coupled with the initial damage due to cracking, may handicap CR2 specimens and ultimately result in lower amounts of self healing when compared to CR1 specimens.

2.4.2 Recovery of Tensile Properties

Uniaxial tensile test was conducted to measure the tensile mechanical properties of ECC specimens after self healing. ECC specimens measuring $230 \times 76 \times 13$ mm were prepared and preloaded to different predetermined strain levels from 0.3% to 3% at 6 months. After straining and unloading, the cracked specimens were exposed to 10 wet–dry cycles (CR1 or CR2). Uniaxial tensile tests were conducted again in the rehealed specimens. In the stress–strain curve of the reloading stage, the permanent residual strain introduced in the preloading stage is not accounted for.

Figures 10 and 11 show the preloading tensile stress–strain curves of ECC specimens as well as the reloading tensile stress–strain curves of rehealed ECC specimens after conditioning cycles CR1 and CR2, respectively. For the CR1 test series, the tensile strain capacity after self healing for these specimens ranges from 1.7% to 3.1%. The tensile strain after self healing for CR2 specimens ranges from 0.8% to

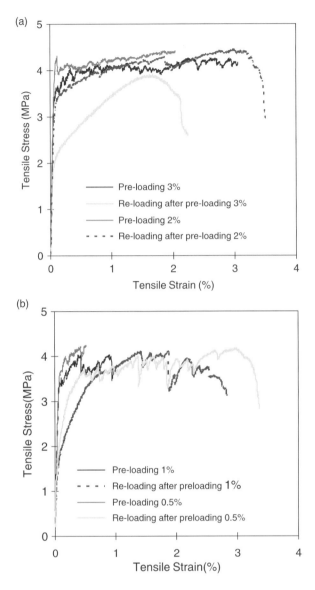

Fig. 10 Preloading and reloading after 10 CR1 (water/air) cycles tensile stress–strain relations of ECC specimens: (**a**) Specimen with preloading above 1%; (**b**) Specimens with preloading to or below 1%

2.2%, slightly lower than that of CR1. However, the ultimate strength after self healing for the CR2 specimens is higher than that of CR1 specimens. The difference of the ultimate tensile strength and tensile strain capacity after self healing can likely be attributed to the different cyclic conditioning regimes. Recall that specimens

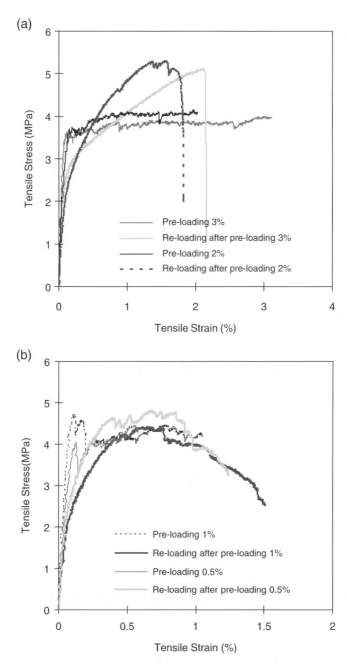

Fig. 11 Preloading and reloading after 10 CR2 (water/hot air) cycles tensile stress–strain relations of ECC specimens: (**a**) Specimen with preloading above 1%; (**b**) Specimen with preloading to or below 1%

subjected to CR2 were submersed in water and then dried in air at 55°C. With this temperature increase, the moisture in the specimens will migrate out and may result in a process similar to steam curing. Therefore, hydration of unreacted cement and fly ash will be accelerated, leading to an increased strength of the ECC matrix and fiber/matrix interfacial bonding (due to a strong hydrophilic nature of PVA fiber). The increase of interface bond strength leads to the increase of fiber bridging strength, and therefore higher composite ultimate tensile strength in CR2. On the other hand, the increase of matrix cracking strength prevents crack initiating and propagation from multiple defect sites and strong interfacial bonding increases the tendency of fiber rupture. Both mechanisms are known [17] to cause a negative impact on the development of multiple microcracking, and therefore lower tensile ductility in CR2 as a consequence.

Figure 12 shows the tensile stress–strain curves of ECC specimens which have been preloaded to 2% or 3% strain levels, then unloaded, and immediately reloaded. Thus these specimens have no opportunity to undergo any self healing. As expected, there is a remarkable difference in stiffness between the virgin specimen and the preloaded specimen under tension. This is due to the reopening of cracks within preloaded specimen during reloading. The opening of these cracks offers very little resistance to load, as the crack simply opens to its previous CW. Once these cracks are completely opened; however, the load capacity resumes, and further tensile straining of the intact material (between adjacent microcracks) can take place. By comparing the material stiffness of self healed specimens in Figures 10 and 11 with that shown for the preloaded specimens without self healing in Figure 12, it can be seen that a significant recovery of the stiffness of ECC specimens after self healing (Figure 13).

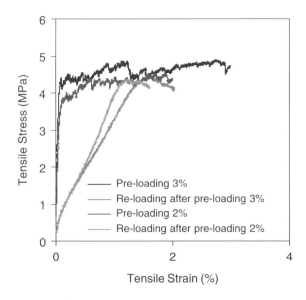

Fig. 12 Preloading and reloading without self healing tensile stress–strain curve of ECC specimens

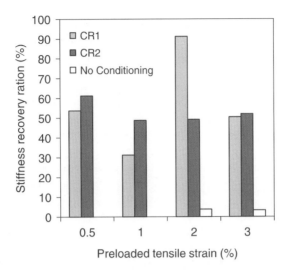

Fig. 13 Stiffness recovery of ECC under different conditioning regime

In other words, self healing of ECC material can result not only in possible sealing of cracks as shown by others, but in true rehabilitation of tensile mechanical properties, in this case the stiffness of the material under tensile load.

This finding is also supported by the rebound of resonant frequency seen in the self healed ECC specimens. As outlined in ASTM C215, resonant frequency is directly related to the dynamic modulus, or stiffness, of a material. The self healing shown through resonant frequency measurements demonstrates the same self healing as that shown in the stiffness gain of tensile coupons. This congruent finding can be used to validate the results of both test series, resonant frequency, and direct tension testing.

Figure 14 shows an ECC specimen subjected to tensile loading after undergoing self healing through the CR1 conditioning regime. This specimen was initially subjected to 2% strain before being exposed to wet–dry cycles. The distinctive white residue, characteristic of crystallization of calcium carbonate crystals, is abundant within the crack and near the crack face on the specimen surface. Further, it can be seen that the majority of cracks which form in self healed specimens tend to follow previous crack lines and propagate through the self healed material. This is not surprising due to the relatively weak nature of calcium carbonate crystals in comparison to hydrated cementitious matrix. The lower first cracking strength in the rehealed specimen (Figures 10 and 11) is also attributed to that the first crack in the rehealed specimen under tension starts from the self healed material (calcium carbonate) which has a lower strength compared to adjacent hydrated cementitious matrix.

However, this is not always the case. As can be seen in Figures 15 and 16, new cracks and crack paths have been observed to form adjacent to previously self healed cracks which now show little or no new cracking. The possibility of this event depends heavily upon the cracking properties of the matrix adjacent to the self

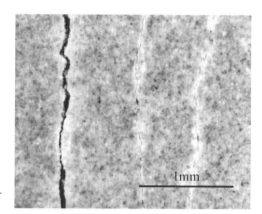

Fig. 14 Cracks through self healed material due to reloading after wet–dry cycles

Fig. 15 Cracks through virgin ECC material adjacent to a self healed crack held tight by self healing material

healing, and the quality of the self healing material itself. However, this phenomenon serves as testament to the real possibilities of mechanical self healing within ECC material.

2.4.3 Recovery of Transport Property

Reduction of permeability coefficient in cracked ECC due to self healing

Permeability specimens were cast into coupon plates with cross-sectional dimensions of 13×76 mm and 305 mm in length. The specimens were preloaded to the

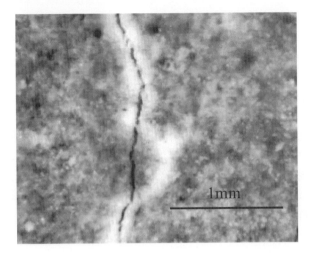

Fig. 16 Meandering new crack path partially deviating from previously self healed crack

predetermined tensile strain. Prior to the permeability testing (CR3), ECC specimens were kept in water for 14 days to ensure complete water saturation. The edges of the coupon specimen were sealed with epoxy to facilitate unidirectional flow through the cross section. Due to the length of time associated with this type of testing, CW permeability measurements were performed in the unloaded state.

Figure 17 shows the rate of permeation through the ECC specimens dropped drastically from the initial values until asymptotically reaching the recorded value, even though the CWs during permeability testing do not change. This phenomenon can be partially attributed to achieving complete saturation and further densification of the matrix throughout the testing period. However, ECC specimens were saturated in water for 14 days prior to permeability testing at an age of 28 days. By the time of testing, the specimens should have been nearly, if not completely, saturated and continuing to undergo little matrix hydration.

Throughout the course of permeability testing, a white residue formed within the cracks and on the surface of the specimens near the cracks. These formations are shown in Figure 18. Figure 18a shows a saturated ECC specimen immediately prior to the beginning of permeability testing, while Figure 18b shows the same specimen after permeability testing. The white residue forms both within the cracks, and within the pores on the surface of the ECC specimen. The effect of self healing of cracks on permeability has been investigated by other researchers [1], and may be significant in the permeability determination of cracked ECC. This can be attributed primarily to the large binder content and relatively low water to binder ratio within the ECC mixture. The presence of significant amounts of unhydrated binders allows for autogeneous healing of the cracks when exposed to water. This mechanism is particularly evident in cracked ECC material due to the small CWs which facilitate self healing. However, this phenomenon is not observed while cracked ECC specimens are

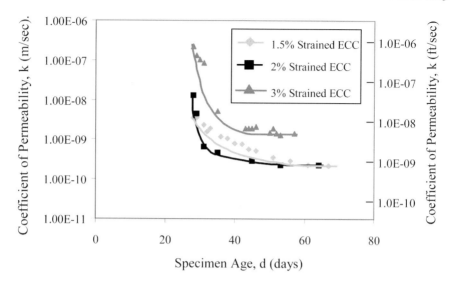

Fig. 17 Development of permeability for ECC strain to 1.5%, 2%, and 3%

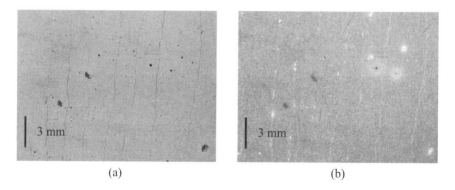

Fig. 18 Appearance of ECC permeability speciemens (**a**) before permeability testing, and (**b**) after permeability testing

simply saturated in water (CR5). During the 14 days of saturation prior to permeability testing, cracked ECC specimens showed no signs of autogeneous healing of the cracks. After only 3 days in the permeability testing apparatus, evidence of self healing became apparent. A similar phenomenon was also seen when cracked ECC specimens were partially submerged in water. Crack healing was only exhibited near the surface of the water, while no healing was observed above or below the water surface.

Surface chemical analysis (XEDS) of the self healing ECC specimens using an environmental scanning electron microscope (ESEM) show that the crystals forming within the cracks, and on the surface adjacent to the cracks, are hydrated cement

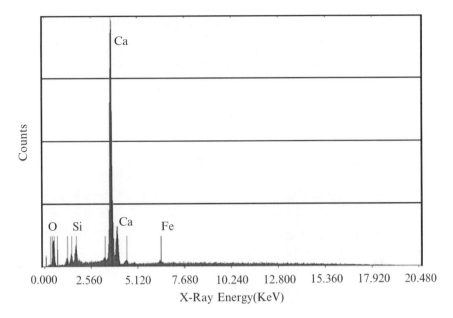

Fig. 19 ESEM surface chemical composition analysis (XEDS) of self healing crack formations

products, primarily calcium carbonate (Figure 19). These crystal formations within the self healed cracks are shown in Figure 20. To facilitate healing of the cracks, and promote formation of calcium carbonate, a flow of water containing carbonates or bicarbonates must be present. Within the permeability testing, these carbonates were introduced by the dissolution of CO_2 in air into the water which flows through the specimens. In the case of the partially submerged specimens, the small amount of carbon dioxide dissolved at the water surface was sufficient to cause limited self healing at that location. However, in the absence of this constant carbonate supply, as in the saturation tanks prior to permeability testing, no self healing of the ECC microcracks can occur. Ultimately, the formation of these crystals slows the rate of permeation through the cracked composite and further reduces the permeability coefficient.

Reduction of diffusion coefficient in ECC due to self healing

Autogenous self healing was also observed in an attempt to measure the diffusion coefficient of damage ECC specimen by means of the chloride-ponding test. Salt-ponding test in accordance with AASHTO T259–80 (Standard Method of Test for Resistance of Concrete to Chloride Ion Penetration) was conducted to evaluate another transport property, effective diffusion coefficient, of material either virgin (uncracked), preloaded (damaged/cracked), or rehealed specimen. After ponding for a certain period (30 days for the preloaded specimen and 90 days for the virgin specimen), the salt solution was removed from the prism surface. Powder samples were

(a)

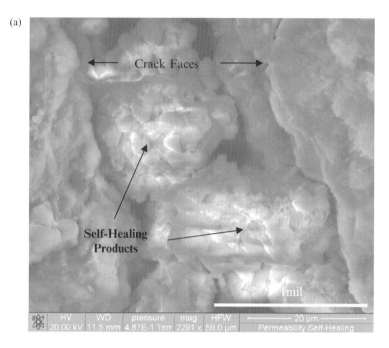

(b)

Fig. 20 Morphology of crack within ECC specimen from ESEM: (**a**) Autogenous self healing crystalline formations in ECC crack after permebaility testing; (**b**) ECC crack before permeability testing

taken from the specimen for chloride analysis at various depths from the exposed surface. Total chloride (acid-soluble) content by weight of material at each sampling point was examined according to AASHTO T 260–97 (Standard Method of Test for Sampling and Testing for Chloride Ion in Concrete and Concrete Raw Materials).

The chloride profiles were then input into statistical and curve-fitting software. Equation 3, Crank's solution to Fick's second law, was fitted to the data. The regression analysis yielded the values of the effective diffusion coefficient (D_e) and surface chloride concentration (C_s) for the specimen.

$$C(x, t) = C_s \left[1 - erf \left(\frac{x}{2\sqrt{D_e t}} \right) \right] \qquad (3)$$

where

$C(x, t)$ = chloride concentration at time t at depth x

C_s = surface chloride concentration

D_e = effective chloride diffusion coefficient

t = exposure time

$erf()$ = error function

In conducting the ponding test, ECC prism specimens measuring $356 \times 76 \times 51$ mm were prepared. In addition to ECC material, mortar prisms were also tested for comparison purpose. The mortar prisms were reinforced with three levels of steel mesh in order to preload the specimen to a predetermined deformation. At the age of 28 days, prisms surface were abraded using a steel brush as required by AASHTO T259– 80. The prisms were preloaded using four-point bending test to a predetermined deformation. The ponding test was then performed in the unloaded state. Plexiglass was used around the side surfaces of the prism to build an embankment for holding chloride solution on the exposed surface of prisms. At 29 days of age, a 3% of NaCl solution was ponded on the cracked surface of prisms. In order to retard the evaporation of solution, aluminum plates were used to cover the top surface of the specimens.

Table 2 shows the preloaded beam deformation (BD) value, their corresponding average CWs, depths and number of cracks for prism specimens. Two virgin prisms from each mixture were tested without preloading for control purpose. Note that preloading of the mortar beams were limited to 0.83 mm due to the large CW ($\sim$400 µm) and crack depth 70 mm generated in these specimens. In the ECC specimens, the CW remains at about 50 µm even after BD at 2 mm. The crack length becomes impossible to measure accurately due to the tight CW. Table 2 also shows the corresponding number of cracks for prism specimens at each BD value. As seen from this table, when the deformation applied to the prism specimens is increased, the number of cracks on ECC is clearly increased but the CW did not change for the different deformation values. Micromechanically designed ECC changes the cracking behavior from one crack with large width to multiple smaller cracks as described in section 1.3.

Table 2 Crack widths, numbers, and depths of preloaded ECC and mortar prisms

Material	Beam Deformation (mm)	Average Crack Width (µm)	Crack Depth (mm)	Crack Number
Mortar	0.5	~50	20	1
	0.7	~150	36	1
	0.8	~300	55	1
	0.83	~400	70	1
ECC	0.5	~0	N/A	0
	1.0	~50	N/A	15
	1.5	~50	N/A	21
	2.0	~50	N/A	35

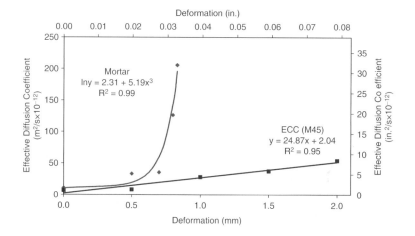

Fig. 21 Diffusion coefficient versus preloading deformation level for ECC and mortar

Figure 21 shows the relationship between the effective diffusion coefficient of chloride ions and the BD level, for mortar and ECC specimens. Despite the same or higher magnitude of imposed overall deformation and higher crack density, the ECC specimens reveal an effective diffusion coefficient considerably lower than that of the reinforced mortar because of the tight CW control. Especially for the higher deformation level, the effective diffusion coefficient of mortar increased exponentially with BD. The effective diffusion coefficient of ECC; however, increased linearly with the imposed deformation value, because the number of microcracks on the tensile surface of ECC is proportional to the imposed BD. The total chloride concentration profiles perpendicular to the crack path indicate no significant chloride penetration even at large imposed deformation (2 mm) for ECC specimens.

The reason for the relatively low diffusion coefficient of cracked ECC specimens is not only due to the tight CW but also the presence of self healing of the microcracks. The self healing of cracks becomes prominent when CW is small. In the case of precracked ECC prisms exposed to salt solution, a distinct white deposit was visible over the crack surface at the end of 1-month exposure period (Figure 22). These

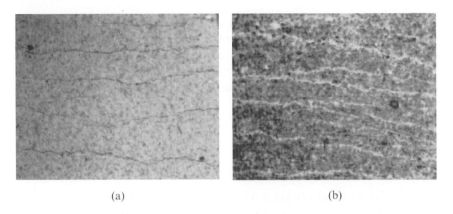

(a)   (b)

Fig. 22 Self healing products in ECC microcracks (**a**) before, and (**b**) after salt-ponding test at 30 days exposure

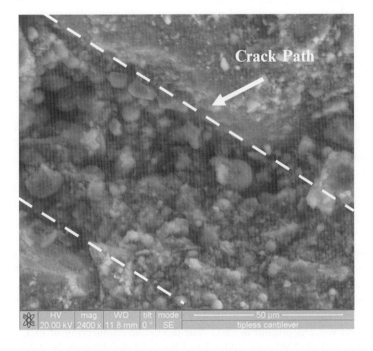

Fig. 23 ESEM micrograph of rehydration products in a self healed crack

deposits were most probably caused by efflorescence due to leaching of calcium hydroxide (CH) into cracks and due to the presence of NaCl ion in solution. This white deposit on the crack surface easily blocked the flow path due to smaller CW of ECC. An ESEM observation of the fractured surface of ECC across a healed crack is shown in Figure 23. The present ESEM observations show that most of the products seen in the cracks were newly formed C-S-H gels. CH and deposition of salts in the crack path were also observed. These observations indicate that microcracks of ECC exposed to NaCl solution healed completely after exposure for 30 days to NaCl solution. This can be attributed primarily to the large fly ash content and relatively low water to binder ratio within the ECC mixture. The continued pozzolanic activity of fly ash is responsible for the self healing of the crack which reduces the ingress of the chloride ions.

3 Conclusions

From the resonant frequency (after wet–dry cycle exposure) and permeability measurements, it appears that a maximum CW of 50 μm is necessary to achieve full recovery of mechanical and transport properties in ECC material. Between 50 and 150 μm, partial recovery can be attained.

Under both water/air cycle and water/hot air cycle environments, a recovery of resonant frequent above 80% is achievable for ECC specimens pretensioned to as much as 3% strain. Most self healing appears to occur within the first 4–5 cycles. Under these same conditions, 100% recovery of the initial elastic modulus can be attained. Further the original ultimate tensile strength and strain capacity are retained after rehealing. However, the first crack strength may be lower especially for those specimens with large prestraining, probably due to the weakness in some partially healed crack planes. In this case, crack formation upon reloading will likely take place where a partially healed crack is present. For specimens with lower amount of prestraining, complete rehealing occurs so that new cracks are observed near rehealed cracks upon reloading. This is in spite of the fact that the products formed inside the cracks are calcites known to be relatively weak. It is possible that the bridging fibers serve as nucleation sites for the calcite crystals while simultaneously act as reinforcements. Further study on the details of reheal product formation process is needed.

Both transport properties, permeability, and chloride diffusion, showed a decrease over time. Precracked ECC specimens exposed to water flow in the permeability experiments showed rehealing through calcite formation, while precracked ECC specimens exposed to saltwater ponding showed rehealing through CH formation.

From these studies, it becomes evident that self healing both in the mechanical and transport sense is present in ECC. The deliberate engineering of ECC to maintain extremely tight CW even under large imposed deformation as carried out in these experiments, is largely responsible for the quality of autogenous self healing in this material. The low water/cement ratio in addition to the large amount of fly ash also aids in promoting self healing via continued hydration and pozzolanic activities. These activities are expected to extend indefinitely since unhydrated cement and unreacted fly ash in concrete materials are known to have a long shelf life.

Similar to all concrete materials, the micro-reservoirs of unhydrated cement and unreacted fly ash in ECC ensures self healing to be pervasive throughout a structure.

The very act of creating a crack is expected to provide advective–diffusive forces that drives free calcium ions towards the crack surface and enhances the tendency to form rehealing products in the crack in the presence of water and dissolved CO_2. This suggests that chemicals are always intrinsically present where it is needed for self healing. (External water and CO_2 in air is expected to be plentiful especially for transportation infrastructure.) Unlike other concrete materials, ECC has the ability to self-control CW down to the 50 μm range and below, which drastically enhances the reliability of self healing, as demonstrated in the test results described for a variety of exposure environments in this article. Also demonstrated is the quality of rehealing especially for ECC specimens preloaded to below 1% tensile strain. Recovery of mechanical and transport properties reaching 80% or above is evident. Additional research is needed to confirm the repeatability of the self healing functionality in ECC subjected to tensile load cycles.

Acknowledgments

This work has drawn upon a body of research with participations from M. Lepech, M. Sahmaran, and Y. Yang. Their contributions are greatly appreciated. The authors would also like to acknowledge helpful comments by K. van Brugel and E. Schlangen at Delft University, and by K. Maekawa at University of Tokyo. This research was funded through an NSF MUSES Biocomplexity Program Grant (Nos. CMS-0223971 and CMS-0329416). MUSES (Materials Use: Science, Engineering, and Society) support projects that study the reduction of adverse human impact on the total interactive system of resource use, the design and synthesis of new materials with environmentally benign impacts on biocomplex systems, as well as the maximization of efficient use of materials throughout their life cycles.

References

1. Edvardsen C. (1999) Water permeability and autogenous healing of cracks in concrete. ACI Mater J 96:448–455
2. Reinhardt H., Joos M. (2003) Permeability and self-healing of cracked concrete as a function of temperature and crack width. J cement concrete Res 33:981–985
3. Pimienta P., Chanvillard G. (2004) Retention of the mechanical performances of Ductal specimens kept in various aggressive environments. Fib Symposium.
4. Granger S., Pijaudier-Cabot G., Loukili A. (2007) Mechanical behavior of self-healed Ultra High Performance Concrete: from experimental evidence to modeling. Accepted for publication in Proc. FRAMCOS 6. Catalina, Italy
5. Ter Heidi N., Schlangen E., van Breugel K. (2006) Experimental study of crack healing of early age cracks. In: Proceedings of Knud Hojgaard Conference on advanced cement-based materials: research and teaching. Lyngby, Denmark
6. White S.R., Sottos N.R., Geubelle P.H., Moore J.S., Kessler M.R., Sriram S.R., Brown E.N., Viswanathan S. (2001) Autonomic healing of polymer composites. Nature 409:794–797

7. Kessler M.K., Sottos N.R., White S.R. (2003) Self-healing structural composite material. Compos Part A: Appl Sci Manuf 34(8):743–753

8. Dry C. (1994) Matrix crack repair and filling using active and passive model for smart time release of chemicals from fibers into cement matrixes. J Smart Mater Struct 118–123

9. Dry C. (1996) Procedures developed for self-repair of polymer matrix composite materials. Compos Struct 35:236–269

10. Li V.C., Lim Y.M., Chan Y. (1998) Feasibility study of a passive smart self-healing cementitious composite. Compos Part B 29B:819–827

11. Dry C. (1996) Release of smart chemicals for the in-service repair of bridges and roadways. In: Proceedings of the Symposium on Smart Materials, Structures and MEMS

12. Lee J.Y., Buxton G.A., Balazs A.C. (2004) Using nanoparticles to create self-healing composites. J Chem Phys 121(11):5531–5540

13. Nishiwaki T., Mihashi H., Jang B.K., Miura K. (2006) Development of self-healing system for concrete with selective healing around crack. J Adv Concrete Techn 4(2):267–275

14. Bang S.S., Galinat J.K., Ramakrishnan V. (2001) Calcite precipitation induced by polyurethane-immobilized Bacillus pasteurii. Enzyme Microb Tech 28:404–409

15. Rodriguez-Navarro C., Rodriguez-Gallego M., Chekroun K.B., Gonzalez-Munoz M.T. (2003) Conservation of ornamental stone by myxococcus santhus-induced carbonate biomineralization. Appl Environ Microbiol 2182–2193

16. Li V.C. (2003) On engineered cementitious composites (ECC) – a review of the material and its applications. J Adv Concrete Tech 1(3):215–230

17. Li V.C., Wu C., Wang S., Ogawa A., Saito T. (2002) Interface tailoring for strain-hardening PVA-ECC. ACI Mater J 99(5):463–472

18. Lepech M.D.A.V.C.L. (2006) Long term durability performance of engineered cementitious composites. Int J Restoration Buildings Monuments 12(2):119–132

19. Kunieda M., Rokugo K. (2006) Recent progress on HPFRCC in Japan, required performance and applications. J Adv Concrete Tech 4(1):19–33

20. Clear C.A. (1985) The effects of autogenous healing upon the leakage of water through cracks in concrete. In: Cement and Concrete Association Technical Report No. 559, p 28

21. Nanayakkara A. (2003) Self-healing of cracks in concrete subjected to water pressure. Symposium on new technologies for urban safety of mega cities in Asia. Tokyo, Japan

22. Institute AC (1995) Building code requirements for structural concrete (ACI318–95) and Commentary (ACI318R-95). American Concrete Institute. Detroit, Michigan

23. Joos M. (2001) Leaching of concrete under thermal influence. Otto-Graf-J 12:51–68

24. Hannant D.J., Keer J.G. (1983) Autogenous healing of thin cement based sheets. Cement Concrete Res 13:357–365

25. Farage M.C.R., Sercombe J., Galle C. (2003) Rehydration and microstructure of cement paste after heating at temperatures up to 300°C. J Cement Concrete Res 33:1047–1056

26. Cowie J., Glassert F.P. (1992) The reaction between cement and natural water containing dissolved carbon dioxide. Adv Cement Res 14(15):119–134

27. Ter Heide N. (2005) Crack healing in hydrating concrete. MSc Thesis, TU Delft, The Netherlands

28. Schiessl P., Brauer N. (1996) Influence of autogenous healing of cracks on corrosion of reinforcement. In: Proceedings on durability of building materials and components 7

29. Mangat P.S., Gurusamy K. (1987) Permissible crack widths in steel fiber reinforced marine concrete. J Mater Struct 20:338–347

30. Jacobsen S., Marchand J., Boisvert L. (1996) Effect of cracking and healing on chloride transport in OPC concrete. J Cement Concrete Res 26:869–881

31. Jacobsen S., Marchand J., Homain H. (1995) SEM observations of the microstructure of frost deteriorated and self-healed concrete. J Cement Concrete Res 25:1781–1790

32. Ismail M., Toumi A., Francois R., Gagne R. (2004) Effect of crack opening on local diffusion of chloride inert materials. Cement Concrete Res 34:711–716

33. Aldea C., Song W., Popovics J.S., Shah S.P. (2000) Extent of healing of cracked normal strength concrete. J. Mater Civil Eng 12:92–96
34. Li V.C. (1997) Engineered cementitious composites – tailored composites through micro-mechanical modeling. In: Fiber reinforced concrete: present and the future., Canadian Society for Civil Engineering, Montreal, pp 64–97
35. Maalej M., Hashida T., Li V.C. (1995) Effect of fiber volume fraction on the off-crack plane energy in strain-hardening engineered cementitious composites. J Am Ceramics Soc 78(12):3369–3375
36. Li V.C., Chan Y.W. (1994) Determination of interfacial debond mode for fiber reinforced cementitious composites. ASCE J Eng Mech 120(4):707–719
37. Yang E., Li V.C. (2007) Numerical study on steady-state cracking of composites. Compos Sci Tech 67:151–156
38. Ramm W., Biscoping M. (1998) Autogenous healing and reinforcement corrosion of water-penetrated separation cracks in reinforced concrete. J Nucl Eng Des 179:191–200
39. Reinharde H., Jooss M. (2003) Permeability and self-healing of cracked concrete as a function of temperature and crack width. J Cement Concrete Res 33:981–985
40. Wang K., Jansen D.C., Shah S.P., Karr A.F. (1997) Permeability study of cracked concrete. J Cement Concrete Res 27:381–393
41. Lepech M.D., Li V.C. (2005) Water permeability of cracked cementitious composites. In: ICF 11. Turin, Italy, Paper 4539 of Compendium of Papers CD ROM.
42. Jacobsen S., Sellevold E.J. (1996) Self healing of high strength concrete after deterioration by freeze/thaw. J Cement Concrete Res 26:55–62
43. Yang Y., Lepech M., Li V.C. (2005) Self-healing of ECC under cyclic wetting and drying. In: Proceedings of international workshop on durability of reinforced concrete under combined mechanical and climatic loads. Qingdao, China
44. Cernica J.N. (1982) Geotechnical engineering. Holt, Reinhart & Winston, New York, pp 97–99
45. Lepech M.D. (2006) A paradigm for integrated structures and materials design for sustainable transportation infrastructure, PhD Thesis. In: Department of Civil and Environmental Engineering, University of Michigan, Ann Arbor

Self Healing Concrete: A Biological Approach

Henk M Jonkers

Delft University of Technology, Faculty of Civil Engineering and GeoSciences/Microlab, Stevinweg 1,
The Netherlands
E-mail: h.m.jonkers@tudelft.nl

1 An Introduction to Concrete

Concrete can be considered as a kind of artificial rock with properties more or less similar to certain natural rocks. As it is strong, durable, and relatively cheap, concrete is, since almost two centuries, the most used construction material worldwide, which can easily be recognized as it has changed the physiognomy of rural areas. However, due to the heterogeneity of the composition of its principle components, cement, water, and a variety of aggregates, the properties of the final product can widely vary. The structural designer therefore must previously establish which properties are important for a specific application and must choose the correct composition of the concrete ingredients in order to ensure that the final product applies to the previously set standards. Concrete is typically characterized by a high-compressive strength, but unfortunately also by a rather low-tensile strength. However, through the application of steel or other material reinforcements, the latter can be compensated for as such reinforcements can take over tensile forces.

Modern concrete is based on Portland cement, a hydraulic cement patented by Joseph Aspdin in the early 19th century. Already in Roman times hydraulic cements, made from burned limestone and volcanic earth, slowly replaced the widely used non-hydraulic cements, which were based on burned limestone as main ingredient. When limestone is burned (or "calcined") at a temperature between 800 and 900°C, a process that drives off bound carbon dioxide (CO_2), lime (calcium oxide; CaO) is produced. Lime, when brought into contact with water, reacts to form portlandite ($Ca(OH)_2$) which can further react with CO_2, which in turn forms back into calcite ($CaCO_3$), or limestone, the pre-burning starting material. However, a major drawback of this non-hydraulic cement is that it will not set under water and, moreover, its reaction products portlandite and limestone are relatively soluble, and thus will deteriorate rapidly in wet and/or acidic environments. In contrast, portland cement produces, upon reaction with water, a much harder and insoluble material that will also set under water. For portland cement production a source of calcium, silicon, aluminum, and iron is needed and therefore usually limestone, clay, some bauxite, and iron ore are burned in a kiln at temperatures up to $1,500°C$. The cement clinker produced is mainly composed of the minerals alite ($3CaO.SiO_2$), belite ($2CaO.SiO_2$), aluminate ($3CaO.Al_2O_3$), and ferrite ($4CaO.Al_2O_3.Fe_2O_3$), which all yield specific hydration products with different characteristics upon reaction with water.

S. van der Zwaag (ed.), *Self Healing Materials. An Alternative Approach to 20 Centuries of Materials Science*, 195–204.
© 2007 *Springer*.

The contribution of these clinker minerals to the composition of general-purpose portland cement in weight percentage is typically 50%, 24%, 11%, and 8% respectively. Important characteristics of clinker minerals are reaction rate and contribution to final strength of the product. For example, of the two calcium silicates, alite is the most reactive and contributes to early strength, while the slower-reacting belite contributes more to longer-term strength. Aluminate contributes to early strength as its hydration reaction is fast but it also generates much heat. The final properties of cement-based materials can thus vary widely as they strongly depend on the mineral composition of the cement used and therefore, different types of cement, each suitable for specific applications, are produced. Quantitatively most important hydration product of general-purpose portland cement is calcium silicate hydrate (C–S–H), an amorphous mineral somewhat resembling the natural mineral tobermorite. A secondary reaction product is calcium hydroxide (portlandite), which together with the very soluble sodium and potassium oxides (Na_2O and K_2O) also present in portland cement, contribute to the high alkalinity of the concrete's pore fluid (pH $\approx$ 13). The high matrix pH is important in structural concrete as it protects the embedded steel reinforcement from corrosion. The protective oxidized thin layer of Fe^{3+} oxides and oxyhydroxides on the reinforcement steel (the passivation film) rapidly degrade when the matrix pH drops below 9, leading to further oxidation and deterioration of the concrete structure due to expansion reactions and loss of strength. Corrosion of the steel reinforcement is in fact one of the major causes limiting the durability, or lifetime, of concrete structures. For further and more detailed information on general concrete properties the reader is referred to Reinhardt (1985) and Neville (1996).

2   Concrete Durability, Deterioration, and Self Healing Properties

A variety of additives or replacements of cement can be applied in order to improve the durability of the final concrete product. Also certain industrial waste or recycled materials can be used to improve the sustainability, or environmental friendliness, of concrete and some even improve certain properties. The production of cement is high-energy consuming as raw materials are burned at $1,500°C$, a process that contributes to a significant amount of atmospheric CO_2 release worldwide. Thus, for both economical and environmental reasons, cement production and use should be minimized. Examples of industrial waste products, which can partly replace and even improve cement properties, are fly ash, blast furnace slag, and silica fume. Fly ash, a waste product from coal-burning power plants, is a source of reactive silica and can substitute 35–75% of cement in the concrete mix. Application of fly ash increases concrete strength as it reduces the required water/cement ratio and also improves resistance against chemical attack as it decreases the matrix permeability. Similarly, silica fume from the silicon industry and blast furnace slag from steel industries can partially replace cement in the concrete mix, as these are sources of reactive silica and both reactive silica and calcium respectively. Other commonly applied additives that improve or change certain concrete characteristics needed for specific applications are air-entraining agents to improve freeze/thaw resistance, setting or retarding agents and plasticizers to enable a lower water/cement ratio to increase concrete strength.

A number of processes negatively affect the durability and result in the unwanted early deterioration of concrete structures. One major cause that initiates various mechanisms of concrete deterioration is the process of cracking what dramatically increases the permeability of concrete. The microstructure of hardened cement paste is porous as it contains isolated as well as interconnected pores. Specifically the connected pores determine permeability, as these allow water and chemicals to enter the concrete matrix. As cracking links both isolated and connected pore systems, this results in a substantially increased permeability. In most concrete-deterioration mechanisms permeability plays a major role. Intrusion of sulfate ions into the matrix may result in ettringite formation, a conversion reaction in which a high-density phase is transformed into a low-density phase, causing expansion and further cracking of the material. Chloride ions penetrating the matrix through the connected pore system will destabilize the passivation film of the steel reinforcement and by doing so accelerate further corrosion. Similarly, in a process called carbonation, CO_2 diffusing through the pore system will react with alkaline pore fluid components such as $Ca(OH)_2$ which will result in a lowering of matrix pH and again depassivation of the protective film on the steel reinforcement. These examples make clear that cracking of concrete should be minimized and that a potential healing mechanism should ideally result in the sealing or plugging of newly formed cracks in order to minimize increases in matrix permeability. An active self healing mechanism in concrete should be ideal as it does not need labor-intensive manual checking and repair what would save an enormous amount of money.

A self healing mechanism or self healing agent in concrete should comply ideally with all, or at least with some, of the following characteristics:

1. Should be able to seal or plug freshly formed cracks to reduce matrix permeability
2. Must be incorporated in the concrete matrix and able to act autonomously to be truly "self-healing"
3. Must be compatible with concrete, i.e. its presence should not negatively affect material characteristics
4. Should have a long-term potential activity, as concrete structures are build to last typically for at least 50 years
5. Should preferably act as a catalyst and not be consumed in the process to enable multiple healing events
6. Must not be too expensive to keep the material economically competitive

Different types of potential self healing mechanisms or agents for autonomous concrete repair can be thought of. One series of mechanisms could involve the secondary formation of minerals which are compatible with the material matrix, i.e. will not negatively affect but rather increase concrete durability by sealing freshly formed cracks and so decrease matrix permeability. A chemical agent such as the inclusion of still nonreacted cement particles in the concrete matrix is feasible as it complies with at least some of the listed self healing properties. Besides this, other agents could work equally well or can contribute to the self healing property of concrete in concert with the previous one. Next to chemicals one could think of an agent of biological origin, and in the next part the possible application of bacteria as healing agent will be considered.

3   The Self Healing Mechanism of Bacterial Concrete

Do bacteria exist which could potentially act as a self healing agent in concrete, and if so, what would be the healing mechanism? From a microbiological viewpoint the application of bacteria in concrete, or concrete as a habitat for specialized bacteria, is not odd at all. Although the concrete matrix may seem at first inhospitable for life, as it is a very dry and extremely alkaline environment, comparable natural systems occur in which bacteria thrive. Inside rocks, even at a depth of more than 1 km within the earth crust, in deserts as well as in ultra-basic environments, active bacteria are found (Jorgensen and D'Hondt 2006; Fajardo-Cavazos and Nicholson 2006; Dorn and Oberlander 1981; DelaTorre et al. 2003; Pedersen et al. 2004; Sleep et al. 2004). These desiccation- and/or alkali-resistant bacteria typically form spores, which are specialized cells able to resist high mechanically and chemically induced stresses (Sagripanti and Bonifacino 1996). A low-metabolic activity and extremely long life-times also characterize spores, and some species are known to produce spores which are viable for up to 200 years (Schlegel 1993).

   In a number of recent studies the potential for application of bacteria in concrete technology was recognized and reported on, e.g. for cleaning of concrete surfaces (DeGraef et al. 2005) as well as for the improvement of mortar compressive strength (Ghosh et al. [53]). Moreover, bacterial treatment of degraded limestone, ornamental stone, and concrete structures for durability improvement has been the specific topic of a number of recent studies (Bang et al. 2001; Ramachandran et al. 2001; Rodriguez-Navarro et al. 2003; De Muynck et al. 2005; Dick et al. 2006). Due to bacterially controlled precipitation of dense calcium carbonate layers, crack-sealing, as well as significant decreases in permeability of concrete surfaces were observed in these studies. In these remediation and repair studies the bacteria and compounds needed for mineral precipitation were brought into contact with the structures' surface after setting or crack formation had occurred, and were not initially integrated as healing agents in the material's matrix. The mechanism of bacterially mediated calcite precipitation in those studies was primarily based on the enzymatic hydrolysis of urea. In this urease-mediated process the reaction of urea $(CO(NH_2)_2)$ and water yields CO_2 and ammonia (NH_3). Due to the high pK value of the NH_3/NH_4^+ system (about 9.2) the reaction results in a pH increase and concomitant shift in the carbonate equilibrium (CO_2 to HCO_3^- and $CO3_2^-$) which results in the precipitation of calcium carbonate $(CaCO_3)$ when sufficient calcium ions (Ca^{2+}) are present.

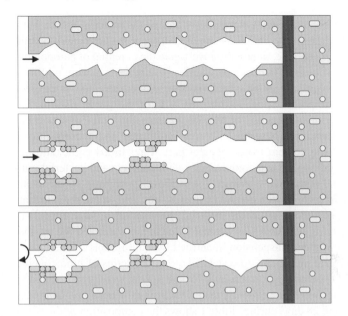

Fig. 1 Scenario of crack-healing by concrete-immobilized bacteria. Bacteria on fresh crack surfaces become activated due to water ingression, start to multiply and precipitate minerals such as calcite ($CaCO_3$), which eventually seal the crack and protect the steel reinforcement from further external chemical attack

Whether this system can be integrated in the concrete matrix in order to act as truly self healing agent or that its application is tailored for remediation purposes remains to be investigated as both enzyme (urease) activity and substrate (urea) may not have a long-term (years) stability when immobilized in the concrete matrix. One drawback of the urease-based system from an environmental viewpoint is the excessive (equivalent) production of ammonium to carbonate ions. Thus ideally, for bacterially based self healing concrete, both the bacteria and environmental friendly biomineral precursor compounds should be embedded in the material. The immobilized bacteria should immediately start to precipitate minerals and seal cracks upon concrete cracking and water entrance. Such a scenario is schematically depicted in Figure 1.

4   Experimental Evidence for Bacterially Controlled Self Healing in Concrete

In a number of published studies the potential of calcite precipitating bacteria for concrete or limestone surface remediation or durability improvement was investigated (Bang et al. 2001; Ramachandran et al. 2001; Rodriguez-Navarro et al. 2003; De Muynck et al. 2005; Dick et al. 2006). However, as the bacteria and mineral precursor compounds were not initially part of the material matrix but rather externally applied, the remediation mechanism in those studies cannot be truly defined as self healing. Therefore, in order to investigate the potential of autonomous bacterially mediated self healing in concrete, a series of experiments were performed. Firstly, a number of potentially suitable bacterial species were selected. Four species of alkali-tolerant (alkaliphilic) spore-forming bacteria of the genus *Bacillus* were obtained from the German Collection of Microorganisms and Cell Cultures (DSMZ), Braun-schweig, Germany. These bacteria were cultivated and subsequently immobilized in concrete and cement stone (cement plus water in a weight ratio of 2:1 without aggregate addition) to test compatibility with concrete and bacterial mineral production potential respectively. As was listed above (paragraph 2), the ideal self healing agent should comply with certain characteristics, and one of them is that its presence should not negatively affect the material characteristics. To test this, a dense culture of *Sporosarcina pasteurii* was washed twice in tap water and the number of bacteria in the resulting cell suspension quantified by microscopic counting before addition to the concrete mix make up water. Two parallel series of nine concrete bars (with and without bacteria) of dimensions $16 \times 4 \times 4$ cm were prepared and triplicate bars of both series were subsequently tested for flexural tensile and compressive strength after 3, 7, and 28 days curing. Table 1 shows the composition of the concrete mix and Figure 2 depicts the strength development of both types of concrete in time.

The results of the concrete compatibility test show that the addition of bacteria to a final concentration of 10^9 cm^{-3} does not affect strength characteristics. Moreover, incubation of cement stone pieces in a medium to which yeast extract and peptone (3 and 5 g L^{-1} respectively) was added as a bacterial food source revealed that on the surface of bacteria-embedded specimen, but not on control specimen, copious amounts of calcite-like crystals were formed (Figure 3). From the latter experiment it can therefore be concluded that suitable bacteria, in this case alkali-resistant spore-forming bacteria, embedded in the concretes' cement paste are able to produce minerals when an appropriate food source is available.

Table 1 Cement, water, and aggregate composition needed for the production of nine concrete bars of dimensions $16 \times 4 \times 4$ cm. The washed cell suspension used for bacterial concrete was part of totally needed makeup water

Compound	Weight (g)
Cement (ENCI CEMI 32.5)	1, 170
Water	585
Aggregate Size Fraction (mm):	
4–8	1, 685
2–4	1, 133
1–2	848
0.5–1	848
0.25–0.5	730
0.125–0.25	396

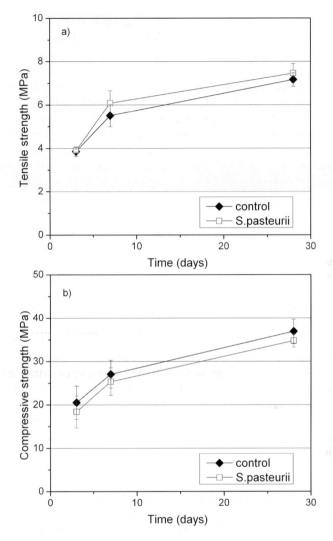

Fig. 2 Flexural tensile (**a**) and compressive (**b**) strength testing after 3, 7, and 28 days curing revealed no significant difference between control and bacterial concrete. The latter contained 1.14×10^9 *S. pasteurii* cells per cubic centimeter of concrete

The mechanism of bacterial mineral production here is likely metabolically mediated. As the bacteria metabolize the available organic carbon sources (yeast extract and peptone) under alkaline conditions, carbonate ions are produced which precipitate with access toward calcium ions present in the concrete matrix. The produced carbonate ions can locally reach high concentrations at the bacterially active "hot

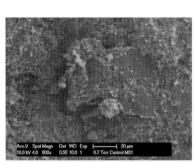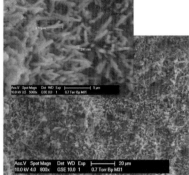

A. Control (no bacteria)   B. Concrete with immobilized bacteria

Fig. 3 Cement stone samples, which were cured for 10 days and subsequently further incubated in yeast extract- and peptone-containing medium. (**A**) Control (cement stone without added bacteria) and (**B**) cement stone containing 10^9 cm^{-3} spores of *B.pseudofirmus*. The inset in Figure 2B ($\times 5000$ magnification) shows a close up of the massive calcite-like crystals formed on the specimen surface

Table 2 Flexural tensile- and compressive-strength characteristics of control and organic carbon-amended concrete bars after a 28 days curing period

Type of Concrete	Tensile Strength (MPa)	Compressive Strength (MPa)
Control	7.78 ± 0.38	31.92 ± 1.98
Na-aspartate	7.33 ± 0.37	33.69 ± 1.89
Na-glutamate	7.16 ± 0.19	28.52 ± 3.56
Na-polyacrylate	6.42 ± 0.47	20.53 ± 4.50
Na-citrate	3.48 ± 1.72	12.68 ± 1.82
Na-gluconate	0	0
Na-ascorbate	0	0

spots" and here calcium carbonate (calcite) crystals form. For autonomous self healing, however, all compounds needed for the healing reaction should ideally be incorporated in the material matrix. As in the previous experiment bacterial food sources were part of the medium but not of the concrete matrix, an additional experiment was done to investigate the compatibility of concrete and various organic compounds. The compounds chosen for this test, all suitable food sources for the applied bacteria, were added to concrete in a concentration of 0.5% of cement weight (see Table 1 for composition of the concrete mix). After 28 days curing, some compounds appeared to be better compatible with concrete then others (Table 2).

No significant difference was found in flexural tensile and compressive strength between control and amino acid (aspartic acid and glutamic acid)-containing concrete bars. Concrete to which polyacrylic acid and citric acid was added suffered significant strength loss, while gluconate- and ascorbic acid-amended concrete did not develop any strength at all. Specific organic compounds such as amino acids appear thus suitable candidates to act as self healing agent in concert with suitable bacteria.

5 Conclusions and Future Perspectives

Some previous studies reported on the successful application of bacteria for cleaning of concrete surfaces as well as concrete-, limestone-, and ornamental-stone crack repair (DeGraef et al. 2005; Dick et al. 2006; Rodriguez-Navarro et al. 2003; Bang et al. 2001; Ramachandran et al. 2001). As the bacteria in these studies were brought into contact with the material only after damage had occurred, these examples cannot be considered as truly, autonomous, self healing mechanisms. The experiments presented in this study, focused on the healing potential of concrete-immobilized bacteria, i.e. bacteria that are part of the concrete matrix. The results of the experiments show that immobilized bacteria mediate the precipitation of minerals and, moreover, the bacteria and certain classes of needed food sources do not negatively affect concrete strength characteristics. It can therefore be concluded that bacterially controlled crack-healing in concrete by mineral precipitation is potentially feasible. The concept, however, needs further developments on some areas. It should still be clarified whether bacterial mineral precipitation effectively seals cracks, i.e. significantly reduces the permeability of cracked concrete in order to protect the embedded reinforcement from corrosion and thus increases the durability of the material. Furthermore, bacterial species must be selected which, when part of the concrete matrix, remain viable for at least the expected lifetime of the construction. If so, the bacterial approach can successfully compete with other (abiotic) self healing mechanisms as such bacteria comply with all the listed characteristics of the most ideal self healing agent.

Acknowledgements Arjan Thijssen is acknowledged for help with ESEM analysis and for providing ESEM photographs. Financial support from the Delft Centre for Materials (DCMat: www.dcmat.tudelft.nl) for this work is also gratefully acknowledged

References

Bang S.S., Galinat J.K., Ramakrishnan V. (2001) Calcite precipitation induced by polyurethane-immobilized Bacillus pasteurii. Enzyme Microb Tech 28:404–409

DeGraef B., DeWindt W., Dick J., Verstraete W., DeBelie N. (2005) Cleaning of concrete fouled by lichens with the aid of Thiobacilli. Mater Struct 38(284):875–882

DelaTorre J.R., Goebel B.M., Friedmann E.I., Pace N.R. (2003) Microbial diversity of cryptoendolithic communities from the McMurdo Dry Valleys, Antarctica. Appl Environ Microbiol 69:3858–3867

De Muynck W., Dick J., De Graef B., De Windt W., Verstraete W., De Belie N. (2005) Microbial ureolytic calcium carbonate precipitation for remediation of concrete surfaces. In: Alexander M., Beushausen H-D., Dehn F., Moyo P. (eds) Proceedings of international conference on concrete repair, rehabilitation and retrofitting. South Africa: Cape Town, pp 296–297

Dick J., DeWindt W., DeGraef B., Saveyn H., VanderMeeren P., DeBelie N., Verstraete W. (2006) Bio-deposition of a calcium carbonate layer on degraded limestone by Bacillus species. Biodegradation 17:357–367

Dorn R.I., Oberlander T.M. (1981) Microbial origin of desert varnish. Science 213:1245–1247

Fajardo-Cavazos P., Nicholson W. (2006) Bacillus endospores isolated from granite: Close molecu-
    lar relationships to globally distributed Bacillus spp. from endolithic and extreme environments.
    Appl Environ Microbiol 72:2856–2863
Ghosh P., Mandal S., Chattopadhyay B.D., Pal S. (2005) Use of microorganism to improve the
    strength of cement mortar. Cement Concrete Res 35(10):1980–1983
Jorgensen B.B., D'Hondt S. (2006) A starving majority deep beneath the seafloor. Science 314:
    932–934
Neville A.M. (1996) Properties of concrete, 4th edn. Pearson Higher Education, Prentice Hall, NJ
Pedersen K., Nilsson E., Arlinger J., Hallbeck L., O'Neill A. (2004) Distribution, diversity and activ-
    ity of microorganisms in the hyper-alkaline spring waters of Maqarin in Jordan. Extremophiles
    8:151–164
Ramachandran S.K., Ramakrishnan V., Bang S.S. (2001) Remediation of concrete using micro-
    organisms. ACI Mater J 98:3–9
Reinhardt H.W. (1985) Beton als constructiemateriaal – Eigenschappen en duurzaamheid. Delftse
    Universitaire Pers, Delft, The Netherlands
Rodriguez-Navarro C., Rodriguez-Gallego M., BenChekroun K., Gonzalez-Munoz M.T. (2003)
    Conservation of ornamental stone by Myxococcus xanthus-induced carbonate biomineraliza-
    tion. Appl Environ Microbiol 69:2182–2193
Sagripanti J.L., Bonifacino A. (1996) Comparative sporicidal effects of liquid chemical agents. Appl
    Environ Microbiol 62:545–551
Schlegel H.G. (1993) General microbiology, 7th edn, Cambridge University Press, Cambridge, UK
Sleep N.H., Meibom A., Fridriksson T., Coleman R.G., Bird D.K. (2004) H-2-rich fluids from ser-
    pentinization: geochemical and biotic implications. PNAS 101:12818–12823

Exploring Mechanism of Healing in Asphalt Mixtures and Quantifying its Impact

Dallas N. Little[1] and Amit Bhasin[2]

[1] *Snead Chair Professor, Zachry Department of Civil Engineering, Texas A&M University,*
Senior Research Fellow, Texas Transportation Institute
E-mail: d-little@tamu.edu
[2] *Associate Research Scientist, Texas Transportation Institute, Texas A&M University*

1 Introduction

Any treatise on healing in asphalt pavements must begin with an answer to the question: "How important is healing in asphalt pavements?" Experience clearly tells us that it is substantially important. Shift factors from laboratory-predicted to field-observed fatigue cracking demonstrate that laboratory data underpredict field observations. A variety of reasons are responsible for the use of shift factors including traffic wander, time of crack propagation, difference in stress states between the laboratory and the field, and crack healing. The shift factor developed by the Asphalt Institute from AASHTO Road Test (laboratory-to-field) data is 18.3.

Lytton et al. (1993) expressed that the shift factor is dominated by healing and is a function of the number of rest periods, time between load cycles, and the rest period duration. Using the relationships developed by Lytton et al. (1993) and considering data from moderately trafficked highways, one can calculate shift factors of between $\sim$3 and $\sim$13.

Damage in asphalt pavements is affected by the quality of the mastic including its cohesive strength, ability to resist fracture damage, and the ability of microcracks in the mastic to heal during rest periods. The fatigue damage process is influenced by fracture in which crack growth is induced and microcrack healing in which microcrack surfaces, at least partially, rebond. The healing process affects the fatigue process most profoundly when the microcracks are small. Therefore, processes that impede the growth of microcracks, and keep them small, affect the fracture and healing processes and, therefore, the damage process. The interruption of microcrack growth due to a dispersed filler that interrupts crack-tip energy is an example of such a process.

2 Organization

This chapter is divided into seven sections. Section 2 simply identifies the organization. It is followed by a brief background of the documented effects of healing on the performance of asphalt mixtures in section 3. Section 4 hypothesizes mechanisms of healing, and section 5 identifies recent methods that have been successfully used to quantify healing in the laboratory and in the field. Section 6 discusses a methodology by which healing data may be used to predict a shift factor and fatigue life in the field. Section 7 is the conclusion.

3 Background

3.1 Laboratory Documentation

Some of the most recent, significant documentation of healing in the laboratory was demonstrated by Kim et al. (2001), by Carpenter and Shen (2006), by Maillard et al. (2004), and by Little et al. (2001). The approaches behind some of these methods are discussed in more detail in section 5.

Data from these laboratory studies clearly demonstrate that healing is real and significant. Little et al. (2001) showed that rest periods (of 24-h duration) applied in traditional flexural beam bending experiments increased the fatigue life by more than 100% depending on the type of binder used. Kim et al. (2001) used torsional loading of asphalt mastics to demonstrate that healing periods of between 30 s and 2 min extend fatigue life and decrease the rate of dissipated damage causing energy (measured as pseudo-strain energy). Their work also showed that the impact of healing is by far the greatest when rest periods are applied before significant damage occurs. Carpenter and Shen (2006) skillfully verified this conclusion by demonstrating that the application of short rest periods between each load cycle not only extends fatigue life but is also responsible for the "endurance limit" of some asphalt mixtures. Carpenter and Shen (2006) used dissipated energy between load cycles to quantify healing.

Bhairampally et al. (2000) demonstrated that the inclusion of rest periods between compressive load cycles extended the time to tertiary damage and that this extension depended on the type of asphalt. Their work concluded that the transition from the secondary phase to the tertiary phase of dynamic compressive creep is related to development and growth of microcracks. Further work by Little and Masad (2006) has shown this to be true using computer assisted tomography. Bhairampally et al. (2000) also demonstrated that a filler (in this case hydrated lime) successfully controlled the rate of damage in the compressive mode of loading because of crack pinning and accentuated the effect of healing of microcracks in the mastic.

3.2 Field Documentation

Some of the most convincing field data regarding healing was reported by Williams et al. (2001). They selected four pavement sections at the Turner Fairbanks accelerated load facility (ALF) considering a full factorial of two thicknesses and two asphalt layer types over a homogeneous subgrade. Surface wave measurements were made to assess pavement stiffnesses before, immediately after, and 24 h after loading passes. Regardless of the pavement type, the trend was that more healing (recovery of stiffness) was recorded closer to the centerline suggesting that more fatigue damage results in a greater potential for and a greater amount of microdamage healing. Williams et al. (2001) reported other convincing support for healing using surface wave analysis of pavements at Mn/ROAD pavement sections and on US Highway 70 in North Carolina using designed experiments.

Nishizawa et al. (1997) used data from four thick pavements to demonstrate that fatigue cracking did not occur because healing effects at the low-strain and low-damage levels compensated for (offset) crack growth.

4 Mechanisms

4.1 *Hypothesis*

Development of a fracture process zone at the crack tip is a precursor to the extension of an existing crack in a viscoelastic material (Schapery 1981). The fracture process zone, Figure 1, can be considered to be comprised of a number of micro or nano cracks. These nanocracks extend and coalesce to extend the crack tip upon the application of external load. An important distinction between the crack and the fracture process zone is that while a crack cannot support external load (considering Mode I failure) the fracture process zone is load bearing.

A common example of a fracture process zone is the formation of craze fibrils in thermoplastic polymers. Craze fibrils form following loss of entanglement of the molecular chains. At low temperatures and high strain rates, crazing occurs primarily due to scission of molecular chains, where as, at higher temperatures it occurs primarily due to disentanglement of the molecular chains (Berger and Kramer 1987).

Immediately after the removal of external load two processes occur. The first is viscoelastic recovery in the bulk of the material and the second is healing in the fracture process zone. Viscoelastic recovery occurs in the bulk of the material even when

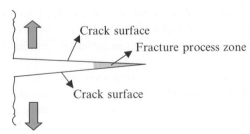

Fig. 1 Crack propagation and fracture process/healing zone in mode I loading

the applied stress or strain is too low to damage the material. In contrast healing occurs only after a stress or strain is induced, which is sufficiently large to generate damage. The phenomenological difference between viscoelastic recovery and healing is that while the former is due to the rearrangement of molecules within the bulk of the material, the latter is due to the wetting and interdiffusion of material between the two faces of a nano crack to achieve properties of the original material.

The three primary steps in the healing process are:

1. Wetting of the two faces of a nanocrack
2. Diffusion of molecules from one face to the other
3. Randomization of the diffused molecules to attempt to reach the level of strength of the original material

De Gennes (1971) proposed a model to explain the movement of a polymer molecule in a worm like fashion inside a cross linked polymeric gel, also known as the reptation model. Berger and Kramer (1987) demonstrated that the disentanglement time for chains in the crazing zone during the crack growth process was in agreement with the reptation model. In a reverse approach, Wool and O'Connor (1981) used the same reptation model to determine the time required for healing caused by the interdiffusion of molecules between crack faces.

Wool and O'Connor (1981) ingeniously described net macroscopic recovery or healing in a material by combining an intrinsic healing function of the material with a wetting distribution function using a convolution integral as follows:

$$R = \int_{\tau=-\infty}^{\tau=t} R_h\,(t-\tau)\,\frac{d\phi\,(\tau,X)}{d\tau}\,d\tau \tag{1}$$

where R is the net macroscopic healing, $R_h\,(t)$ is the intrinsic healing function of the material, $\phi\,(t,X)$ is the wetting function, and τ is the time variable.

The wetting distribution function $\phi\,(t,X)$, defines wetting at the contact of the two-crack surfaces on a domain X over time t. From a material property point of view, an asphalt binder with a higher surface free energy promotes wetting. For analytical purposes, the wetting function can be simplified by considering instant wetting or constant rate wetting. The domain in which wetting occurs can be obtained from geometric considerations for the crack growth in the material.

The intrinsic healing function, $R_h\,(t)$, defines the rate at which two-crack faces that are in complete contact with each other ("wet") regain strength due to interdiffusion and randomization of the molecules from one face to the other. From a material property point of view, an asphalt binder that has a molecular structure that favors interdiffusion of molecules will promote healing.

An important consideration in this model is that surface energy is also a function of the molecular structure of the material. It is possible, that certain molecular structures that promote interdiffusion may also reduce surface energy of the binder. Therefore, any effort to relate material properties to the healing of asphalt binders must be based on the combined effect of wetting (dependent on surface energy) and intrinsic healing (dependent of molecular structure).

Kim et al. (1990) and Little et al. (2001) conducted mechanical tests to determine the macroscopic healing (cumulative effect of wetting and interdiffusion) of different asphalt binders. Healing was achieved by applying rest periods of different durations at prespecified intervals in a cyclic load test. The pseudo-stain energy, which is the energy obtained from the area in the stress-pseudo-strain regime, was used to measure damage accumulated by the sample in a particular cycle. The use of pseudo strain in lieu of strain was used to eliminate the contribution of viscoelastic recovery in a stress-strain hysteresis loop. The healing index, H, for each rest period was quantified as follows:

$$H = \frac{\phi_A^R - \phi_B^R}{\phi_A^R} \tag{2}$$

where ϕ_A^R and ϕ_B^R are the pseudo-strain energy after and before a rest period measured in a cyclic load test. The healing index plotted against the duration of the rest period was used to derive the intrinsic healing function of different asphalt binders (Figure 2). Section 5 presents several other methods by which to quantify healing in asphaltic materials.

Little et al. (2001) also reported that the short term healing rate in asphalt binders (the rate at which healing occurs during the first 10 s of the rest period) was inversely proportional to the Lifshitz-van der Waals or nonpolar component of surface tension of the asphalt binder. Similarly, the long-term healing rate (the rate at which healing occurs after the first 10 s) was directly proportional to the acid-base component of surface tension of the asphalt binder.

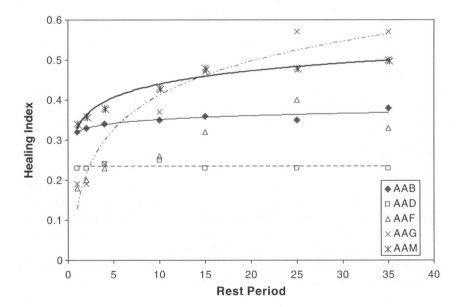

Fig. 2 Healing of mixtures with different asphalt binders (Little et al. 2001)

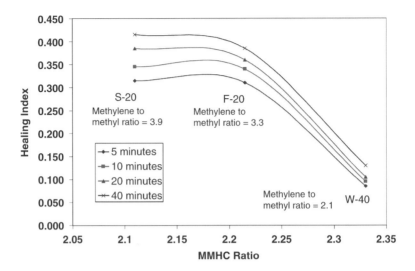

Fig. 3 Correlation between healing index to MMHC and methylene to methyl ratio (Kim et al. 1990)

Kim et al. (1990) investigated the characteristics of asphalt binders that promote the healing. They used FTIR measurements to obtain two parameters: methylene to methyl ratio and MMHC ratio (methyl and methylene hydrogen atoms to methyl and methylene carbon atoms) in independent aliphatic molecules and in aliphatic molecules attached to aromatics. A higher methylene to methyl ratio indicates a longer chain and a lower MMHC indicates a chain with fewer branches. These two parameters or indices were derived based on FTIR evaluation of the bulk asphalt binders. Figure 3 illustrates the correlation between healing index and these two parameters for three asphalt binders designated as S-20, F-20, and W-40. It can be concluded from Figure 3 that the presence of longer aliphatic molecules and fewer branches facilitate the healing process. Molecules with these characteristics also facilitate the interdiffusion process as proposed by Wool and O'Connor (1981).

Kim et al. (1990) correlated net healing of asphalt mixtures with the molecular structure (intrinsic healing function) of the asphalt binder. In contrast Little et al. (2001) correlated net healing of asphalt mixtures with the surface energy (wetting function) without accounting for the differences in the intrinsic healing function of the asphalt binders. As described in Equation (1), the correlation of net healing must be made with the wetting function by taking into account the cumulative effect of intrinsic healing and wetting. Current efforts at the Texas A&M University are focused on the development of a methodology by which to evaluate the net healing of asphalt materials based on their material properties.

Another characterization of healing was presented by Schapery (1989) and subsequently adopted by Lytton (2001) in the form of the following basic equation:

$$2\gamma_h = E_R D_h \left(t_a \right) H_V \tag{3}$$

where γ_h is the surface energy, E_R is the reference elastic modulus (used in the correspondence principle to convert the viscoelastic case to an equivalent elastic case with a typical numerical value of 1.0). D_h (t_a) is the compressive creep compliance of the material corresponding to the time, t_a, that is required for a crack to heal through a distance, a, which is the length of the fracture process zone. H_V is the viscoelastic integral similar to the J-integral except that is corresponds to the change in dissipated pseudo strain energy per unit of crack healing area from one compressive load cycle to the next. In this equation, the total surface energy is directly related to healing. According to the classic Condom–Morse diagram, the interactions due to the acid-base surface energy components (van Oss et al. 1988) are inversely related to the distance between them while the Lifshitz-van der Waals components are inversely related from the 6th to the 9th power of the distance between them. As a result, a large Lifshitz-van der Waals component, whether associated with separate aliphatics or napththenics or as appendages to aromatics may initially resist wetting due to an "aggregate association" of these interactions in the plane of the crack face. However, with time and reptation action, both the nonpolar (Lifshitz-van der Waals) and polar (acid-base) components contribute to healing as Schapery (1981, 1989) suggested. In other words, the short-range interactions may be limited to the plane parallel to the crack face. However, with time and reptation and wetting action, both nonpolar and polar interactions occur across the interface. This example reinforces the use of a combination of wetting and intrinsic healing as described in Equation (1) to explain and characterize healing in asphaltic materials.

5 Quantification of Healing

5.1 Review of Methods

Several different approaches have been used by different researchers to quantify the magnitude of healing in different types of asphalt binders, mastics, or mixtures. The commonality among most of these approaches is that healing is quantified by measuring the response from a cyclic load test by allowing intermittent periods of recovery. Some of these methods are briefly discussed in this section.

Carpenter and Shen (2006) used the ratio of dissipated energy change (RDEC) in a cyclic load test to quantify the fatigue-cracking resistance of the asphalt mixture. Simply defined, RDEC is a measure of the energy that causes incremental damage from one load cycle to the next and can be easily computed at any given cycle, a, as follows:

$$RDEC_a = \frac{DE_a - DE_b}{DE_a \, (b - a)} \tag{4}$$

where DE_a and DE_b are the dissipated energy (area within the stress-strain hysteresis loop) at cycles a and b, respectively. The fatigue resistance of the mixture is quantified as the plateau value (PV), which is the magnitude of the $RDEC$ in the

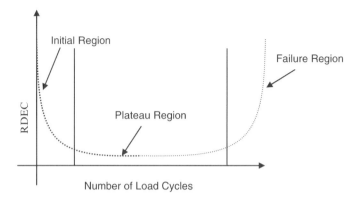

Fig. 4 Plateau value of RDEC used to determine crack-growth resistance in asphalt mastic (Carpenter and Shen 2006)

steady-state response regime of the cyclic load test (Figure 4). A higher magnitude of the PV indicates greater incremental damage energy and shorter fatigue-cracking life. To account for healing, they introduced rest periods after each load cycle. In this approach, the effect of healing can be quantified as the reduction in the *PV* due to the application of rest periods in the cyclic load test.

Carpenter and Shen (2006) demonstrated that the *PV* and the rest period have a straight-line relationship on a log–log scale. For a given strain level, this relationship can be extrapolated to determine the rest period that will yield a *PV* that corresponds to the fatigue-endurance limit of the material. Kim and Roque (2006) used a similar approach to quantify the healing characteristics of asphalt binders used in different asphalt mixtures. They quantified healing in terms of the recovered dissipated creep strain energy per unit time.

Maillard et al. (2004) conducted tensile tests on films of asphalt binder lodged between glass spheres to simulate an asphalt film that is bound by aggregates. They measured the rate of healing in the asphalt film by transmitting ultrasonic waves through the sample. A decrease in the amplitude of the ultrasonic signal corresponds to damage in the film. An increase in the amplitude of the ultrasonic signal for an undisturbed sample after applying a tensile load corresponds to the healing process. Figure 5 illustrates healing after each cycle of tensile load on an asphalt film measured using the ultrasonic waves.

5.2 Application of a Dynamic Mechanical Analyzer to Quantify Healing

Kim et al. (2003) used the dynamic mechanical analyzer (DMA) to measure the fatigue-cracking characteristics of fine aggregate and asphalt binder matrix. They quantified healing as the increase in fatigue life of the mixture due to the application

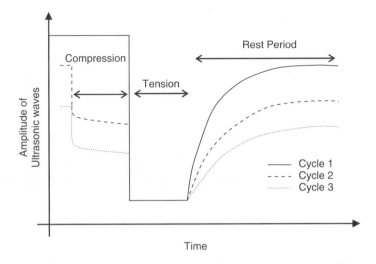

Fig. 5 Healing after the application of a tensile load measured using ultrasonic waves (Maillard et al. 2004)

of a specified number of rest periods during the test. The following section describes in further detail the use of DMA to quantify healing in the fine aggregate matrix.

Fatigue cracking and healing in asphalt mixtures is a process that is concentrated mostly in the fine aggregate and asphalt binder matrix (FAM) of the mix. Therefore, by evaluating the mechanical properties of the FAM, one is able to improve the sensitivity of measurement of the property of interest and also avoid the interaction effect of other parameters such as coarse aggregate gradation. The DMA is an effective tool by which various different mechanical properties of the FAM can be determined.

Preparation of specimens for testing with the DMA is very similar to the procedure followed for preparing asphalt mixtures. The only difference is that the asphalt binder is mixed and compacted with only fine aggregates (material passing #16 sieve). The cylindrical test specimens for the DMA are 12 mm in diameter and 50 mm in height. About 20 specimens can be obtained by sawing and coring a single 150 mm diameter and 75 mm height sample compacted using the Superpave gyratory compactor (SGC). Figure 6 illustrates the test setup along with a test specimen. Detailed methodology for the mixture design procedure and sample preparation for the DMA is available in the literature (Kim and Little 2005).

Figure 7 illustrates the sequence of load cycles that are applied, typically in a controlled strain mode, in order to determine a number of important properties of the FAM. There are two methods to assess fatigue-cracking life of the FAM using the DMA: (1) the loss of dynamic modulus (normalized with respect to the undamaged modulus); and (2) the rate of change of dissipated pseudo-strain energy. The dissipated pseudo-strain energy is the amount of energy dissipated in each load cycle due to the damage incurred by the specimen. As described previously, dissipated pseudo-strain energy is computed as the hysteresis area within the stress-pseudo-strain curve.

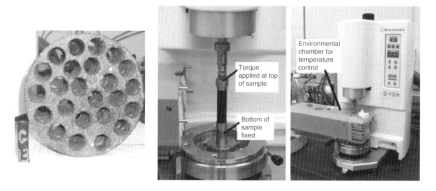

Fig. 6 Specimen extracted from SGC sample (left) and test setup (right) for DMA

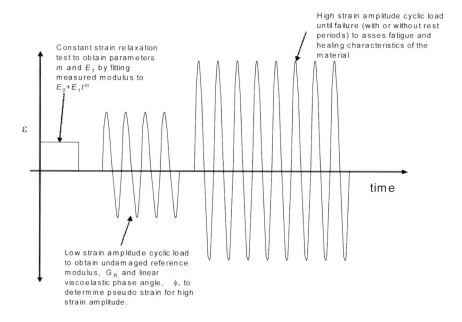

Fig. 7 Sequence of loading for a typical DMA test to obtain various material properties for FAM

Kim et al. (2003) used the DMA to determine the healing properties for two different asphalt binders AAD and AAM. They included 2 min rest periods 10 times during each test. The rest periods were introduced at equal damage levels for each mixture. For example, the fatigue life of AAM was shorter than the fatigue life of AAD. Therefore, the rest periods for AAM were applied at proportionately shorter intervals for AAM as compared to AAD. Figure 8 illustrates the increase in fatigue life due to inclusion of the rest periods for both asphalt binders. It is evident from the figures that the improvement in fatigue life due to the introduction of rest periods is small for

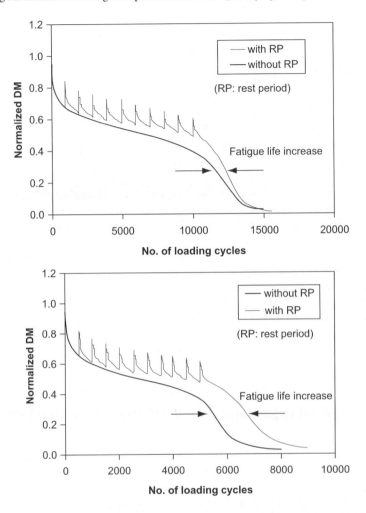

Fig. 8 Impact of rest period on fatigue life and of asphalt using binders AAD (left) and AAM (right) (Kim et al. 2003)

AAD and significantly high for AAM. This suggests that the asphalt binder AAM has better healing characteristics as compared to AAD. Sufficient replicate tests verified statistical differences between the healing properties of AAD and AAM.

Little and Bhasin (2006) used a similar approach to quantify the healing propensity of binders in different FAM. They computed the dissipated pseudo-strain energy, W_R, as the area in the stress-pseudo-strain hysteresis curve for any given cycle. The parameter b which represents the rate of accumulation of damage was obtained from the relationship of the dissipated pseudo-strain energy with number of load cycles, N, from experimental data using the following form:

$$W_R = a + b \ln(N)$$

Table 1 Healing quantified as the reduction in the rate of change of dissipated pseudo-strain energy

FAM	Parameter b		Change in b
	No Rest	With Rest	
RA ABD	231	181	1.3
RL ABD	314	174	1.8
RK ABD	272	177	1.5
RA AAD	126	139	0.9
RL AAD	202	200	1.0
RK AAD	293	318	0.9

In order to quantify healing for different materials, they applied nine rest periods of 4 min each during the cyclic test at high strain amplitude. The rest periods were applied at 2.5%, 5%, 10%, 15%, 20%, 25%, 30%, 40%, and 50% of the fatigue life value of that particular material measured without any rest period. When rest periods were applied, the parameter b was computed by fitting the dissipated pseudo-strain energy measured immediately after the rest period to the corresponding number of load cycles using the aforementioned equation. A relative decrease in the parameter b was then used to quantify the healing potential for each material. Table 1 presents the healing potential for FAM using two different types of asphalt binders and three different types of aggregates. From Table 1 it is evident that the asphalt binder type and the aggregate type are important in determining the healing potential of the asphalt mastic within a FAM.

6   Incorporating Healing to Predict Fatigue Cracking Life

Lytton et al. (1993) made one of the earliest attempts to incorporate healing in the evaluation of fatigue life of pavements based on material properties. Fatigue life determined in the laboratory, $N_{f(lab)}$, is typically multiplied by a shift factor, SF, in order to estimate fatigue performance of the mixtures in the field, $N_{f(field)}$, as follows:

$$N_{f(field)} = N_{f(field)} \times SF \qquad (5)$$

Large variations in the value of shift factors have been reported in the literature. Lytton et al. (1993, 2003) proposed that healing of asphalt mixtures contributed to the shift factor. They also proposed that the total shift factor was a geometric combination of the shift factor due to healing, SF_h, shift factor due to residual stresses, SF_r, and shift factor due to dilation, SF_d, as follows:

$$SF = SF_h + SF_r + SF_d = \left[1 + a\,(t_r)^b\right] + SF_r + SF_d \qquad (6)$$

where the shift factor due to healing is expressed as a function of the rest period between load cycles recorded in seconds, and a and b are healing coefficient and exponent, respectively. This model was calibrated using laboratory and field data to obtain values of a and b for asphalt mixtures from different climatic zones. The shift factor for healing obtained from these parameters varied from 1.09 to 2.7 for a 30 s rest period. This approach demonstrated the importance of incorporating healing in prediction of pavement performance based on laboratory data.

Little et al. (2001) and Lytton and coworkers (2001) further improved the above methodology by determining the healing potential of individual asphalt binders using methods previously described. The measured healing potential was then used to represent the rate of healing in an asphalt binder as:

$$\frac{dh}{dN} = fn\left(t_r, h_1, h_2, h_\beta\right) \tag{7}$$

where t_r is the rest period, and h_1, h_2, and h_β are the short-term healing rate, long-term healing rate, and maximum healing achieved by the binder, respectively. The three terms related to healing were found to be empirically correlated to the different surface free energy components of the asphalt binder. The rate of healing per load cycle, $\frac{dh}{dN}$, which is analogous to the rate of crack growth, $\frac{dc}{dN}$, allows one to include healing in the determination of fatigue-cracking life of the material based on principles of fracture mechanics. The improvements in this methodology over the previous approach are:

1. The contribution of healing was incorporated in a fatigue-cracking model based on principles of fracture mechanics instead of the use of a shift factor.
2. The propensity of different asphalt binders to heal at different rates was accounted for by measuring the healing index of the binders individually in laboratory experiments.

Researchers at the Texas A&M University have developed a model to quantify the fatigue crack growth in asphalt mastics (fine aggregate matrices) at a given number of load cycles as a function of properties such as surface free energy, rate of dissipation of the pseudo-strain energy per load cycle, and creep compliance of the material. Ongoing work at the Texas A&M University is focused on the measurement of healing as an intrinsic material function of time and temperature, and the inclusion of this healing function in the crack growth model.

7 Conclusions

Microdamage healing exists in asphalt mixtures. Its impact is substantial to the prediction of fatigue life of asphalt mixtures. Further investigation is required to clearly define the healing process. However, it is apparent that the two important steps in healing phenomenon are wetting of the crack faces and interdiffusion of molecules between the wetted surfaces to gain strength approaching that of the undamaged material. Therefore, the surface energy components, morphology, and nature of the generic components of the asphalt binder affect wetting and bonding.

Several reasonable approaches are available by which to quantify healing. The most reliable are based on recovered energy methods.

References

Berger L.L., Kramer E.J. (1987) Chain Disentanglement during High Temperature Crazing of Polystyrene. Macromolecules 20:1980–1985

Bhairampally R.K., Lytton R.L., Little D.N. (2000) Numerical and Graphical Method to Assess Permanent Deformation Potential for Repeated Compressive Loading of Asphalt Mixtures. *Transportation Research Record: J Transportation Res Board* 1723:150–158

Carpenter S.H., Shen S. (2006) A dissipated energy approach to study HMA healing in fatigue. 85th annual meeting of the Transportation Research Board, Washington DC

de Gennes P.G. (1971) Reptation of a polymer chain in the presence of fixed Obstacles. J Chem Phys 55(2):572–579

Kim B., Roque R. (2006) Evaluation of healing property of asphalt mixture. 85th annual meeting of the Transportation Research Board, Washington, DC

Kim Y., Little D.N. (2005) Development of specification type tests to assess the impact of fine aggregate and mineral filler on fatigue damage. 0-1707-10, Texas Transportaion Institute, Texas

Kim Y.R., Lee H., Little D.N. (2001) Microdamage healing in asphalt and asphalt concrete, vol IV: a viscoelastic continum damage fatigue model of asphalt concrete with microdamage healing. Texas Transportation Institute, College Station, Texas

Kim Y.R., Little D.N., Benson F.C. (1990) Chemical and mechanical evaluation on healing mechanism of asphalt concrete. Proc AAPT 59:240–275

Kim Y.R., Little D.N., Lytton R.L. (2003) Fatigue and healing characterization of asphalt mixes. J Mater Civil Eng (ASCE) 15:75–83

Little D.N., Bhasin A. (2006) Using surface energy measurements to select materials for asphalt pavement. Final Report for Project 9–37, Texas Transportation Institute, Texas

Little D.N., Lytton R.L., Williams A.D., Chen C.W. (2001) Microdamage healing in asphalt and asphalt concrete, vol I: microdamage and microdamage healing, project summary report. FHWA-RD-98-141, Texas Transportation Institution, College Station, Texas

Lytton R., Uzan J., Fernando E.G., Roque R., Hiltunen D. (1993) Development and validation of performance prediction models and specifications for asphalt binders and paving mixes, Report No. SHRP-A-357. Strategic Highway Research Program

Lytton R.L., Chen C.W., Little D.N. (2001) Microdamage healing in asphalt and asphalt concrete, vol III: a micromechanics fracture and healing model for asphalt concrete. FHWA-RD-98-143, Texas Transportation Institution, College Station, Texas

Maillard S., de La Roche C., Hammoum F., Gaillet L., Such C. (2004) Experimental investigation of fracture and healing at pseudo-contact of two aggregates. 3rd Euroasphalt and Eurobitume Congress, Vienna

Nishizawa T., Shimeno S., Sekiguchi M. (1997) Fatigue analysis of asphalt pavements with thick asphalt mixture. layer. 8th International Conference on Asphalt Pavements, Seattle, Washington, pp 969–976

Schapery R.A. (1981) Non linear fracture analysis of viscoelastic composite materials based on a generalized. J Integral Theory

Schapery R.A. (1989) On the mechanics of crack closing and bonding in linear viscoelastic media. Int J Fract 39:163–189

van Oss C.J., Chaudhury M.K., Good R.J. (1988) Interfacial Lifshitz-van der Waals and polar interactions in macroscopic systems. Chem Rev 88:927–941

Williams A.D., Little D.N., Lytton R.L., Kim Y.R., Kim Y. (2001) Microdamage healing in asphalt and asphalt concrete, vol II: laboratory and field testing to assess and evaluate microdamage and microdamage healing. FHWA-RD-98-142, Texas Transportation Institution, College Station, Texas

Wool R.P., O'Connor K.M. (1981) A theory of crack healing in polymers. J Appl Phys 52(10): 5953–5963

Self Healing in Aluminium Alloys

Roger Lumley

CSIRO Manufacturing and Materials Technology, Clayton South MDC, Victoria, Australia
E-mail: Roger.Lumley@csiro.au

1 Introduction

Self healing is a key property of biological materials, examples being the autonomous repair of fractured bones or torn skin tissue, as is discussed in other chapters of this book. More than a century ago, principles were developed defining the reconstitution of fractured bone *in vivo* which stated that, in essence, material dissolved from where it was not required was redeposited to where it was required as a response to mechanical stimuli and damage. As far as metals and other inanimate materials are concerned, it is well known that damage to oxide films, which normally protect the surfaces of metals such as aluminium (Al) and titanium (Ti) from corrosion, can be repaired by reoxidation in air, which can be seen as a form of self-repair. Now attention is being directed to processes that may possibly heal defects, such as cracks, which can develop in the interior of materials during manufacture, or when they are in service. Such self healing processes may then allow failures to be averted and the useful lives of components and structures to be extended.

Self healing of materials can be considered on macro-, micro- or nano/atomic scales. Sintering of metal powders to form a solid body is perhaps the best examples of the mechanisms that may be used for self healing at the macro level. At the other end of the scale are processes that involve only the localized movement of atoms, such as dynamic precipitation, which can immobilize dislocations and other defects in an alloy when it is loaded. At the intermediate or micro level, evidence is accumulating that self healing of small, internal cracks is feasible. Self healing on all three of these scales is considered in separate sections in this chapter.

Damage is defined by the Oxford Dictionary as "physical harm, which reduces the value, operation, or usefulness of something". It can take different forms with materials and can arise for many reasons including wear, corrosion, overload, impact, fatigue, and creep. Since this chapter will be concerned primarily with Al alloys, it is desirable first to review the factors that influence the strength and fracture characteristics of these materials. Moreover, many commercial Al alloys respond to precipitation hardening and some basic principles of this phenomenon that are relevant to self healing need to be considered. All processes of self-repair involve movement of atoms, and it is also desirable to give a brief review of the underlying principles of mass transfer of atoms by solid state diffusion.

1.1   Aluminium as an Engineering Material

Aluminium is now second only to iron as the most commonly used industrial metal. From its very first utilization, Al has competed with, and indeed replaced many more established materials due to its favorable properties. In particular, Al is only one third the density of steel, and is resistant to corrosion in most environments due to the highly tenacious oxide film that covers the metal. Aluminium can be readily fabricated into a wide range of consumer products, ranging from packaging and foil through to high-strength engineering products for automotive and aerospace applications. It can be readily cast into intricate shapes and conducts heat and electricity nearly as well as copper. Aluminium's excellent recyclability is also another factor since the remelting of scrap normally requires only about 5% of the energy needed to extract the same weight of primary metal from bauxite ore. In general, the desirable aspects of Al and its alloys may be summarized according to Table 1 and the range of alloys produced commercially are summarized in Figure 1. Many textbooks describe Al and its alloys in greater detail, and, as such, the reader is referred to these for additional information [e.g. 1, 2].

1.2   Damage in Aluminium Alloys

Damage that influences the mechanical behavior of metals may range from the macroscale right down to the nano- or atomic scale. In the context of self healing, it is also quite clear that, as a crack or defect feature approaches the macroscale, the ability to repair the feature by self healing becomes dramatically reduced. Therefore, self healing in metals is generally considered as being applicable at the microscale and below. Design issues are often a controlling factor that may limit the service life of components and structures and the following equation is the key to defining the fracture conditions for a large cracked plate [3]

$$K_c = \sigma \sqrt{\pi a}$$

where K_c is the critical stress intensity factor or fracture toughness of the material and is a function of the basic material properties, σ is the design stress (normally being a nominated fraction of the yield stress of the material) and a is the maximum allowable flaw size.

This relationship may be used in several ways to design against a component failure, and the utility of this equation is that once any two key variables are defined, the third factor is fixed. Typically therefore, such design criteria also tend to dictate the choice of material selected for the application since if the fracture toughness (K_c) is reached, then rapid failure will result. Factors governing K_c for Al alloys are well documented, and the various microstructural features affecting those that respond to age hardening are clearly depicted by the "Staley Toughness Tree" (Figure 2) [4].

Table 1 Characteristics of aluminium and relative importance in different products

Use	Characteristics			Type of product						
	Density	Conductivity	Corrosion resistance	Decorative	Castings	Forgings	Sheet	Extrusions	Cable	Foil
Transport	↑↑↑	↑↑	↑↑↑	↑↑↑	↑↑↑	↑↑↑	↑↑	↑↑	–	–
Machinery	↑↑↑	↑↑	↑↑	↑	↑↑↑	↑↑↑	↑	↑	–	–
Building	↑↑	–	↑	↑↑↑	–	–	↑↑	↑↑	–	–
Household	↑↑	↑↑↑	↑↑↑	↑↑	–	–	↑	–	–	↑
Chemicals and Food	↑↑	↑↑	↑↑↑	↑	↑	↑	↑	↑	–	↑
Packaging	↑	↑	↑↑↑	↑↑↑	–	–	–	–	–	↑
Electrical	↑	↑↑↑	↑	–	↑	↑	–	↑	↑	↑

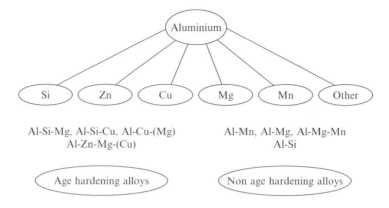

Fig. 1 A summary of the range of aluminium alloys produced commercially

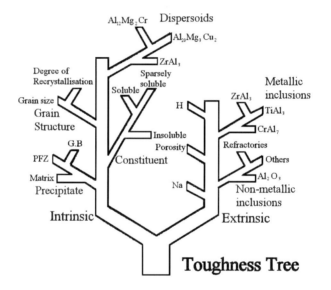

Fig. 2 The "Staley Toughness Tree", a representation of microstructural features that affect fracture toughness in age-hardenable aluminium alloys. (From Starke and Staley [4])

Fracture toughness properties of some of these Al alloys are also given in Table 2 [5].

If it is considered that design of a component or structure is such that the critical flaw size "a" may be readily exceeded during service, then failure is unlikely to be averted by self healing because there are over-riding factors which can cause failure. What self healing may achieve, however, is to slow down or prevent the growth of these defects so that they do not reach their critical flaw size and cause failure during their expected lifetimes.

Table 2 Representative fracture toughness of a range of plate aluminium alloys [5]

Alloy and Temper	Typical Plane Strain Fracture Toughness, K_{Ic}		
	L-T (MPa$\sqrt{m}$)	T-L (MPa$\sqrt{m}$)	S-T (MPa$\sqrt{m}$)
2090-T81	38	–	–
2014-T651	24	22	19
2024-T351	36	33	26
2024-T851	24	23	18
2419-T851	43	38	30
6013-T651	42	–	–
6061-T651	39	–	–
7050-T7651	34	31	26
7075-T651	29	25	20
7475-T651	43	37	32
7475-T7351	55	45	36

1.3 Diffusion Phenomena in Metals and Alloys

The atomic transport of matter by diffusion is represented ideally by the net flux, J, of atoms per second per unit area of reference plane in opposite directions ($\pm x$) in the presence of a concentration gradient, dc/dx, as given by Fick's first law:

$$J = -D(dc/dx)$$

where D is the diffusion coefficient, given by:

$$D = Do \exp(-Q/RT),$$

Do (the frequency factor in cm^2/s), Q (the activation energy for diffusion in kJ/mol) and R (the gas constant, 8.314510 J/Kmol) are all constants, so the only variable is the temperature T, in Kelvin.

For Al alloys, the diffusion rates of key elements are compared to the self diffusion rate of Al over a range of temperatures in Figure 3 [6].

Vacant lattice sites or other inhomogeneities within a metal are essential in facilitating atomic diffusion and four basic modes are recognized which are depicted in Figure 4. Each assumes that there is a certain degree of imperfection within a solid. Because these diffusion processes have different activation energies, each will have a characteristic diffusion coefficient at any given temperature. An example of the roles of the different diffusion processes is shown in Figure 5a for silver [7] and it is particularly interesting to note that, whereas vacancy diffusion (D_v) decreases dramatically as temperature decreases, the changes to pipe diffusion along dislocations (D_p), grain boundary diffusion (D_{gb}), and surface diffusion (D_s) are much less. Additional data showing hetero-diffusion associated with dislocations is presented in Figure 5b, and it is important to note that, again with Ag as the example, self-diffusion in the presence of dislocations is up to 10^8 faster than self-diffusion through the lattice. For example, atomic diffusion along a dislocation line is actually faster at 0.5Tm than

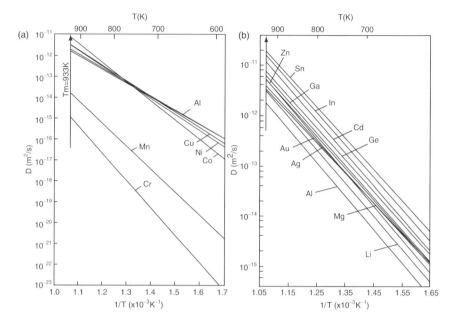

Fig. 3 (**a**) Diffusion coefficients for slow diffusing impurity elements in aluminium vs. reciprocal of temperature. (**b**) Diffusion coefficients for fast-diffusing impurity elements in aluminium vs. reciprocal of temperature. Self-diffusion of aluminium is also shown in both plots (From Landolt–Börnstein [6])

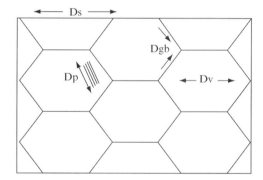

Fig. 4 The basic modes of atomic diffusion in polycrystalline metals. D_s is surface diffusion, D_{gb} is grain boundary diffusion, D_p is pipe diffusion associated with dislocations, and D_v is vacancy diffusion

is vacancy-assisted diffusion at a temperature close to Tm. What this means is that the mass transfer occurring via pipe diffusion along dislocations is more like that occurring in the material at much higher temperatures. For Al, it has been estimated that pipe diffusion of copper atoms along dislocations at ambient temperature occurs at a rate 10^6 times greater than that for bulk diffusion of vacancies [8]. However, it should be noted that rates of vacancy and solute diffusion at ambient temperatures can also be much changed if a metastable supersaturation of vacancies is retained within the material by quenching from a high temperature [1]. Furthermore, rates are

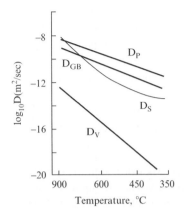

(a)

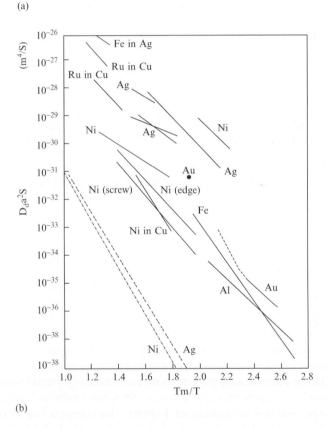

(b)

Fig. 5 (**a**) Shows the variation of diffusion coefficients with temperature in polycrystalline Ag. (**b**) Shows the self-diffusion and heterodiffusion (associated with dislocation structures) in a range of metals and alloys. Different values for the same elements (e.g. Ag) are for different dislocation forms. Broken lines are for the lattice self-diffusion coefficient in Ni and Ag multiplied by a^2, where $a = 5 \times 10^{-10}$ m (From Murr [7], and Landolt–Börnstein [6])

also higher if vacancies trapped or generated by some microstructural features can be released [9]. In these ways, the enhanced vacancy and solute diffusivity characteristics at higher temperatures may be maintained at a substantially lower temperature. The fact that high levels of atomic mobility are possible at relatively low temperatures is fundamental to some of the processes of self healing that will be discussed.

2   Solid-state Sintering Processes

The primary aim of sintering powder bodies is to bond particles and to increase density via the removal of porosity. The driving force for sintering processes depends on the surface energy or chemical potential of the body being sintered [10]. The mechanisms of mass transport of material causing densification are all that differs between various sintering processes. In terms of the potential for self healing behavior, it is evident that some or all of these mechanisms may be utilized, and hence they deserve further scrutiny.

Sintering processes provide opportunities to remove or modify features that, in a solid material, would be considered to be stress raisers and defects such as cracks. The major issue is one of mass transport, which is generally negligible below 0.5 Tm and exceptions have been discussed in the previous section. Therefore, unless the conditions governing mass transfer of material at elevated temperature can be retained and exploited at lower temperature, self healing within reasonable time frames will not be expected to occur by normal thermally activated processes. There are however exceptions to this, such as chemically activated processes, which will be discussed in the next section.

2.1   Solid-state Mass Transfer

Sintering without a liquid phase is limited to mass transport by the vapor phase or by various types of solid state diffusion. These processes, which are summarized in Table 3 and Figure 6 (with reference to Figure 4), are governed by rate-determining equations that are based on the work of Ashby [11–13]. With the notable exception of ice, vapor phase sintering is rare at close to the normal temperatures to which materials are exposed, and it is not considered here in further detail. It is worth noting, however, that for micron-range particles with correspondingly small bonds between them, the minimum vapor pressure required for mass transfer and shape modification of the open space is of the order of 1–10 Pa. This represents the lower limit for bonding of free surfaces.

Solid-state sintering involves mass transport by diffusional means whereby atoms and/or vacancies move through the material [14] and material is transported in local regions by the most thermodynamically favorable route or routes [12]. These diffusional events result in the (centre–centre) approach of adjacent particles or surfaces, a

Table 3 Sintering mechanisms and associated neck growth rate equations [11, 12]

Mechanism #	Transport Path	Associated Diffusion Coefficient	Source of Matter Transported to Neck	Associated Neck Growth Rate Equation
1	Surface diffusion	D_s	Surface	$\chi_1' = 2D_s\delta_s FK_1{}^3 \equiv dx_1/dt$
2	Lattice diffusion	D_v	Surface	$\chi_2' = 2D_v FK_1{}^2$
3	Vapor transport	None	Surface	$\chi_3' = PvF(\Omega/2\pi\Delta_0 kT)^{0.5}K_1$
4	Boundary diffusion	D_{gb}	Grain boundary	$\chi_4' = 4D_{gb}\delta_{gb}FK_2{}^2/x$
5	Lattice diffusion	D_v	Grain boundary	$\chi_5' = 4D_v FK_2{}^2$
6	Lattice diffusion	D_p	Dislocations	$\chi_6' = 0.444K_2 Nx^2 D_p F$ $(K_2 - 1.5Gx/F_s r)$

Where:

χ_i' is the sintering neck growth rate

δ_s is effective surface thickness, δ_{gb} is effective grain boundary thickness and E_{Ad} (energy of adhesion) $\approx F_s/10$; $F = F_s\Omega/kT$;

(where Ω is the atomic or molecular volume, k is Boltzmanns constant, T is temperature (K)),

Δ_0 is the theoretical compact density,

K_i's represent the curvature differences (driving mechanisms),

P_v in mechanism 3 is the is vapour pressure ($P_v = P_o \exp(-Q_{vap}/kT)$),

N in mechanism 6 is the dislocation density (cm^{-2}),

G is shear modulus,

r is particle radius

and x is the half width of the neck.

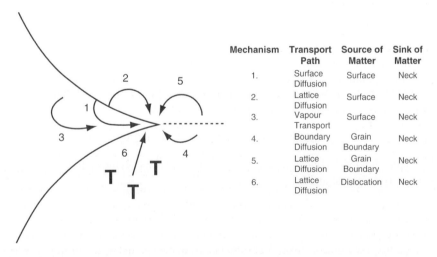

Fig. 6 Alternate paths for matter transport to the neck region during solid-state sintering. See Table 3 for additional details

process that is driven by the differences in free energy or chemical potential between the surfaces, and the point of contact (the "neck"). Because, in practice, the geometry of a neck also approximates to the geometry of a crack, it is possible to use the rate-determining equations for sintering as a means for evaluating the potential for crack closure in solid materials. To a first approximation, this is achieved by utilizing data in Table 3 and Figure 6 although it must be noted that the complexity and, usually, rate of matter transport to the neck is increased in the following cases:

1. In systems of two or more components where, for example, the diffusion rate of one species in the other may be faster than the respective self-diffusivities
2. When new or additional (diffusive) mechanisms are introduced, (e.g. see Figure 6)
3. Existing mechanisms are accelerated by the application of an external stress [11]

An example of complex atomic diffusion is in ternary Al–Zn–Mg alloys, wherein the diffusivity of Zn atoms accelerates as the concentration of Mg in the alloy increases [15]. In this instance, it may be that the inherently high diffusivity of Mg in Al has the effect of conveying vacant lattice sites to Zn atoms that would work to accelerate the jump frequency of Zn as it diffuses.

2.2 Solid-state Sintering in Aluminium Alloys

Bonding or sintering of Al powder alloys has traditionally been thought to be a difficult process, due to the tenacious oxide film present on the surfaces of Al powders. Generally in non-Al systems, the activity of the sintering atmosphere is used to reduce the oxide layer. In Al it is possible to modify the oxide layer on Al powders by the use of nitrogen gas to assist or facilitate sintering [16, 17]. However, a simpler way to promote sintering of pure Al powders is to use small additions of Mg, with quantities of less than 0.2 wt% [18] being all that is required. This process is facilitated by the solid-state diffusion of Mg through the Al and becomes evident at temperatures as low as 275°C. As the Mg diffuses through the Al matrix to the Al/Al_2O_3 interface, the following partial reduction reaction occurs to produce nanoscale crystallites of cubic spinel:

$$3Mg + 4Al_2O_3 \rightarrow 3MgAl_2O_4 + 2Al$$

At magnesium (Mg) concentrations between 4% and 8% [19] the complete reduction reaction is dominant and is:

$$3Mg + Al_2O_3 \rightarrow 3MgO + 2Al$$

The complete reduction reaction can occur directly, or proceed via the partial reduction reaction mentioned above. As a direct consequence of the action of Mg in reducing the alumina film on the Al powders, solid-state sintering and bonding of Al powder bodies with >99% purity is possible. Additions of Mg also enhance liquid-phase sintering in more complex alloys by the same means.

2.3 Transient Liquid Phase Sintering in Aluminium Alloys

Transient liquid phase sintering (TLPS), which is a widely used process in powder metallurgy, occurs when the liquid phase exists for a finite period only, then dissipates due to solubility effects or secondary reactions. TLPS has particular relevance to self healing in Al and other materials, as it involves bonding of component materials by the presence of liquid phases which are only temporarily present. Dental amalgams are a good example of TLPS that occurs at close to ambient temperatures.

Three distinguishable variants of TLPS are possible [14] which are:

1. The powders undergo exothermic reactions that are responsible for liquid-phase formation (e.g. Al–Ni, Al–Ti). This includes such processes as in situ microfusion [20].
2. Starting powders pass through a liquid-forming stage, but the liquid is absorbed by the solid phase (e.g. Al–Zn, Al–Cu, Al–Mg) [21–23]
3. Cooling from liquid-phase sintering temperatures leads to the formation of a glassy phase in boundary regions, which subsequently crystallizes. This mechanism is predominantly found in ceramics.

2.4 Non-thermally Activated Sintering in Multicomponent Alloys

An additional means of initiating solid-state sintering has also recently been reported that is not thermally activated, and is termed precipitation-induced densification. This process is particularly significant when considering self healing because it causes the closure of porosity and crack-like geometries at reduced temperatures through the decomposition of a solid solution. Lumley and Schaffer [24] have shown that substantial pore closure may be initiated by the precipitation of particles of a second phase into the void space of porous materials by slowly cooling from elevated temperature, and this is found to apply to a range of Al alloys as shown in Table 4 [25]. In particular, the propensity for an alloy to undergo precipitation, and its solute content, are both very important in this process. During slow cooling from an elevated sintering temperature, the alloy behaves according to its phase diagram, in that the *decrease* in temperature forces the heterogeneous precipitation of a second phase as the alloy moves towards its equilibrium, two-phase state (Figure 7). Figure 8 shows a backscattered SEM image of the surface of an internal pore, demonstrating precipitation into the open pore space, and Figure 9 presents an example of the efficacy of the process in closing pore structures compared to longer sintering times at higher temperatures. It is important to note that the diffusing solute elements are strongly attracted to the highest free energy surfaces, namely that the driving force for precipitation is proportional to the reduction in Gibbs free energy, ΔG, as expressed by:

$$\Delta G = \Delta Gs + \Delta G + \Delta G\phi$$

Table 4 Volume changes in different aluminium alloy systems by the process of precipitation induced densification

Alloy	Density After Sintering (% theoretical)	Density After Slow Cooling (% theoretical)	Densification (+/−)	Solute Content (atomic %)	Predicted Precipitate Species	Propensity for Precipitation
Al–8Zn	89	88	−	3.5	γ	Very low
Al–2.5Mg	90.5	91	+	2.9	$\beta\,Al_3Mg_2$	Very low
Al–2.5Mg–1Cu	88	90	+/−*	3.4	S Al_2CuMg	Moderate
Al–8Zn–2.5Mg	90	93	+	6.2	η $MgZn_2$	High
Al–8Zn–2.5Mg–1Cu	90	98	+	6.6	η $MgZn_2$	Very high
Al–4.5Cu–1.6Mg	89	93	+	3.9	S Al_2CuMg	Very high

*Test samples showed a mix of expansion and shrinkage

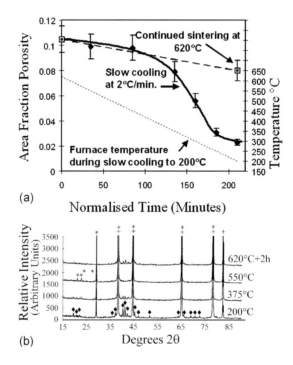

(a)

(b)

Fig. 7 (**a**) The volume changes to sintered Al–8Zn–2.5Mg–1Cu alloy during slow cooling; (**b**) XRD scans tracking the cooling process. During slow cooling, significant densification (pore closure) occurs simultaneously to precipitation of η phase (from Lumley and Schaffer [24]). For (**b**), diamonds are for the η phase $MgZn_2$, crosses for the matrix aluminium, and asterisks for the T phase, $Mg_{32}(AlZn)_{49}$

where ΔGs is the surface free energy term, $\Delta G\varepsilon$ is the strain energy term, and $\Delta G\phi$ is the chemical free energy change.

Solute segregation to any interface such as a grain boundary, free surface or dislocation will reduce the total free energy of the matrix. This causes the chemical

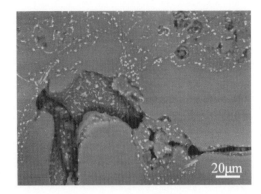

Fig. 8 SEM image showing precipitation onto a pore surface occurring during precipitation induced densification. (From Lumley and Schaffer [24])

free energy term to dominate, resulting in heterogeneous interface (precipitate) nucleation.

Such chemically driven pore closure can move the system back to one whereby sufficient mass transfer of *species different from the base material* may occur, and the diffusivity may already be naturally anomalous because solute elements often diffuse at a rate faster than the self-diffusivity of Al (Figure 3). Because solute segregation to interfaces, and subsequent heterogeneous precipitation, are both predicted and observed experimentally, it is also the case that self healing may be initiated by chemical potential. Since the alloy wants to equilibrate towards its two-phase condition, away from its metastable state, it will do so by precipitating material into any open void, as an energetically preferred site for precipitation to occur.

3   Precipitation Hardening in Aluminium Alloys

Age hardening was first reported in a ternary Al–Cu–Mg alloy in 1911 by Alfred Wilm [26]. This phenomenon is associated with alloy systems in which there is a decreasing solid solubility of solute elements with decreasing temperature, as shown in Figure 10. Quenching from a high temperature (Ta) suppresses the equilibrium separation of a second phase and results in the formation of a metastable, supersaturated solid solution (SSSS) [27] at a lower temperature. Ageing the quenched alloy in the range of ambient temperature up to around 250°C, for a sufficient time, causes a decomposition of the SSSS resulting in precipitation of submicroscopic particles of one or more second phases, which serve as obstacles to crystallographic slip during deformation [28], thereby strengthening the alloy.

Historically, precipitation hardening in the Al–Cu system has received special attention and the complete precipitation sequence has been identified as:

$$S.S.S.S \rightarrow GP\ zones \rightarrow \theta'' \rightarrow \theta' \rightarrow \theta$$

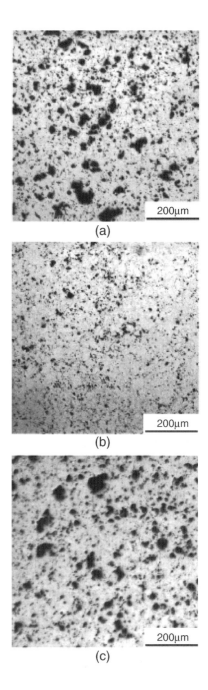

Fig. 9 (**a**) Al-8Zn–2.5Mg–1Cu alloy sintered 2 h at 620°C, (**b**) subsequently slow cooled as in Figure 7, and (**c**) continued sintering at 620°C for an extended timeframe (see Figure 7a) (From Lumley and Schaffer [24])

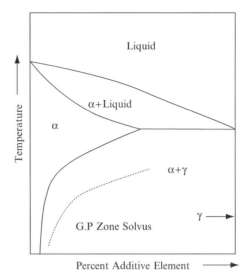

Fig. 10 Model phase diagram for an alloy capable of undergoing precipitation strengthening. Solution treatment is conducted within the single-phase region of the phase diagram, prior to quenching. Upon heat treatment, the alloy is again heated into the $\alpha + \gamma$ phase region, where γ is typically an intermetallic phase (e.g. A_2B)

GP (Guinier–Preston) zones are ordered, solute-rich clusters of atoms that are coherent with the parent lattice and are of only one or two atomic planes in thickness. Their population densities may be as high as $10^{18}/\mathrm{cm}^3$. Although they are readily sheared by mobile dislocations, they are able to cause high strengthening because of the localized compressive strain field each precipitate imparts into the surrounding lattice. The phase θ'' is also coherent with the Al lattice, whereas the intermediate phase θ' is semi-coherent and occurs as plates which form on the $\{001\}_\alpha$ planes and are typically much larger than the θ''. θ' also has a very high tendency to nucleate heterogeneously on defects, such as dislocations. The equilibrium precipitate $\theta(Al_2Cu)$ is incoherent with the Al lattice and, because it is normally coarsely dispersed, it causes little strengthening. Generally, θ forms when the alloy is averaged at relatively high temperatures, and often appears on grain boundaries. The hardness-time curve for an Al–4Cu alloy aged at 150°C is shown in Figure 11, together with the peak aged microstructure which contains both the θ'' and θ' phases as plates viewed edge-on, on the $\{100\}_\alpha$ planes.

Quenching from a high temperature also results in a supersaturation of vacancies that are available to facilitate solute diffusion during ageing. For example, with Al alloys that are amenable to age hardening at ambient temperature, it has been estimated that diffusion rates are up to 10^{10} faster than expected because of the presence of vacancies retained by quenching from high temperature [29]. Vacancies also promote clustering of solute atoms and this phenomenon also contributes to hardening in some alloys. Although it is not possible to observe vacancies directly, positron annihilation spectroscopy (PALS) now allows quantitative measurements to be made of

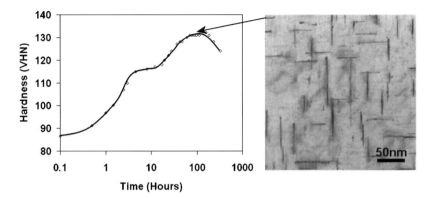

Fig. 11 Transmission electron micrograph showing precipitate structure, and hardening diagram of Al–4Cu aged at 150°C viewed in an [001]α crystallographic orientation. Fine particles of θ″ and θ′ are shown within the material edge on, that provide strengthening to the alloy

solute atom/vacancy interactions [30], and is a particularly powerful tool when used in conjunction with 3D atom probe field ion microscopy.

Wrought Al alloys that are amenable to precipitation strengthening are classed according to the major alloying elements that are present and for wrought products comprise the 2000 series alloys (Al–Cu–(Mg)), the 6,000 series (Al–Mg–Si), and the 7,000 series (Al–Zn–Mg–(Cu))[31]. For cast products, equivalent age hardenable alloy systems are the 200 series alloys (Al–Cu–X), the 300 series alloys (Al–Si–X) and 700 series alloys (Al–Zn–Mg–(Cu)). For all alloy types, solute contents and ratios of constituent elements are the keys to their ageing behavior because the nanoscale precipitates that form are mixtures of phases based on intermetallic compounds (e.g. Al_2Cu, Al_2CuMg, Mg_2Si, $MgZn_2$) and the proportion of the precipitate phases, depends on the alloy composition [32].

It is the ageing heat treatment given to Al alloys that controls the precipitate size, distribution and morphology, which then control mechanical behavior. For example, the precipitation process occurring in an alloy that is solution treated, quenched, and aged at ambient temperature (T4 temper) is much less advanced than that for the same alloy aged to peak hardness at 150°C (T6 temper). Many different means have been investigated by which heat treatment of Al alloys may modify precipitation processes in order to benefit their mechanical, thermal or chemical behavior [33, 34]. All have the common theme of improving the range and application of alloys in engineering service through control of their microstructures. It should also be noted that cold working an Al alloy after solution treatment and quenching, but prior to ageing, invariably accelerates the precipitation processes. Current theories suggest that this follows because vacancy contents are increased and the higher dislocation content both facilitates pipe diffusion of solutes, and provides favorable sites where heterogeneous nucleation of precipitates can occur.

3.1 Secondary Precipitation

One recent development in the heat treatment of Al alloys has revealed that precipitate nucleation may be stimulated to occur more than once by a process referred to as secondary precipitation [35]. In this process, a solution treated and quenched alloy is initially aged for a relatively short period at an elevated temperature to produce an underaged microstructure which still retains a substantial amount of solute in solid solution. If the alloy is then quenched again, secondary precipitation is observed to occur at a lower temperature. The preexisting precipitates from the initial ageing period may be chemically modified [36], and new GP zones may form. An illustration of this effect in an Al–Cu alloy is shown by comparing the microstructures in Figure 12. In particular, whereas the initial precipitate species formed by underageing at the higher ageing temperature are quite coarse, the secondary precipitates are much finer and form between the primary precipitates.

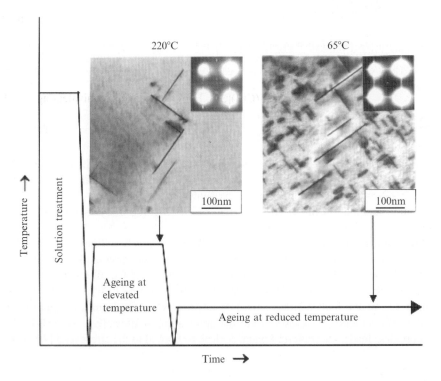

Fig. 12 The process of secondary precipitation in a model Al–4Cu alloy. Following initial ageing at 220°C, θ' precipitates within the alloy are relatively coarse. During subsequent ageing, a much finer dispersion of GP zones form within the matrix aluminium. (From Lumley et al. [35])

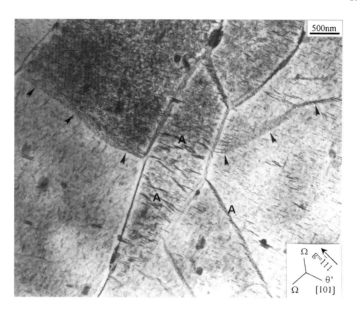

Fig. 13 Dynamic precipitation in an underaged Al–Cu–Mg–Ag alloy following 500 h creep at 300 MPa and 150° C. Bands of dynamically precipitated particles are formed from solute present in association with mobile dislocations (arrowed). Examples of dynamically precipitated θ' phase are marked "A." (From Lumley et al. [37])

3.2 Dynamic Precipitation

Dynamic precipitation is a form of precipitation that occurs while the alloy is under load, and occurs as a response to the generation of moving dislocations during service. An example of this phenomenon is shown in Figure 13, in which extensive dynamic precipitation has occurred in an Al–Cu–Mg–Ag alloy in the process of creep.

Dynamic precipitation is particularly prevalent in alloys based on the Al–Cu–Mg system [e.g. 37–43], and is another form of heterogenous precipitation controlled by interfacial free energy. It has been shown in binary Al–Cu alloys that dynamic precipitation causes preferential precipitation on appropriate $\{001\}_\alpha$ planes, that lie within 45° of the tensile stress axis, and effectively results from a reorganisation of the microstructure as a response to the application of stress [e.g. 39–41]. Under compressive loads, the preferred plane is close to normal to the stress axis [40, 41].

3.3 Dynamic Precipitation and Self Healing during Creep

Some effects of dynamic precipitation that occur during creep of the solution treated and quenched, but underaged alloy 2024, have been described by Kloc et al. [42] who found that creep behavior was modified by continuous precipitation during testing. Recently Wang et al. [43] observed that dynamic precipitation also occurred during creep of two Al–Cu–Mg–Ag alloys. In one alloy, precipitation of the S phase, (Al$_2$CuMg) was observed during creep loading, and in the other alloy, both the S phase and the θ' phase were precipitated during the creep process. Where θ' was dynamically precipitated in conjunction with S, the creep performance was improved despite the fact that room temperature yield stress of the alloy was lower.

Arumulla and Polmear [44] showed earlier that a substantial reduction in secondary creep rate (65%) could be achieved if an experimental, high purity Al–Cu–Mg–Ag alloy was tested in the lower strength underaged condition (proof stress 313 MPa) rather than the peak aged, T6 condition (proof stress 363 MPa). These results were obtained at 125°C and at a stress of 265 MPa. In addition, the time to failure for the underaged condition was also increased. Effects of underageing have been investigated further [37] using a higher strength experimental composition (Al–5.6Cu–0.45Mg–0.45Ag–0.3Mn–0.18Zr) and comparative creep curves are shown in Figure 14a. This alloy is hardened by a combination of the precipitates Ω and θ' which form as thin plates on the $\{111\}_\alpha$ and $\{001\}_\alpha$ planes respectively. Both have the composition Al$_2$Cu although plates of Ω contain thin surface layers of Ag and Mg atoms that partition there during the ageing process [45]. The secondary creep rate for the underaged (UA) condition (2 h at 185°C) was found to be approximately one third of that for the fully hardened T6 alloy. A similar result was obtained for the commercial Al alloy 2024 in the UA and T6 conditions (Figure 14b) for which the time to failure in accelerated tests was also increased from $\sim$260 h in the T6 condition to $\sim$480 h by underageing prior to testing.

Changes in precipitate lengths for the T6 and UA conditions during creep at 150°C are shown in Figure 15 for the same Al–Cu–Mg–Ag alloy. It is clear that the Ω phase increases in length at a faster rate in the T6 condition. What is more significant is that whereas the θ' decreases in length and appears to be dissolving in the T6 condition, this phase is retained and increases in length in the UA alloy. Figure 16 shows a comparison between the initial microstructures at the beginning of creep at 300 MPa and 150°C and then after 500 h creep. It is apparent the θ' phase is actually being redissolved during 500 h creep from the T6 condition, whereas for the UA case, the structure retains a high proportion of the θ' phase within the microstructure following creep. The dissolution of the θ' phase in the T6 material is not simply due to the extended time of exposure at 150°C as this phase is still present in large quantities when there is no load applied.

It has been proposed that the beneficial effect of underageing on creep properties may arise because uncommitted solute is available in solid solution and can participate in further precipitation [37]. Such solute may also be expected to interact with mobile dislocations which will impede their motion and effect deformation behavior [46]. The beneficial effect of underageing in reducing secondary creep rates would appear to arise because of five main factors:

1. The presence of solute in solid solution that causes solute atmospheres around dislocations and impedes their motion

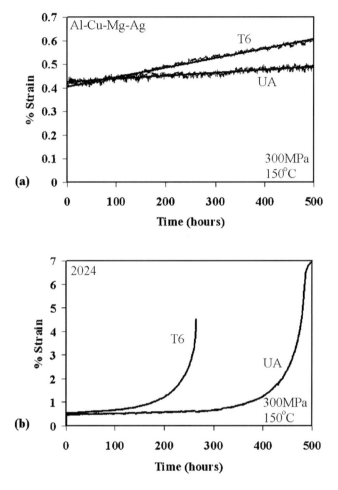

Fig. 14 Creep curves for the T6 and UA condition in (**a**) Al–Cu–Mg–Ag alloy, and (**b**) 2024 alloy. (From Lumley et al. [37])

2. Retention of the θ' precipitates within the matrix of the UA alloy whereas this phase mostly dissolves and is incorporated into other precipitate phases, subgrain boundaries and grain boundaries during creep of the peak aged alloy
3. Additional dynamic precipitation of θ' that occurs during creep of the UA alloy
4. Retention of a finer dispersion of the Ω phase in the UA alloy
5. Retention of a fine subgrain structure in the UA alloy during creep due, in part, to the more complex distribution of precipitates in the matrix and at subgrain boundaries that apparently impede boundary migration

All of these features provide more obstacles to the movement of dislocations in the UA alloys, and are associated with dynamic changes occurring during secondary

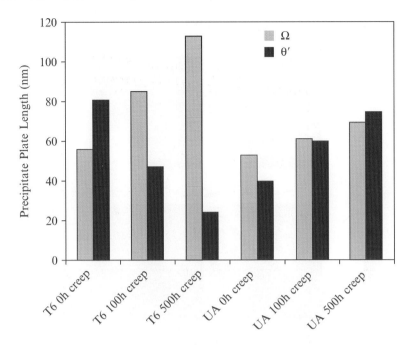

Fig. 15 Changes in precipitate lengths during creep of material aged to the UA (2 h) and T6 conditions (From Lumley et al. [37])

creep in both the UA and T6 alloys. Each is associated with the retained solute content of the alloy, and is assumed to act together in providing enhanced creep resistance. In total, they provide a form of self healing of the microstructure of the UA alloy during creep.

It is important to note from the work of Lumley et al. [37] that the creep stress exponent, n, for both the UA and T6 alloys was close to 5 within the test parameters of the study (5.9 and 4.3 respectively). The creep stress exponent is derived from the equation:

$$n = (\log \varepsilon_1 - \log \varepsilon_2)/(\log \sigma_1 - \log \sigma_2)$$

at a constant temperature, and creep stress exponent values of close to 5 are typically observed in metals and alloys that are solid solutions where creep involving dislocation glide and climb is the dominant deformation mechanism, with the activation energy close to that for lattice diffusion [47]. The observation that five-power-law creep was dominant supports the conclusion that the presence of "free" solute in solid solution in the UA alloy was the key to providing enhanced creep resistance because of the effects of dynamic precipitation that arise. This proposal is also supported by consideration of the motion of dislocations through the microstructure of an alloy containing elements both in solid solution and in precipitates [48]. Here it is known that:

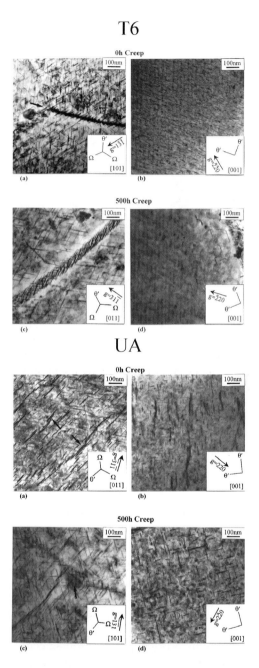

Fig. 16 Microstructure of Al–Cu–Mg–Ag alloy in either the T6 or UA condition, at 0 h creep and 500 h creep. For the T6 material, the microstructure becomes depleted of the θ′ precipitate during the creep process by dissolution caused by the cutting action of dislocations. For the UA material, dynamic precipitation bands form immediately, as the sample is loaded, and after 500 h creep, still display a high proportion of θ′ precipitates within the microstructure. These θ′ particles are continually precipitated into the microstructure (From Lumley et al. [37])

1. Moving dislocations will attract clouds of solute atoms both to their cores and to the associated strain fields
2. The binding energies of the solute atoms to the dislocation are inversely proportional to both the size of the solute cloud and to the solute concentration in the surrounding matrix
3. When a dislocation interacts with a precipitate, solute atoms will be transferred to the precipitate, causing Ostwald ripening, if their binding energy to the precipitate exceeds that to the dislocation. Conversely, solute atoms will be removed from the precipitate if their binding energy to the dislocation is greater

The Al matrix in the fully hardened, T6 condition is solute depleted with respect to the UA matrix, where precipitation is less advanced and more solute remains in solid solution. Therefore, in the peak-aged T6 case, mobile dislocations attract relatively small solute clouds from the matrix during creep and the binding energy of these elements to the dislocation cores will be relatively high. This implies that, when dislocations interact with precipitates having lower binding energies (and hence, reduced stability), these precipitates will be more readily dissolved because the solute atoms may diffuse away to dislocation cores. (If the binding energies of the precipitate solute atoms are less than the binding energy between the solute atom and the dislocation, solute elements may be removed from the precipitate and be incorporated into the mobile dislocation.) This explains why θ', which is known to be less stable than Ω [49], redissolves preferentially in the T6 alloy during creep (Figures 15 and 16).

Dissolution of θ' during creep of the UA alloy might also be expected to occur. However, because the solute content of the matrix is higher than in the T6 condition, solute clouds existing around mobile dislocations remain significant, and the binding energy of most solute atoms (present in precipitates) to the dislocation will be reduced. If the solute concentration in the dislocation core and the surrounding solute cloud is effectively saturated, dissolution of precipitates will be minimized. It is proposed that this excess solute within the solute cloud facilitates dynamic precipitation at mobile dislocations, primarily of the θ' phase for which dislocations are preferred nucleation sites.

In the case of dynamic precipitation during creep, there are several aspects of self healing behavior that require consideration. These are:

1. Deterioration of the microstructure in the form of θ' dissolution occurs during secondary creep of the T6-treated alloy, which leaves the alloy with fewer impediments to dislocation glide and climb. In the underaged material, this damage is largely avoided by the dynamic precipitation of θ' particles.
2. The comparative creep curves for T6 and UA 2024 alloy provided in Figure 14, show that secondary creep rate is reduced and the onset of tertiary creep, (where cracks form and propagate through the microstructure), is delayed. The role of dynamic precipitation on processes of tertiary creep require further study, but it may be hypothesized that tertiary creep cracking may be influenced by pipe diffusion, grain boundary diffusion, and subgrain boundary diffusion of solute elements. Tertiary creep cracking typically involves the formation of voids on grain

boundaries, often in triple points. These voids coalesce, forming cracks that the propagate through the material more rapidly. If void formation and growth can be slowed or altered by precipitating material into the void in the same means as precipitation induced densification (Figures 8 and 9), then the onset of tertiary creep must as a result be offset to longer time frames, and the subsequent cracking that occurs will proceed more slowly.

3.4 Dynamic Precipitation and Self Healing during Fatigue

In parallel to the creep study of the Al–Cu–Mg–Ag alloy, the behavior of the UA and peak aged (T6) material has been compared during fatigue testing at room temperature over a range of stresses [50] (Figure 17). Trends observed in creep were replicated during fatigue in that the UA alloy showed a superior performance. Microstructural evidence again showed that dissolution of θ' occurred in the T6 condition during testing whereas evidence of dynamic precipitation was observed in the matrix of the UA alloy (Figure 18). In effect, the damage normally caused by dislocations in the UA alloy was also being reversed by the dislocations, when they are saturated with solute elements. This is a form of bulk self repair since it retains or replenishes the durability of the underlying microstructure. Earlier work by Garrett and Knott [51] on fatigue of Al–Cu alloys also found a substantial improvement in the lifetimes of underaged material compared to peak aged or overaged material (Figure 19). This earlier work showed that crack propagation during fatigue of the underaged material was being constrained to the $\{001\}_\alpha$ planes of the Al alloy, and that this cracking was accompanied by substantial slip band formation adjacent to the propagating crack (Figure 20a). This behavior was not observed in the peak aged or overaged conditions. These workers proposed that the observed improvement for the underaged material was due to a process of restricted crystallographic slip which, in light of recent findings, could be expected to result from the formation of dynamic precipitation bands of $(001)_\alpha$ type precipitates (θ'), on dislocations associated with slip bands. Thus it appears likely that heterogeneous precipitation associated with the crack tip and its plastic zone may also be influencing fatigue crack propagation.

Because a crack has high free energy at its tip and a high dislocation density within its plastic zone (e.g. Figure 20b), it is a preferred nucleation site for heterogenous precipitation, as was discussed in section 2.4. The movement of dislocations and the formation of slip bands within the microstructure will enhance diffusivity and assist the delivery of solute elements to the crack tip, as an analogous mechanism to the process of precipitation induced densification of sintered alloys. Precipitation into a sharp angle by the process of precipitation induced densification for a geometry that approximates to a crack, is shown in Figure 21. It is therefore suggested that a process of crack closure or crack modification may also occur during fatigue, in addition to the other microstructural effects already discussed. What is most important is that enough solute atoms can be rapidly delivered to the crack site, which seems inherently practical by pipe diffusion along dislocation lines.

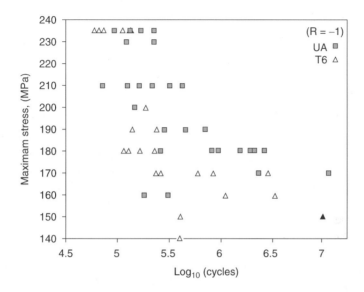

Fig. 17 Results of fatigue resistance in (**a**) the T6 condition, and (**b**) the UA condition for an Al–Cu–Mg–Ag alloy (from Lumley et al. [50]). Despite the lower 0.2% proof strength, the UA material displays improved fatigue life compared to the T6-treated alloy. The 0.2% proof strength in the UA condition was 412 MPa and in the T6 condition 470 MPa. $R = -1$ (zero mean stress). 10^7 data points are run-outs

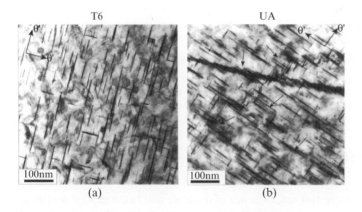

Fig. 18 Microstructures of the T6 and UA material following 100,000 fatigue cycles at ±160 MPa. Similar to the examples for creep shown in Figure 16, the T6 microstructure is now depleted of θ' precipitates in specific orientations. The UA material however retains precipitates in all orientations due to the action of dynamically precipitated θ' (examples arrowed). (From Lumley et al. [50])

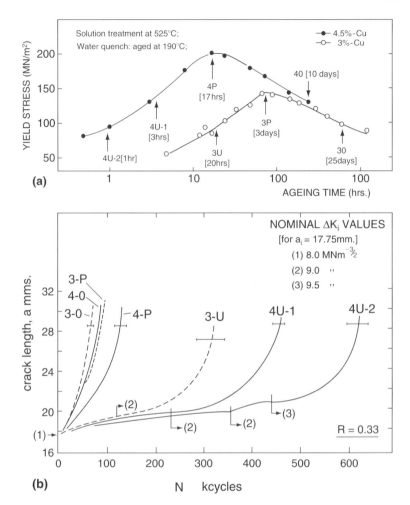

Fig. 19 Fatigue crack growth for underaged, peak–aged, and overaged binary Al–Cu alloys: (a) shows the condition from which the material was subjected to fatigue, and (b) shows the crack growth rates of the different conditions. (From Garrett and Knott [51])

3.5 Dilatometry Studies

Another interesting consequence of dynamic and heterogenous precipitation associated with creep or fatigue is that there is potential for changes in the volume of the material that is facilitated during the decomposition of the solid solution. This can be either expansion or contraction depending on the precipitate species formed [53, 54].

Solute associated with mobile dislocations has differing effects on local volume changes, depending on the size of the particular solute atom. However, it is important to note that, where heterogeneous precipitation of θ' occurs, the localized volume

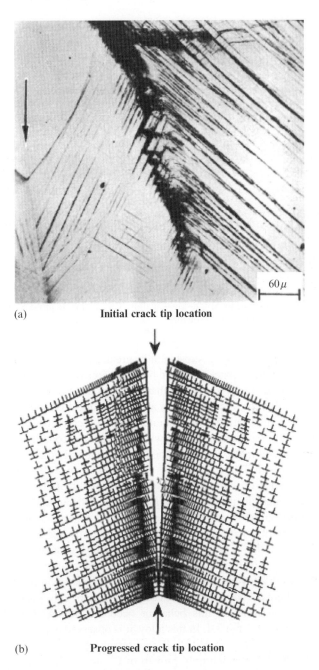

(a)   **Initial crack tip location**

(b)   **Progressed crack tip location**

Fig. 20 (**a**) Optical microstructure of a fatigue crack in UA Al–Cu alloy, showing propagation of the crack along an $(001)_\alpha$ plane, and shear banding associated with dislocation structures nearby the crack (from Garrett and Knott [51]). (**b**) A model of the dislocation arrangement around a fatigue crack tip. (From Riemelmoser and Pippan [52])

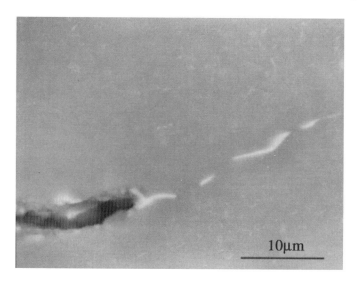

Fig. 21 Backscattered electron micrograph showing a crack-like feature with heterogeneous precipitation at the tip of the defect in an Al–Zn–Mg–Cu alloy. The precipitate phase $MgZn_2$ is observed as the white phase present in the microstructure

expansion of the region may be substantially greater than that of the surrounding matrix region where precipitates are more homogeneously distributed. In particular, abnormally dense dynamic precipitation that occurs in bands along dislocations may be expected to result in local compression around a dislocation source. The result of this will be that, as dislocations are generated around a crack tip (e.g. Figure 20b), and heavy dynamic precipitation occurs within the plastic zone in quantities much higher than the general matrix material, the local region will be compressed thereby restricting crack growth. Figure 22a shows the volume changes that occur in association with precipitation of θ' in the Al–Cu–Mn–Si alloy 2025[53], and it is observed that linear dimensional changes of approximately 0.16% occur to the material during ageing over a range of temperatures. The dimensional changes associated with precipitation however differ with each alloy system. Generally, the growth of Al alloys as a result of precipitation is greatest in alloys containing substantial amounts of Cu, but is reduced progressively with increasing Mg content in alloys such as 2014 and 2024. The Al–Mg–Si alloy 6061 displays effectively no change in dimensions during heat treatment, whereas 7000 series alloys contract (Figure 22b) [54].

An interesting variation on this behavior occurs in the Al–Cu–Mg–Ag alloy referred to in sections 3.3 and 3.4. In this alloy, it is observed experimentally that the Ω phase (approximately two thirds of all precipitates) ceases to grow after 2–2.5 h of ageing at 185°C, after which only growth of θ' (approximately one third of all precipitates), occurs (Figure 15). Differential dilatometry at 185°C $\pm$ 1.5°C for the Al–Cu–Mg–Ag alloy using pure Al as a reference is shown in Figure 23. In this case, precipitation of different phases corresponds to bulk contraction and expansion

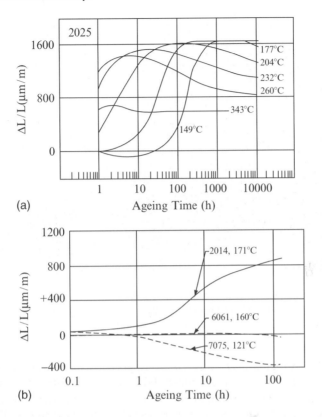

(a)

(b)

Fig. 22 (**a**) Dimensional changes occurring during ageing of the Al–Cu–Mn–Si alloy 2025, aged at different temperatures. (**b**) Dimensional changes during ageing of different aluminium alloys during artificial ageing. (From Hunsicker [53])

events of the test sample. Initially as the sample reaches approximately 140°C, there is a contraction occurring that would appear to be associated with the rapid formation of the Ω phase. In the interval between when the sample finishes this event, and the time Ω finishes precipitating (~120–150 min), a more moderate contraction is observed that may correspond to competitive shrinkage due to Ω phase formation, and expansion due to θ' formation. Expansion due to θ' formation then appears to dominate after 2.5 h (including the time to heat to temperature), continues rapidly for ageing times up to 8 h, and then more slowly at times beyond 10 h (the peak aged T6 condition). In total, from the UA condition (2–2.5 h), the sample then experiences a growth of close to 40 μm (0.16%) across the sample length (25 mm). Consequences of these volume changes are the subject of current research, but it appears possible that the compression of a crack should not only slow crack growth but also change the interfacial free energy through geometrical effects. This follows because, at lower interfacial angles, there is a higher likelihood that heterogeneous precipitation will occur rapidly in this region, of the form discussed in sections 2.4 and 3.4.

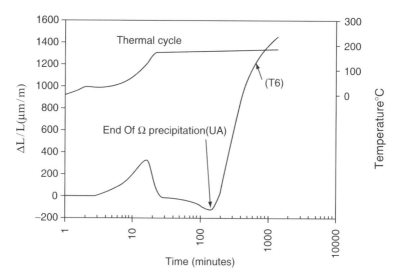

Fig. 23 Dimensional changes associated with ageing Al–Cu–Mg–Ag alloy at 185°C. Sample length 25 mm. See text for details. (Differential dilatometry courtesy Wright W. CSIRO)

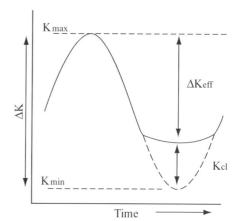

Fig. 24 A simple representation of plasticity-induced crack closure in metals and alloys during fatigue

The outcomes of the localized volume changes that may occur during fatigue can be best explained by the same principles as those used to describe plasticity-induced crack closure [55], oxide-induced crack closure and surface roughness-induced crack closure [56, 57]. A simple representation for plasticity-induced crack closure is provided in Figure 24. Plasticity-induced crack closure is reported to modify fatigue crack growth rates through altering the relationship between the applied stress intensity factor range at the crack tip such that:

$$(\Delta K_{eff} = K_{max} - K_{cl}) \leq (\Delta K = K_{max} - K_{min}).$$

where ΔK is the stress intensity factor range, ΔK_{eff} is that actually experienced at the crack tip, K_{max} and K_{min} are the maximum and minimum stress intensities, and K_{cl} is the stress intensity factor at closure.

In effect, it would appear to be the case that if the local volume of the material surrounding the crack increases due to dynamic precipitation of θ' phase, the value of ΔK_{eff} should decrease, hence, working to minimize crack propagation and further improve fatigue life.

Whether or not crack modification is occurring by a process of dynamic precipitation, it seems clear that for both creep and fatigue, dynamic precipitation tends to compensate for microstructural damage in Al alloys, with the result that a higher integrity structure is maintained for an extended lifetime. Although dynamic precipitation has the potential to delay both crack initiation and crack propagation in Al alloys, it is also necessary to consider some of the macroscopic limitations of such a process, as mentioned in section 1.2.

4   Analogies to Other Materials

In the above sections, it is shown how metallurgical processes such as sintering and dynamic precipitation in response to loading conditions may initiate different levels of self healing in Al alloys. It is now interesting to consider if these processes may have parallels with the behavior of natural materials. Certainly, precipitation of crystalline materials into cracks does occur on a wide scale in geological processes (e.g. Figure 25), but biological processes governing the repair of bone and tissue are more relevant because of their inherent efficiency, as discussed later in this text. In particular, the Wolff–Roux law provides some interesting analogies. This relationship defines the repair of bone *in vivo* in particular, and, in summary, may be considered to state that: "material is dissolved from where it is not required and redeposited to where it is required, as a response to mechanical stimuli and damage," or alternately:

$$\Delta \text{Bone Structure} = f(\Delta \text{Mechanical Stimulus, Physiologic Mechanisms})$$

In the case of bone, the physiological mechanisms are effectively the actions of osteoblasts, and osteoclasts associated with bone remodeling.

Although its possible relationship to inanimate materials may at first appear to be tenuous, an attempt can be made to modify the Wolff–Roux law so that it applies to metallic materials. If it is assumed that the service life of an Al alloy is related to its microstructure, as well as the changes that may occur by the dynamic processes that occur during service, then it follows that:

$$\Delta \text{Service life} = f(\text{Service Environment}, \Delta \text{Microstructure})$$

If the microstructure responds or adapts unfavorably to its conditions of loading and accumulates damage, then the service life will be reduced. Examples of this were provided earlier in terms of both creep and fatigue of T6-treated Al alloys wherein one

Fig. 25 Precipitation of crystalline minerals (dendrites) into cracks within rocks. The mineral is a magnesium compound and is a constituent of the parent rock. (Images courtesy Lumley GI Ground Breaking Innovations)

precipitate species is dissolved during service loading. As a result, the microstructure is deleteriously affected and the service life is potentially less when compared to what it may have been had the damage not occurred. On the other hand, if the microstructure of an alloy responds favorably to its service conditions, then the useful life has the potential to be increased because damage is mitigated through microstructural rejuvenation. Examples of this were provided earlier whereby creep and fatigue behavior are improved when the alloy is tested in the underaged condition. For example, the dynamic formation of bands of precipitates produces efficient obstacles to the propagation of subsequent dislocations that move through the material. As a result, the microstructure resists or slows the damage by the continual dynamic precipitation of new particles as free solute atoms in the solid solution precipitate on mobile dislocations.

It therefore follows that changes in microstructure provided by dynamic precipitation arise as a response to mechanical stimulus. However, since these processes also involve heterogenous precipitation, the Gibbs free energy indicates that such dynamic precipitation is very favorable when the free energy of the interface to which it is precipitating is relatively high. This condition addresses both the cases where dynamic precipitation occurs within the microstructure, since the dislocation itself has a high free energy, and also at the preferred nucleation sites of a free surface such as that produced when a short crack or pore structure is present. The degree to which dynamic precipitation can influence the microstructure is also a function of the uncommitted solute concentration within the alloy. That is, the active uncommitted diffusing species within the Al alloy must be of sufficient quantity to have an effect.

For a responsive dynamic process to occur, the solute elements that facilitate the healing behavior must also have sufficient mobility to reach their required destination. Several means have been suggested by which this may occur in other materials, but for metals, it is the diffusive mechanisms shown in Figure 4 that are the only viable means for mass transfer. In particular, pipe diffusion, wherein solute elements travel along dislocation lines and defect structures, provides an opportunity for this to occur. In this case, the feature associated with damage (dislocations), also provides the mechanism (pipe diffusion of solute atoms) to effect repair. Diffusion along grain boundaries may also facilitate mass transfer, especially for nanocrystalline microstructures in which the grain boundary volume is relatively high.

Therefore in summary for metals it can be stated that:

$$\Delta \text{ Microstructure} = f(\text{Mechanical stimulus}, \gamma, c, d^*)$$

Where mechanical stimulus = applied stress, strain rate; γ = free energy of interfaces; c = uncommitted solute concentration; d^* = net diffusivity (atomic mobility of free species and is the sum of vacancy diffusion, pipe diffusion, grain boundary diffusion, and surface diffusion).

Now when we consider this analogy to the Wolff–Roux law, the final aspect that must be considered is the size of the defect, or its effective atomic volume. This is important because if the defect is bigger than what is reasonably able to be addressed, self healing will have little or no effect in eliminating the damage which has occurred.

Although it is not always necessary to address the full volume of the defect, obviously the scale must be comparable.

5   Summary

The framework for initiating self healing in Al alloys has been presented here with some practical examples showing how self healing may manifest itself in Al alloys. Because the studies conducted thus far have been relatively limited, the information contained within this chapter should not be considered exhaustive, and is open to improvement, modification and discussion. However, it does appear clear that self healing in Al alloys in service, and potentially other metals, does seem feasible.

In the past, alloys have generally been designed to be stable and resist change when in service. Now an exciting aspect of the concept of self healing is that this earlier design paradigm may need to be modified to allow metals and alloys to undergo controlled microstructural changes so that they can respond to service conditions. This concept will provide challenges both with respect to design principles and in being accepted in engineering practice.

Acknowledgements The author would like to thank Professor Ian Polmear for assistance in preparing this chapter, and CSIRO for supporting this work.

References

1. Polmear I.J. (2006) Light alloys, from traditional alloys to nanocrystals, 4th edn. Butterworth-Heinemann, Oxford, UK
2. Altenpohl D.G. (1998) Aluminium: technology, applications, and environment, 6th edn. The Aluminum Association, Washington DC
3. Hertzberg R.W. (1996) Deformation and fracture mechanics of engineering materials, 4th edn. Wiley, NY, pp 333–334
4. Starke E.A., Staley J.T. (1996) Application of modern aluminum alloys to aircraft. Prog Aerospace Sci 32:131–172
5. Kaufman J.G. (2001) Fracture resistance of aluminum alloys. ASM International, The Aluminum Association, p 102–103
6. Landolt–Börnstein numerical data and functional relationships in science and technology (1991) In: Mehrer H (ed) Diffusion in metals and alloys, vol 3/26. Springer, Berlin, pp 195–196
7. Murr L.E. (1975) Interfacial phenomena in metals and alloys. Addison-Wesley Publishing Company, Reading, MA, p 292
8. Jannot E.J. (private communication), Institut für Metallkunde und Metallphysik. RWTH Aachen, 52074 Aachen, Germany.
9. Somoza A., Macchi C.E., Lumley R.N., Polmear I.J., Dupasquier A., Ferragut R. (2007) Role of vacancies during creep and secondary precipitation in an underaged Al-Cu-Mg-Ag alloy. 14th ICPA – International Conference on Positron Annihilation, 2006. Hamilton, Canada, Physica Status Solidi (in press)
10. Courtney T.H. (1984) Densification and structural development in liquid phase sintering. Metall Trans A 15A:1065–1074

11. Murr L.E. (1975) Interfacial phenomena in metals and alloys. Addison-Wesley Publishing Company, Reading, MA, p 404

12. Kingery W.D., Bowen H.K., Uhlman D.R. (1976) Introduction to ceramics, 2nd edn. Wiley, New York

13. Ashby M.F. (1974) A first report on sintering diagrams. Acta Metall 22:275–289

14. Richerson D.W. (1982) Modern ceramic engineering, properties, processing and use in design, vol.1. Marcel Dekker, New York

15. German R.M. (1985) Liquid phase sintering. Plenum Press, New York

16. Nakao Y., Sugaya K., Seya S., Sakima T.(1996) Process for producing aluminium sintering. US Patent #5525292

17. Schaffer G.B. and Hall B.J. (2002) The influence of the atmosphere on the sintering of aluminum. Metall Trans A 33A:3279–3284

18. Lumley R.N., Sercombe T.B., Schaffer G.B. (1999) Surface oxide and the role of magnesium during the sintering of aluminium. Metall Mater Trans A 30A:457–463

19. Levi C.G., Abbaschian G.J., Mehrabian R. (1978) Interface interactions during fabrication of aluminium alloy-alumina fiber composites. Metall Trans A 9A:697–711

20. Hu C., Baker T.N. (1993) Microstructure of Al-Fe alloys sintered via in-situ microfusion process. Mater Sci Tech 9:48–56

21. Lumley R.N., Schaffer G.B. (1996) The effect of solubility and particle size on liquid phase sintering. Scripta Materialia 35:589–595

22. Savitskii A.P. (1993) Liquid phase sintering of the systems with interacting components. Russian Academy of Sciences, Tomsk, Russia (translated by Kokunova MV)

23. Lumley R.N., Schaffer G.B. (1996) The particle size effect in multicomponent aluminium alloys. Proceedings of the 1996 World Congress on powder metallurgy and particulate materials. Adv Powder Metall Particulate Mater MPIF 11:185–196

24. Lumley R.N., Schaffer G.B. (2006) Precipitation induced densification in a sintered Al-Zn-Mg-Cu Alloy. Scripta Materialia 55:207–210

25. Lumley R.N. (1996) Unpublished research

26. Wilm A. (1911) Physical-metallurgical experiments on aluminium alloys containing magnesium. Metallurgie 8:223. (Translated to English). In: Martin JW (1968) Precipitation hardening. Pergamon Press, Oxdford, pp 103–111

27. Merica P.D., Waltenberg R.G., Scott H. (1919) Heat treatment of duralumin. Bull. AIME, 9, Chem Metall Eng 21:551

28. Jeffries Z., Archer R.S. (1921) The slip interference theory of the hardening of metals. Chem Metall Eng 24:1057

29. Altenpohl D.G. (1998) Aluminium : technology, applications and environment, 6th edn. The Aluminium Association. Washington DC, P. 150

30. Dupasquier A., Mills A.P. (eds) (1995) Positron spectroscopy of solids. IOS, Amsterdam

31. International alloy designations and chemical composition limits for wrought aluminium and wrought aluminium alloys (1998) The Aluminum Association, Washingtom DC, pp 2–9

32. Polmear I.J. (2006). Light alloys, from traditional alloys to nanocrystals, 4th edn. Butterworth-Heinemann, Oxford UK. 57–59

33. Brandes A., Brook G.B. (eds) (1998) Smithells metals reference book 7th edn. Butterworth-Heinmann, Oxford, pp 22–2

34. Aluminum Standards and Data (1998) Metric SI edn., The Aluminum Association, Washington DC, pp 3–12 to 3–17

35. Lumley R.N., Polmear I.J., Morton A.J. (2003) Interrupted ageing and secondary precipitation in aluminium alloys. Mater Sci Tech 19:1483–1490

36. Buha J., Lumley R.N., Crosky A.J., Hono K. (2007) Secondary precipitation in an Al-Mg-Si-Cu alloy. Acta Materialia 55:3015–3024

37. Lumley R.N., Morton A.J., Polmear I.J. (2002) Enhanced creep performance in an Al-Cu-Mg-Ag alloy through underageing. Acta Materialia 50:3597–3608

38. Lumley R.N., Morton A.J., Polmear I.J. (2000) Enhanced creep resistance in underaged aluminium alloys. Mater Sci Forum 331–337:1495–1500
39. Skrotzki B., Shiflet G.J., Starke E.A. (1996) On the effect of stress on nucleation and growth of precipitates in an Al-Cu-Mg-Ag alloy. Metall Trans A 27A:3431–3444
40. Hosford W.F., Agrawal S.P. (1975) Metall Trans A 6A:487–491
41. Zhu A.W., Chen J., Starke E.A. (2000) Precipitation strengthening of stress aged Al-XCu alloys. Acta Materialia 48:2239–2246
42. Kloc L., Cerri E., Spigarelli S., Evangelista E., Langdon T.G. (1996) Significance of continuous precipitation during creep of a powder metallurgy aluminium Alloy. Mater Sci Eng A216:161–168
43. Wang N., Kazanjian M., Starke E.A. (1998) Microstructure and creep behavior of Al-Cu-Mg and Al-Cu-Mg-Ag alloys. 3rd Pacific Rim international conference on advanced materials and processing. The minerals, metals Mater Soc, Warrendale, PA, USA, pp 713–718
44. Arumulla S.R., Polmear I.J. (1985) Fatigue and creep behaviour of aged alloys based on Al-4Cu-0.3Mg. Proceedings of the 7th international conference on strength of metals and alloys. Montreal, Canada, Pergamon Press, Oxford 1:453–458
45. Hutchinson C.R., Fan X., Pennycook S.J., Shiflet G.J. (2001) On the origin of the high coarsening resistance of Ω plates in Al–Cu–Mg–Ag alloys. Acta Materialia, 49:2827–2841
46. Hull D., Bacon D.J. (1997) Introduction to dislocations, 3rd edn. Butterworth-Heinemann, Oxford
47. Wang J., Wu X., Xia K. (1997) Creep behavior at elevated temperatures of an Al-Cu-Mg-Ag alloy. Mater. Sci. Eng A234–236, 287–290
48. Smallman R.E. (1985) Modern physical metallurgy, 4th edn. Butterworth, London, pp 192–193, 243–258
49. Ringer S.P., Yeung W., Muddle B.C., Polmear I.J. (1994) Precipitate stability in Al-Cu-Mg-Ag alloys aged at high temperatures. Acta Metallurgica Materialia 42:1715–1725
50. Lumley R.N., O'Donnell R.G., Polmear I.J., Griffiths J.R., (2005) Enhanced fatigue resistance by underageing an Al-Cu-Mg-Ag alloy. Mater Sci Forum 29:256–261
51. Garrett G.G., Knott J.F. (1975) Crystallographic fatigue crack growth in aluminium alloys. Acta Metallurgica 23:841–848
52. Riemelmoser F.O., Pippan R. (1998) Mechanical reasons for plasticity-induced crack closure under plane strain conditions. Fatigue Fract Eng Mater Struct 21:1425–1433
53. Hunsicker H.Y. (1967) In: Van Horn KR (ed) Aluminum, properties, physical. Metall Phase Diag 1:158–159
54. Hunsicker H.Y. (1980) Dimensional changes in heat treating aluminium alloys. Metall Trans A 11A:759–773
55. Elber W. (1970) Fatigue crack closure under cyclic tension, engineering. Fract Mech 2:37–45
56. Suresh S., Ritchie R.O. (1984) Propagation of short fatigue cracks. Int Metals Rev 29:445–476
57. Suresh S., Ritchie R.O. (1982) A geometrical model for fatigue crack closure induced by fracture surface morphology. Metall Trans 13A 1627–1631

Crack and Void Healing in Metals

Hua Wang[1], Peizheng Huang[2], and Zhonghua Li[1]

[1] *School of Naval Architecture, Ocean and Civil Engineering, Shanghai Jiaotong University, 200240 Shanghai, P. R. China. E-mail: whxiansh@sjtu.edu.cn*
[2] *College of AeroSpace Engineering, Nanjing University of Aeronautics and Astronautics, 210016 Nanjing, P. R. China*

1 Introduction

Metals are known to degrade, or even worse, lose functionality during service [1, 2]. The major factors shortening the service life of these materials are internal defects, such as cracks and voids [3, 4]. These internal defects are very difficult to detect, and even more difficult to repair [5–7]. During service, they might coalesce into a major crack, leading to failure [8]. When damage defects are generated in materials, the internal energy of the material increases together with some entropy increment, that is, the system is in a metastable state of the thermodynamic equilibrium. If some energy is imported from the environment, the system can overcome the energy barrier and automatically evolve along the way of minimizing the total Gibbs free energy of the system, so that the defects could be self healed [9–11] and the material performance can be partially restored [12–14]. It is therefore essential to understand the evolution of defects so that the healing mechanisms can be understood and employed to achieve the desired specific engineering requirement.

In this chapter, we will first discuss the details of the healing processes of cracks by finite element (FE) modeling. Next, void healing processes will be analyzed and some analytical solutions are presented. From the numerical and analytical analyses, the parameters controlling the rate of the healing processes are derived.

2 FE-modeling of Crack Healing

Experiments have revealed that there are several geometrically distinct stages during crack healing [3, 15, 16]. The initial stage is characterized by crack-tip blunting and regression; in the second stage one or more arrays of "cylindrical" pore channels form upon crack splitting. They then evolve into discrete spherical voids via Rayleigh instability [17–19].

Meanwhile, it has been realized that many factors can affect the healing behavior significantly, such as crack geometry [20], crack arrangement [17], crack face microstructure [8], etc. Here, based on developed FE methods [21–23], several procedures are analyzed to understand the crack healing processes and mechanisms in metals.

S. van der Zwaag (ed.), *Self Healing Materials. An Alternative Approach to 20 Centuries of Materials Science*, 255–277.
© 2007 *Springer*.

2.1 Governing Equations and FE Methods

To formulate the FE model, a weak form of the variational formulation for combined surface diffusion, grain boundary (GB) diffusion and evaporation–condensation has been developed [24, 25]. This weak statement is applicable for tracing arbitrarily large shape change in two (2D) or three dimension (3D). Numerical examples have demonstrated that the methods are accurate and robust, and can capture intricate details of the crack healing [26–28].

Based on Herring's classical theory [29], the flux of surface diffusion, J, is proportional to driving force F

$$J = MF, \tag{1}$$

where $M = \Omega D_s \delta_s / kT$ is the diffusion mobility of atoms on the surface, and is assumed to be isotropic, Ω is the atomic volume, D_s the self-diffusivity on the surface, δ_s the effective thickness of atoms participating in matter transport, k Boltzmann's constant, and T the absolute temperature. The driving force F is defined by the amount of free energy decrease associated with one unit volume of matter moving one unit distance on the surface.

As matter is deposited onto (or removed from) a free surface, a normal velocity V_{ns} of the free surface results. Conservation of matter requires that

$$V_{ns} = -\nabla \cdot J \tag{2}$$

where V_{ns} is taken as positive when matter is added to the surface, $\nabla \cdot J$ stands for the surface divergence of the flux.

Let δI be the virtual mass displacement and δr_{ns} the virtual normal displacement on the surface. We then have

$$\delta r_{ns} = -\nabla \cdot (\delta I) \tag{3}$$

Following the principle of virtual work [30], we obtain the integral form

$$\int F \cdot \delta I \, dA = -\delta G \tag{4}$$

where δG is the increment of free energy, and dA is the element area.

Substituting Equation (1) into (4), we have

$$\int \frac{J \cdot \delta I}{M} dA = -\delta G \tag{5}$$

Once J is solved from Equation (5), the surface normal velocity is obtained from Equation (2), which then updates the shape for a small time increment.

However, it should be noted that the surface velocity V_{ns} due to matter redistribution is not a velocity of any material element. V_{ns} represents the "thickening" rate of the surface as matter is deposited onto the surface [31].

Hence a surface diffusion problem that conserves solid mass as formulated above is difficult to implement in a FE setting. However, if an evaporation–condensation process is introduced into microstructure evolution, the overall mass conservation is easier to treat by FE method [32, 33]. The kinetic law for evaporation–condensation is

$$V_{nv} = mp \tag{6}$$

where m is the special evaporation-condensation rate, p the free-energy reduction associated with per unit volume of matter deposited per unit surface area to the solid, which relates to the surface curvature by Laplace–Young relation. Furthermore, the explicit expression of p is unnecessary in formulating the FE method, and therefore not listed here.

The resulting equations of surface normal velocity and surface virtual displacement relative to the combined action of evaporation–condensation and surface self-diffusion are

$$\begin{cases} V_n = V_{ns} + V_{nv} \\ \delta r_n = \delta r_{ns} + \delta r_{nv} \end{cases} \tag{7}$$

where δr_{nv} is surface virtual displacement of evaporation–condensation process.

Associated with the virtual motion, the free energy changes by δG. According to the definition of the two kinds of driving force, F and p, matter redistribution and exchange on the surface area element δA, reduce the free energy by an amount $(F \cdot \delta I + p \delta r_{nv}) dA$. Thus

$$\int (F \cdot \delta I + p \delta r_{nv}) dA = -\delta G \tag{8}$$

Substituting Equations (1)–(6) into Equation (8), we have

$$\int \left\{ \frac{J \cdot \delta I}{M} + \frac{(V_n + \nabla \cdot J)[\delta r_n + \nabla \cdot (\delta I)]}{m} \right\} dA = -\delta G \tag{9}$$

Let L be a representative length in the problem. A dimensionless ratio mL^2/M measures the relative rate of evaporation–condensation and surface diffusion [33].

When $mL^2/M \ll 1$, the effect of evaporation–condensation can be neglected and mass conservation is preserved in the processes of dynamic evolution [34, 35]. Based on the above weak statement, a FE method is constructed, the details of which can be found [21–23, 33].

2.2   FE-Modelling Results and Discussion

2.2.1   Healing of 2D Crack within a Grain

The shape of the cross section of a cylindrical crack is assumed to elliptical with major and minor axes W and D_0 (Figure 1), and an aspect ratio is defined as $\beta = W/D_0$.

The simulations on cracks of different aspect ratios prove that there exists a critical aspect ratio, $\beta = 133$, below which the crack will directly evolve into a cylindrical pore channel, out of which two or more cylindrical pore channels will be formed

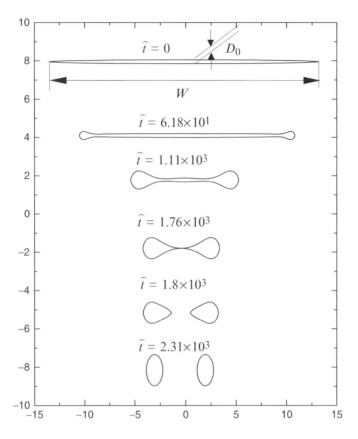

Fig. 1 Shape evolution of a crack with aspect ratio β = 133

for $133 \leq \beta < 275$ and $\beta \geq 275$, respectively. Figure 1 shows the splitting and cylinderization processes of a typical crack with aspect ratio $\beta = 133$ during healing. Similar behavior was reported on the study of the instability for rod-shaped particles of finite length [36].

If the crack intersects multiple grains, a thermal groove will develop along the GB to establish static equilibrium between surface tension and the boundary tension [37], when such a system is heated sufficiently to allow appreciable surface diffusion. Two shoulders or hillocks will develop adjacent to the groove. These hillocks, if they become sufficiently large, can bridge the crack and lead to split as observed by Rödel et al. [3]. Figure 2a shows a crack running through grain I and grain II.

For simplicity, the GB of the two grains is assumed to be perpendicular to the center of initial crack surfaces and is isotropic. As shown in Figure 2b at $\widehat{t} = t/\left(D_0^4/M\gamma_s\right) = 18.2$, the groove is bordered by two ridges. This process takes place on upper and lower crack surfaces. The upper and lower ridges move toward each other as the grooves deepen. As they overlap, the crack is split into three segments as

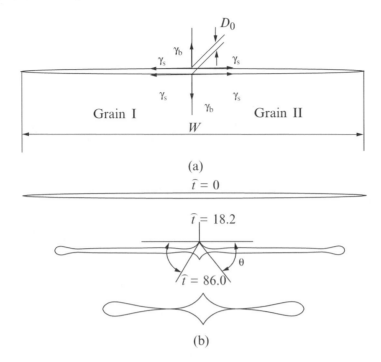

Fig. 2 Healing process of a crack ($\beta = 60$) perpendicular to a GB. (**a**) Initial crack profile; (**b**) shape evolution process

shown in Figure 2b at $\hat{T} = 86.0$. After splitting, a cylinderization process follows as shown in Figure 1.

The grooving behavior depends on the ratio between the GB and surface tensions, $\hat{\gamma} = \gamma_b/\gamma_s$. For given aspect ratio β, there is a critical value of $\hat{\gamma}_c$ at which the crack splits. Based on a number of FE analyses, an approximate formulation of $\hat{\gamma}_c$ is found as follows

$$\hat{\gamma}_c = -0.0349 + e^{-(\beta-43.8743)/68.9126} \tag{10}$$

2.2.2 Healing of Penny-shaped Crack within a Grain

The healing of a penny-shaped crack within a grain is also controlled by surface diffusion alone, which is analyzed by the developed axisymmetric FE method [21]. Similarly, the shape of the crack is characterized by an aspect ratio $\beta = W/D_0$, here W is the diameter and D_0 is the thickness of the crack (Figure 3a).

The FE modelling is performed for cracks with different aspect ratio's β. It is found that the penny-shaped crack, depending upon various initial aspect ratio β, will evolve into a sphere (Figure 3b), or a ring (Figure 3c), or a ring with a center void (Figure 3d), or multiple rings with or without a center void (Figure 3d). A critical

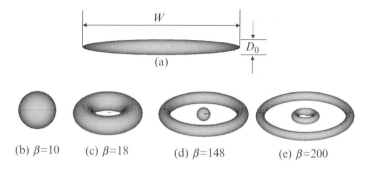

(a)

(b) β=10   (c) β=18   (d) β=148   (e) β=200

Fig. 3 Shape evolution of a penny shaped crack. (**a**) Initial shape; (**b**) $\beta = 10$; (**c**) $\beta = 18$; (**d**) $\beta = 148$; (**e**) $\beta = 200$

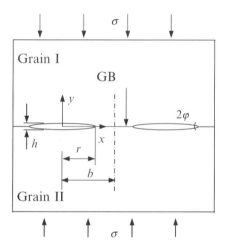

Fig. 4 Schematic illustration of cracks on the GB

aspect ratio of $\beta = 18$ is obtained for a penny-shaped crack, below which a crack will form a spherical void and above which it will split into a doughnut-like channel pore or more complicated ring-void system.

2.2.3 Healing of 2D Grain Boundary Cracks

Figure 4 shows 2D-GB cracks arranged at regular intervals $2b$ under pressure. The sizes of the microcracks along and perpendicular to the grain boundary are represented by r and h, respectively. The subscript o attached to r and h denotes their initial values. For a GB crack, the GB diffusion is coupled by boundary conditions at the triple point of the crack surface and the GB. The excess chemical potential of the atoms on the GB, $\Delta\mu = -\Omega\sigma_n$ induced by normal pressure σ_n, will derive an atom flux from the GB to the crack surfaces. As a result, the crack tips recede and blunt, and the crack gradually shrinks with time as shown in Figure 5. The GB flux is given by $J_b = (D_b\delta_b/kT)\partial\sigma_n/\partial x$, where D_b is the GB diffusivity and δ_b is the

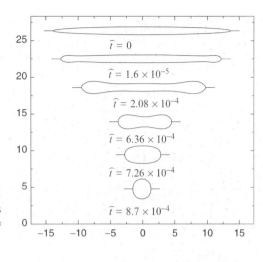

Fig. 5 Healing process of a typical GB crack ($\hat{b} = 20$, $f = 0.01$, $\hat{\sigma} = 10$, $\varphi = 70°$, $\beta = 20$)

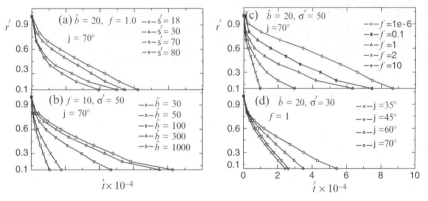

Fig. 6 Healing of a GB crack ($\beta = 20$) depends on: (**a**) stress; (**b**) spacing; (**c**) diffusivity ratio; (**d**) equilibrium dihedral angle

GB thickness. The coupling of the diffusion on the crack surface and on GB requires continuity in the chemical potential and the flux at the triple point. Using the continuity conditions, together with the equilibrium angle condition $\cos\varphi = \gamma_b/2\gamma_s$, the healing process of a GB crack can be analyzed [23].

The healing rate of the GB cracks depends on: (1) The ratio of the GB diffusion and surface diffusion coefficients $f = D_b\delta_b/D_s\delta_s$; (2) the dimensionless crack spacing $\hat{b} = b/r_0$; (3) the dimensionless applied pressure $\hat{\sigma} = \sigma r_0/\gamma_s$; (4) the equilibrium dihedral angle $\varphi = \arccos(\gamma_b/2\gamma_s)$. The equilibrium dihedral angle φ at the triple point is related to the curvature difference near the crack tip. It influences the excess chemical potential of the atom at the triple point, and therefore the healing rate.

Figure 5 displays a typical healing process of a GB crack with the normalized timescale $\hat{t} = t/(r_0^4/M\gamma_s)$. Figure 6 plots the normalized $\hat{r} = r/r_0$ as a function of time $\hat{\tau}$ for four groups of stress $\hat{\sigma}$, crack spacing $\hat{b}$, diffusivity ratio f, and equilibrium dihedral angle φ. The slope of these curves at any point represents the velocity of the crack tip along the GB, indicating the healing rate of the crack.

The initial healing rate is fast because of the pronounced difference in curvature along the crack surfaces. The healing time decreases with increasing $\hat{\sigma}$ and f, or with decreasing $\hat{b}$, and φ.

3   Void Healing

In section 2, we analyzed the features of crack-healing processes. The analyses were performed up to the formation of cylindrical or the doughnut pore channels. However, neither the cylindrical nor the doughnut pore channel is stable. They will split and evolve into discrete voids in the grain or on the GB via Rayleigh instability. On the other hand, for metals, voids can be directly introduced mainly on grain boundaries by either creep or fatigue at elevated temperature [38–40]. And it has been realized that the recovery of metal properties is positively attributed to the healing of voids [38].

To study void healing, the fate of these voids first need to be clarified [41–44]. Will a void relax to a circular void and shrink, or will it collapse to a narrow slit? Here, we will discuss the conditions under which voids can be healed.

3.1   Energetics

The instability of the void shape is an outcome of the competition between the variation in the elastic strain energy and the interface energies.

The appropriate thermodynamic potential, consisting of both elastic energy, surface, and GB energies, is a function of the void shape and volume. The work done by the load either varies the energy in the solid, or produces entropy in the diffusion process.

Denote w as the strain energy per volume, γ_s the surface energy per area on void surface and γ_b the energy per area on the GB. They are taken to be independent from each other for practical purpose. That is, γ_b and γ_s are independent of the applied stress, and the strain field in the body is determined by the elasticity theory neglecting the effect of interface energies.

The total elastic energy and interface energies are

$$U_e = \int w \, dV, \quad U_s = \int \gamma_s \, dA_s, \quad U_b = \int \gamma_b \, dA_b \qquad (11)$$

where V is the volume of the solid, A_s the area of void surface, and A_b the area of the GB. Under the fixed mechanical load, the suitable potential is

$$\Phi = U_e + U_s + U_b - (\text{load} \times \text{displacement}) \qquad (12)$$

Furthermore

$$U_e = (\text{load} \times \text{displacement})/2 \qquad (13)$$

for linear elastic solids. Thus, the thermodynamic potential for the linear elastic solid under constant load is

$$\Phi = -U_e + U_s + U_b \qquad (14)$$

The potential is a function of void shape and volume. For a given void, U_e is determined by the elasticity problem, U_s is integrated over the area of the void surface and U_b along the GB.

The first law of the thermodynamics requires that

$$\frac{d\Phi}{dt} + (\text{dissipation rate}) = 0 \qquad (15)$$

The second law of thermodynamics requires the dissipation to be positive when atoms diffuse. That is, atoms diffuse to reduce the potential of the system.

For simplicity, symmetrical problems are considered. In the following, the potential energy is calculated for 2D ellipses having varying area and shape, and 3D ellipsoids having varying volume and shape. The shape that minimizes the potential is taken to be in equilibrium.

3.2 Thermodynamic Potential of 2D Model

Figure 7 illustrates the 2D model of a small cylindrical void in an elastic solid under biaxial stresses. As diffusion varies the void shape, the solid minimizes its energy by either varying the elastic field or creating the interfaces.

The morphological change, including change of void volume and shape, is an outcome of the competition between the variation in the elastic and interfacial energies.

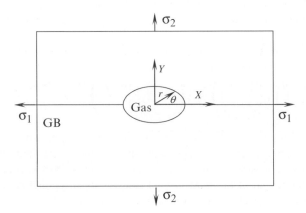

Fig. 7 A gas-filled cylindrical GB void in an elastic solid under biaxial stresses

The shape for an ellipse having the same area as a circular void of radius ρ can be described by

$$X = \rho\sqrt{\frac{1+m}{1-m}}\cos\theta \quad Y = \rho\sqrt{\frac{1-m}{1+m}}\sin\theta \tag{16}$$

The circle corresponds to $m = 0$, the X-direction slit to $m \to +1$, and the Y-direction slit to $m \to -1$. The ellipse has perimeter

$$L_s = \frac{\rho}{\sqrt{1-m^2}}\int_0^{2\pi}\left(1+m^2-2m\cos 2\theta\right)^{1/2}d\theta \tag{17}$$

The integral is evaluated numerically.

For an infinite body with an elliptic hole having the same area as a circle of transient radius ρ and the body with a circular hole of initial radius ρ_0, the elastic energy differs by [45]

$$\Delta U_e = \frac{\pi\rho_0^2(\sigma+p)^2}{2E}\left(\alpha^2\left(\frac{3-m}{1+m}\omega^2 - 2\omega + \frac{3+m}{1-m}\right) - \left(3\omega^2 - 2\omega + 3\right)\right) \tag{18}$$

where p is the internal pressure, being positive for pressure; σ_1, σ_2 are the applied stresses, positive for tension; $\alpha = \rho/\rho_0$, $\omega = (\sigma_1+p)/(\sigma_2+p)$, being the area parameter and the stress bias parameter, respectively. In equations of this section, we use σ to identify σ_2; Equation (18) is derived for plane stress condition. For plane strain condition, replace Young's modulus E by $E/(1-\nu^2)$, where ν is the Poisson's ratio.

And the GB energy and surface energy differ by

$$\Delta U_b = 2\rho_0\gamma_b\left(1 - \alpha\sqrt{\frac{1+m}{1-m}}\right) \quad \Delta U_s = \gamma_s\left(L_s - 2\pi\rho_0\right) \tag{19}$$

Combining Equations (18) and (19), the total difference in the potential is given by

$$\frac{\Delta\Phi}{2\pi\rho_0\gamma_s} = -\frac{\Lambda}{4}\left(\alpha^2\left(\frac{3-m}{1+m}\omega^2 - 2\omega + \frac{3+m}{1-m}\right) - \left(3\omega^2 - 2\omega + 3\right)\right)$$

$$+ \left(\frac{L_s}{2\pi\rho_0} - 1\right) + \lambda\left(1 - \alpha\sqrt{\frac{1+m}{1-m}}\right), \tag{20}$$

where $\Lambda = (\sigma+p)^2\rho_0/\gamma_s E$ is the dimensionless loading parameter; $\lambda = \gamma_b/\pi\gamma_s$ is the interface energy ratio. For a given Λ, Φ is a function of the shape parameter m, area parameter α, stress ratio ω, and interface energy ratio λ. As atoms diffuse to reduce the potential of the system, void changes its shape and area. For all void shapes, the equilibrium shape minimizes Φ.

3.2.1 Critical Load of 2D Void in a Grain

When the interface energy ratio $\lambda = 0$, Equation (20) reduces to the thermodynamic potential of a 2D void in grain

$$\frac{\Delta\Phi}{2\pi\rho_0\gamma_s} = -\frac{\Lambda}{4}\left(\alpha^2\left(\frac{3-m}{1+m}\omega^2 - 2\omega + \frac{3+m}{1-m}\right) - \left(3\omega^2 - 2\omega + 3\right)\right)$$
$$+ \left(\frac{L_s}{2\pi\rho_0} - 1\right).$$
(21)

And if $\omega = 1$, Equation (21) becomes the potential of a 2D void in an infinite isotropic under an equal biaxial stresses

$$\frac{\Delta\Phi}{2\pi\rho_0\gamma_s} = -\Lambda\left(\alpha^2\frac{1+m^2}{1-m^2} - 1\right) + \left(\frac{L}{2\pi\rho_0} - 1\right)$$
(22)

Figure 8a displays the function $\Phi(m)$ at several constant loading levels for $\omega = 1$. Each minimum and maximum represents a stable and unstable state respectively. From Figure 8a, the following behaviors can be found:

1. When $\alpha\Lambda = 0$, the stress vanishes. Φ reaches a minimum at $m_e = 0$; and maxima at $m = \pm 1$. The circular void is stable and the two slits are unstable: any ellipse will relax to a circle.
2. When $\alpha\Lambda \in (0, 3/8)$, the surface energy dominates; Φ reaches a local minimum at $m_e = 0$, two maxima at some m_1 and m_2, and two minima at $m = \pm 1$. The maxima act as energy barriers: an ellipse of $m_1 < m < m_2$ will relax to the circle, but an ellipse of $m > m_2$ or $m < m_1$ will collapse to the slits.
3. When $\alpha\Lambda \in (3/8, \infty)$ the strain energy dominates; Φ reaches the maximum at $m_c = 0$, and minima at $m = \pm 1$. The circle is unstable but the slits are stable: any elliptic void will collapse to the slits.

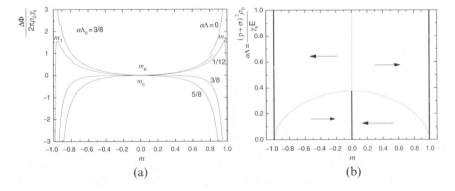

Fig. 8 (a) The potential as a function of void shape parameter at several load levels; and (b) stability conditions projected on the (m, aΛ) plane for a 2D void under biaxial stress state $\sigma = \sigma_1 = \sigma_2$

In Figure 8b, the values for m_e, m_1, m_2, and m_c are projected onto the $(m, \alpha \Lambda)$ plane. The values of m_e minimizing Φ are the heavy solid lines, and values of m_1, m_2, and m_c maximizing Φ are the dotted lines, corresponding to the stable and unstable equilibrium states, respectively. These lines divide the $(m, \alpha \Lambda)$ plane into four regions. A point in each region corresponds to an ellipse under a constant level, evolving toward a stable equilibrium state, either an ellipse of m_e, or the slits of $m = \pm 1$. The evolution direction in each region is indicated by an arrow. From Figure 8, it can be seen that the void will relax to a rounded void if the applied stresses is below the critical value, i.e. $\alpha \Lambda < \alpha \Lambda_c$, but collapse to a slit of vanishing thickness when $\alpha \Lambda > \alpha \Lambda_c$.

A void in a real material differs from the idealised case in many ways. Surface energy is anisotropic in crystals; for aluminum, the {111} planes have the lowest surface energy, and are the preferred void surfaces. The elastic modulus misfit causes asymmetry. The stresses in two directions are not exactly the same. Given these imperfections, the circular symmetry breaks even at vanishing stress levels.

If an imperfection is not too large in magnitude, it only changes the potential slightly, and therefore the locations of its minima and maxima. The lines in Figure 8 will bend somewhat, but the essential features should remain unchanged. To illustrate the general idea, we study a void in an isotropic crystal under a biased stress state, i.e. $\sigma_1 \neq \sigma_2$. Figure 9a is for $\omega = 0.8$, a representative for any stress ratios in the interval $0 < \omega < 1$. The values of minimizing Φ are the heavy solid lines, and the ones of maximizing Φ are the dotted lines. The heavy solid curve ends at $\alpha \Lambda_c$, and is continued by the dotted curve. Several asymmetries are noted when comparing Figures 8b and 9a. For small $\alpha \Lambda$, the local minimum no longer occurs at $m = 0$, nor do the two maxima at the same value of $|m|$. At a critical value $\alpha \Lambda_c$, the minimum and the maximum on the right-hand side annihilate, but the maximum on the left-hand side persists. As expected, under the biased stress, the equilibrium shape is noncircular even for a small value of $\alpha \Lambda$. Figure 9b is the corresponding diagram under uniaxial stress state $\omega = 0$, where $m = -1$ is an unstable equilibrium state, and $m = +1$ is stable. The critical value is reached at about $m = 0.5$, corresponding to an ellipse with axis ratio of 3.

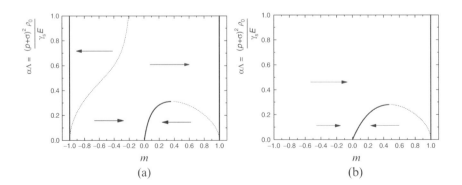

Fig. 9 Stability conditions projected on the $(m, \alpha \Lambda)$ plane. (**a**) Biased biaxial stress state $\sigma_1/\sigma_2 = 0.8$; (**b**) uniaxial stress state $\sigma_1 = 0$, $\sigma_2 = \sigma$

An inspection shows that Figure 9a will degenerate to Figure 8b as $\omega \to 1$, and to Figure 9b as $\omega \to 0$. Although the three figures look very different, the essential feature remains unchanged: an initial void will, depending to the biased stress state, relax to a more or less rounded void as $\alpha \Lambda < \alpha \Lambda_c$, but collapse to a slit as $\alpha \Lambda > \alpha \Lambda_c$.

3.2.2 Critical Load of 2D Void on the GB

Another imperfection studied here is $\lambda \neq 0$, corresponding to a void on GB. To visualize the effect of λ on the critical value $\alpha \Lambda_c$ and the equilibrium shape of the void, Figure 10 displays equilibrium states on the $(m, \alpha \Lambda)$ space for $\lambda = 0.116$ under several remote stress states ($\omega = 1, 0.8, 0$). As before, the figures feature the same: the void will relax to a more or less rounded void if $\alpha \Lambda < \alpha \Lambda_c$, but collapse to a slit if $\alpha \Lambda > \alpha \Lambda_c$. However, the stable equilibrium shape of the GB void, even at stress free state, not a circle but an ellipse, with elongation in the plane of the GB as observed by Gittins [38].

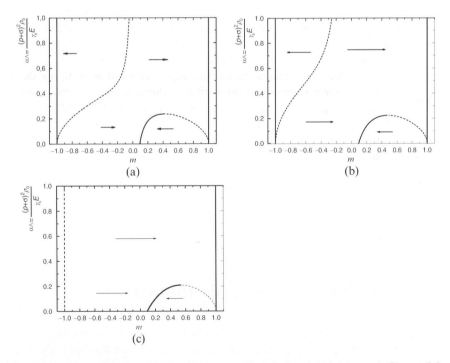

Fig. 10 Stability conditions projected on the (m, a Λ) plane for λ = 0.116: (**a**) ω = 1; (**b**) ω = 0.8; (**c**) ω = 0

3.2.3 Approximate Equation of the Critical Load and the Equilibrium Shape

Both the critical value $\alpha \Lambda_c$ and the equilibrium shape parameter m_e are a function of α, ω, and λ. They should be determined before we discuss the healing process of the void. Although they can be sought by detailed numerical analysis, it is not convenient for practice. An approximately explicit expression for $\alpha \Lambda_c$ and m_e can be obtained by expanding Equation (20) in powers of m

$$\frac{\Delta \Phi}{2\pi \rho_0 \gamma_s} = \psi_0 + \alpha \psi_1 m - \frac{\alpha \psi_2}{4} m^2 + \frac{\alpha \psi_3}{2} m^3 + \cdots , \tag{23}$$

where

$$\left. \begin{array}{l} \psi_0 = (1 - \alpha)(\lambda - 1) + \frac{\Lambda}{4}\left(1 - \alpha^2\right)\left(3 - 2\omega + 3\omega^2\right) \\ \psi_1 = \alpha \Lambda \left(\omega^2 - 1\right) - \lambda \\ \psi_2 = 4\alpha \Lambda \left(1 + \omega^2\right) + 2\lambda - 3 \\ \psi_3 = 2\alpha \Lambda \left(\omega^2 - 1\right) - \lambda \end{array} \right\} \tag{24}$$

Only three leading terms of m are retained for small m. Equilibrium requires that $d\Phi/dm = 0$, i.e.

$$3\psi_3 m^2 - \psi_2 m + 2\psi_1 = 0 \tag{25}$$

From Equation (25), the critical value $\alpha \Lambda_c$ can be approximately expressed by

$$\alpha \Lambda_c \approx \frac{(13\lambda - 3)\omega^2 - (7\lambda + 3) + \sqrt{27\omega_1 + 18\omega_2 \lambda + 3\omega_3 \lambda^2}}{8\omega_4} \tag{26}$$

where

$$\left. \begin{array}{l} \omega_1 = 1 - 2\omega^2 + \omega^4 \\ \omega_2 = 1 + 4\omega^2 - 5\omega^4 \\ \omega_3 = 3 + 2\omega^2 + 27\omega^4 \\ \omega_4 = 1 - 4\omega^2 + \omega^4 \end{array} \right\} \tag{27}$$

When $\alpha \Lambda \leq \Lambda_c$, the equilibrium state parameter m_e of the ellipse is the real root of Equation (25), i.e.

$$m_e = \frac{\psi_2 + \sqrt{\psi_2^2 - 24\psi_1 \psi_3}}{6\psi_3} \tag{28}$$

3.3 Thermodynamic Potential of the 3D Model

Figure 11 shows the 3D model of a gas-filled oblate ellipsoidal void in infinite elastic solid. Here, an axisymmetric problem will be analyzed in which the void evolves as a sequence of oblate ellipsoidal voids. The semi-axes of the void on xy plane are equal under the remote triaxial stresses $\sigma_1 = \sigma \leq \sigma_3$ and an internal pressure p, such that as diffusion changes the shape of the cross section of the void parallel to the xz plane, the shape in the cross-section parallel to the xy plane maintains circular. Hence, the

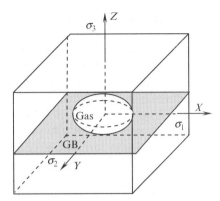

Fig. 11 A gas-filled ellipsoidal void in elastic solid under triaxial stresses $\sigma_1 = \sigma_2 \leq \sigma_3$ and internal gas pressure ρ

void can be represented by its xz cross section. The shape of ellipsoidal void having the same volume as a spherical one of radius ρ can be described by

$$a_x = a_y = \rho\sqrt{\frac{1+m}{1-m}}\cos\theta \qquad a_z = \rho\frac{1-m}{1+m}\sin\theta \qquad (29)$$

where m is the shape parameter, $0 \leq m < 1$. The spherical void corresponds to $m = 0$, the crack to $m \rightarrow 1$.

Following a similar procedure, the total difference in the potential can be formulated as [45]

$$\frac{\Delta\Phi}{4\pi\rho_0^2\gamma_s} = -\frac{\Lambda}{3}(a^3A - A_o) + (a^2B - 1) + \lambda\left(1 - a^2\frac{1+m}{1-m}\right) \qquad (30)$$

where $\lambda = \gamma_b/4\gamma_s$ is interface energy ratio; $\Lambda = \rho_0(\sigma_3 + p)^2/\gamma_s E$ is a dimensionless loading parameter that describes the relative importance of the elastic energy and the void surface energy; A and B are dimensionless numbers listed in Appendix, A_o is A of a spherical void with radius ρ_0. For a given Λ, the thermodynamic potential Φ for 3D model is also a function of the shape parameter m, volume parameter a, stress ratio ω and interface energy ratio λ.

3.3.1 Critical Load of 3D Void in Grain

When $\lambda = 0$, Equation (30) represents a void within grain. Through numerical computations, the stable and unstable equilibrium states under three typical stress states $\omega = 0, 0.5, 1$ are plotted on the $(m, a\Lambda)$ plane (Figure 12). The solid and dotted lines correspond to the stable and unstable equilibrium states, respectively. The line for $\omega = 0.5$ is a representative for any stress ratios in the interval $0 < \omega < 1$. The solid and dotted lines divide the $(m, a\Lambda)$ plane into two regions for $\omega = 1$ and three regions for

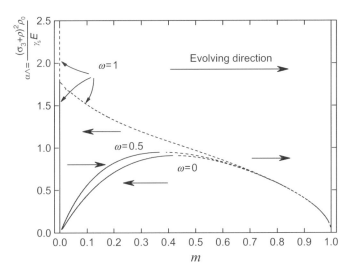

Fig. 12 Stability conditions for a void in grain projected on the $(m, a\Lambda)$ plane

$\omega = 0, 0.5$. A point in each region corresponds to a cavity under a constant level of $a\Lambda$, evolving toward a stable equilibrium state, either an ellipsoid cavity or a crack.

Although the lines look different, they mean practically the same thing, as the feature shown in 2D model: A cavity will relax to a more or less rounded shape if $a\Lambda < a\Lambda_c$, but collapse to a crack if $a\Lambda > a\Lambda_c$.

3.3.2 Critical Load of 3D Void on the GB

Figure 13 displays the stability conditions for a GB void ($\lambda = 1/11$) on the $(m, a\Lambda)$ plane under several remote stress states. The applied load $a\Lambda$ reaches its critical value at the joint point of the heavy solid lines and the dotted lines. Again, a void settles to a rounded shape when $a\Lambda$ is below the critical value, but collapses to a crack when $a\Lambda$ is above it.

3.3.3 Approximate Equation of the Critical Load and the Equilibrium Shape

Following a similar procedure as 2D model, we can obtain an approximate equation to predict the critical load

$$a\Lambda_c \approx \frac{\omega_5 + \omega_6\lambda - \left(\omega_7 + \omega_8\lambda + \omega_9\lambda^2\right)^{1/2}}{\omega_{10}} + \omega_{11} \tag{31}$$

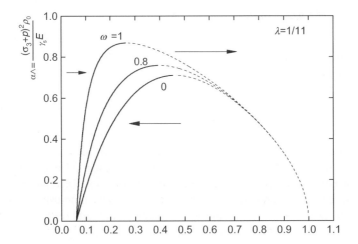

Fig. 13 Stability conditions for a void on GB projected on the $(m, a\Lambda)$ plane

where

$$
\begin{aligned}
\omega_5 &= 2.991 + 1.5151\omega + 1.254\omega^2 \\
\omega_6 &= 11.959 - 1.55\omega + 0.9\omega^2 \\
\omega_7 &= 48.386 - 11.042\omega - 7.75\omega^2 - 7.79\omega^3 - 21.6\omega^4 \\
\omega_8 &= 43.367 + 41.32\omega + 43.245\omega^2 + 7.3\omega^3 + 18.8\omega^4 \\
\omega_9 &= 19.8 + 25.75\omega + 78.9\omega^2 + 34\omega^3 + 73\omega^4 \\
\omega_{10} &= -3.84153 + 1.96341\omega + 1.714\omega^2 + 1.151\omega^3 + 2.263\omega^4 \\
\omega_{11} &= -0.12013 + 0.07652\omega + 0.22715\omega^2 - 0.17983\omega^3
\end{aligned}
\tag{32}
$$

where ω_{11} is introduced to correct the error caused by the truncation of the expansion. And the shape parameter at equilibrium state can be formulated as

$$
m_e = \frac{1.6 - 2\lambda + a\Lambda\omega_{12} + \left(\lambda_1 + a\Lambda\omega_{13} + a^2\Lambda^2\omega_{14} + a\Lambda\lambda\omega_{15}\right)^{1/2}}{6\lambda - 2.2857 + 3a\Lambda\omega_4}
\tag{33}
$$

where

$$
\begin{aligned}
\lambda_1 &= 2.56 - 1.828\lambda - 8\lambda^2 \\
\omega_{12} &= -0.8755 - 0.1347\omega + 0.1102\omega^2 \\
\omega_{13} &= -1.4955 - 0.7576\omega - 0.627\omega^2 \\
\omega_{14} &= 0.936 + 0.491\omega + 0.428\omega^2 + 0.288\omega^3 + 0.5658\omega^4 \\
\omega_{15} &= -5.98 + 0.7752\omega - 0.453\omega^2
\end{aligned}
\tag{34}
$$

3.4   Void Healing Rate

In this section, we will derive the healing rate of an array of voids located on a planar GB under hydrostatic pressure. As shown in Figure 14, the initial radius of the voids is ρ_0, and the void spacing $2b$. Let r be the distance along the GB to the center of a void. The profile represents the cross section of cylindrical void in 2D model, and of ellipsoidal void in 3D model. Here we will derive the healing rate of 2D model [46, 47], and as a reference, that of 3D model may be similarly obtained [48, 49].

The following analyses are based on the assumption that the pressure is below the critical load, such that the void is stable during healing. Also, creep is assumed to be slow compared to surface diffusion and GB diffusion, therefore neglected. As such, surface diffusion and GB diffusion are the dissipative processes included in the analyses. It should be noted that the basic Equations (20) and (30) are strictly valid only for an infinite solid containing a single void. For the model shown in Figure 14, it can be approximately used if the initial size of voids is small compared to the void spacing.

In 2D model, the excess chemical potential on the GB is given by [6, 50]

$$\Delta\mu = -\sigma_n(r)\Omega \tag{35}$$

where $\sigma_n(r)$ is the net normal stress acting on the GB, positive for tensile stress.

The atom flux along the GB is expressed by

$$J = -\frac{D_b}{kT\Omega}\nabla(\Delta\mu) \tag{36}$$

where $D_b = D_{0b}\exp(-Q/RT)$ is assumed to be isotropic. Here, D_{0b} is the frequency factor, R the gas constant and Q the activation energy for the GB diffusion.

For the GB diffusion, the steady state can be expressed by [51, 52]

$$div J = \eta \tag{37}$$

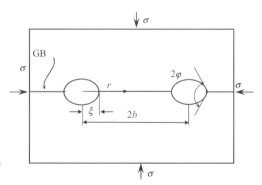

Fig. 14 An array of voids on a planar GB interface

where η is the number of atoms removed per unit time and per unit volume of the GB, independent of the position on the boundary because of the steady state.

Substituting Equation (36) into (37), we obtain the differential equation

$$\nabla^2 (\Delta\mu) + \frac{\eta k T \Omega}{D_b} = 0 \tag{38}$$

The excess chemical potential $\Delta\mu$ on the void surface is expressed as

$$\Delta\mu_s = \gamma_s \kappa \Omega \tag{39}$$

where κ is the curvature of the void surface, taken to be positive for a convex surface. $\Delta\mu$ must be continuous at the cavity apex $(r = \xi)$ where it meets the GB. For the elliptic void described by m_e, the curvature at the apex, κ_{tip}, is given by

$$\kappa_{tip} = -\tau^{3/2}/\rho \quad \tau = \frac{1 + m_e}{1 - m_e} \tag{40}$$

where ρ is the radius of a circle with the same area of an ellipse of m_e. Thus, the continuous condition of the excess chemical potential at the void tip is

$$-\frac{\gamma_s \tau^{3/2}}{\rho}\Omega = -\sigma_n(r = \xi)\Omega \tag{41}$$

where $\xi = \rho\tau^{1/2}$. It is assumed that the surface diffusion is rapid enough for voids to reach its equilibrium shape m_e. Thus, the period of the transient surface diffusion is rather short and can be ignored.

Because of symmetry, the matter flux vanishes at $r = b$, i.e. the boundary condition is

$$(\partial\Delta\mu/\partial r)_{r=b} = 0 \tag{42}$$

Finally, the condition of mechanical balance is given by

$$\int_{\xi}^{b} \frac{\Delta\mu}{\Omega}dr = \int_{\xi}^{b} \sigma_n(r)dr = \sigma b \tag{43}$$

Here and later we set $\sigma = \sigma_2$, being the pressure normal to the GB interface. The equilibrium angle φ (see Figure 14), as section 2, defined as $\cos\varphi = \gamma_b/2\gamma_s$ [1], can be satisfied only when the gradient in excess chemical potential along the void surface and the GB is small enough, so that the (quasi-) equilibrium state is approximately valid. The true equilibrium state is satisfied only when no atom flux occurs. On the other hand, the diffusion process is essentially dynamic. So, the diffusion process may change the angle from the equilibrium one [51]. We can, therefore, assume the equilibrium angle is free to change during the void-healing process.

The healing rate of a cylindrical void can be given as

$$\frac{dV}{dt} = -4b\left(\frac{\delta_b}{2}\right)\eta\Omega \tag{44}$$

Thus, solving the differential Equation (38) for $\Delta\mu$, subject to the boundary conditions of Equations (41), (43). Then, substitute the obtained η into Equation (44), we have the normalized rate of void healing

$$\frac{d\rho^*}{dt} = \frac{d\rho}{dt}\frac{kTb^3}{6D_b\Omega\gamma_s} = \frac{\frac{\rho\sigma}{\gamma_s} - \tau^{3/2}(1 - x\tau^{1/2})}{x^2(2 - 6x - 3x^3\tau^{1/2} + x^3\tau^{3/2} + 6x^2\tau^{1/2} - 3x^2\tau + 3x^2)} \tag{45}$$

And similarly, the normalized healing rate of 3D model can be formulated as [18]

$$\frac{d\rho^*}{dt} = \frac{d\rho}{dt}\frac{kTb^3}{2D_b\Omega\gamma_s} = \frac{\rho\sigma\sqrt{\tau}/\gamma_s + x^2\rho p_0\tau^{3/2}/(a^3\gamma_s) - (1+\tau^3)(1-x^2\tau)}{3 - 4x^2\tau + x^4\tau^2 + 4\ln(x\sqrt{\tau})} \tag{46}$$

where $x = \rho/b$.

Equations (45) and (46) exhibit the dependence of the void healing rate on the external stress, the ratio ρ/b, τ (related to shape parameter m_e) and material parameters related to surface and the GB diffusion.

In the following computations, date for pure iron with bcc lattice (α-Fe) are quoted: $\gamma_s = 2.2\,J/m^2$, $\gamma_b = 0.8\,J/m^2$, $E = 2.1 \times 10^{11}\,N/m^2$ [1]. For cylindrical voids with initial radius $\rho_0 = 2\,\mu m$ and void spacing $b = 20\,\mu m$, the interfacial energy ration is $\lambda = \gamma_b/\pi\gamma_s = 0.116$, the hydrostatic pressure σ should be less than the critical value $\sigma_c = -\sqrt{0.2367\gamma_s E/\rho_0} = -237\,MPa$, as analyzed in section 3.2. For ellipsoidal voids with initial radius $\rho_0 = 2\,\mu m$ and the spacing $b = 20\,\mu m$, the interface energy ratio $\lambda = \gamma_b/4\gamma_s = 1/11$. And the critical compressive stress $\sigma_c = -\sqrt{0.86\gamma_s E/\rho_0} = -478\,MPa$, below which a void maintains its stable shape as it shrinks, as discussed in section 3.3. Figure 15 shows the normalized shrinkage rate at different external pressure levels. Under fixed external pressure, the void shrinkage rate increases monolonously as the voids become small. The void shrinkage rate will be accelerated as the external pressure increases.

Figure 16 shows the relative dependence of the healing rate on the void spacing b. The rate decreases significantly with increasing void spacing.

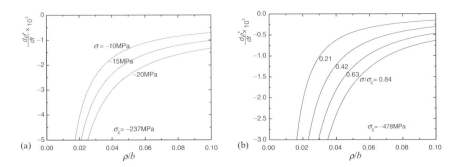

Fig. 15 Normalized healing rate of GB void under several pressure levels below the critical load. (a) 2D model; (b) 3D model

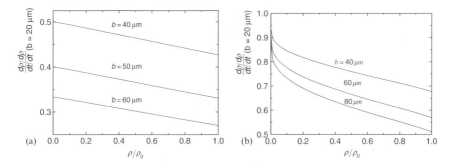

Fig. 16 Relative dependence of the healing rate on various void spacing. (**a**) 2D model; (**b**) 3D model

   The results are well matched with that the growth rate of the voids is approximately inversely proportional to the void spacing for a purely diffusive creep [54]. Hence, void spacing is one of the important factors that control the healing rate of the GB voids.

4   Outlook

Obviously, it is vital to understand the healing features and mechanisms of materials and to determine the physical parameters that control the processes of the healing [55], and some of which for crack and void healings in metals were analyzed in this chapter. Due to the many factors, complexity of the healing mechanism, and processes, crack and void healings are not yet completely understood. However, these discussions will provide deeper insights into healing mechanism in metals in the future.

   The restoration of the material integrity of cracked metallic materials can take place passively under suitable conditions of temperature and pressure; or actively by building elaborately designed healing mechanism into metals [56, 57]. We restricted ourselves to aspects of metal healing processes and conditions, about which there should be much concern. There is a long way to go to realise complete and repeated healing in metallic materials. Still, we hope that the concepts on which we base our considerations may help to advance the theoretical understanding and design of self healing mechanisms.

Appendix

The increase of the strain energy upon introducing an ellipsoidal void filled with gas into an infinite solid was computed by following the method of Eshelby [45]. The coefficient in Equation (30) is

$$A = \frac{1}{2}\left(2\omega^2 C_{11} + 2\omega C_{13} + \omega C_{31} + C_{33}\right)$$

where

$$C_{11} = \frac{(1 - S_{11})(1 - v) - 2v\,S_{13}}{\Delta},$$

$$C_{13} = \frac{S_{13} - (1 - S_{33})\,v}{\Delta},$$

$$C_{31} = \frac{2S_{31}(1 - v) - 2v(1 - S_{11} - S_{12})}{\Delta},$$

$$C_{33} = \frac{1 - S_{11} - S_{12} - 2v\,S_{31}}{\Delta},$$

$$\Delta = (1 - S_{33})(1 - S_{11} - S_{12}) - 2S_{31}\,S_{13}.$$

All S_{ij} can be found in [45].

The surface energy was calculated by integrating the surface tension over the spheroid surface. The coefficients is

$$B = \frac{\varsigma^2}{2} + \frac{\ln\left(\sqrt{\varsigma^6 - 1} + \varsigma^3\right)}{2\varsigma\sqrt{\varsigma^6 - 1}}$$

where $\varsigma = \sqrt{(1 + m)/(1 - m)}$.

Acknowledgements The authors thank Professor Sun J and his research group in Xi'an Jiaotong University, China, who provided valuable discussions. Financial supports from the National Natural Science Foundation of China (Grant No. 10572088 and 10602034) are acknowledged.

References

1. Chuang T.J., Kagawa K.I., Rice J.R., Sills L.B. (1979) Acta Metall 27:265
2. Chuang T.J. (1982) J Am Ceram Soc 65:93
3. Rödel J., Glaeser A.M. (1990) J Am Ceram Soc 73:592
4. Carter W.C., Glaeser A.M. (1987) Acta Metall 35:237
5. Hull D., Rimmer D.E. (1959) Phi Mag 4:673
6. Raj R., Ashby M.F. (1975) Acta Metall 23:653
7. Thouless M.D., Liniger W. (1995) Acta Metall Mater 43:2493
8. Singh R.N., Routbort J.L. (1979) J Am Ceram Soc 62:128
9. Wanamaker B.J., Wong T.F., Evans B. (1990) J Geophys Res 95:1563
10. Rödel J., Glaeser A.M. (1988) In: Yoo MH, Clark WAT, Braint CL (eds) Interfacial structures properties and design. Mater Res Soc. Pittsburgh, PA, p 485
11. Svoboda J., Riedel H. (1995) Acta Metall Mater 43:499
12. Park S.M., Oboyle D. (1977) J Mater Sci 12:840
13. Evans A.G., Charles E.A. (1977) Acta Metall 25:919
14. Suo Z., Wang W. (1994) J Appl Phys 76:3410
15. Powers J.D., Glaeser A.M. (1993) J Am Ceram Soc 75:2225

16. Scott C., Tran V.B. (1985) Am Ceram Soc Bull 64:1129
17. Gupta T.K. (1978) J Am Ceram Soc 61:191
18. Rödel J., Glaeser A.M. (1990) Sintering of advanced ceramics. The American Ceramic Society, Westerville, OH, p 243
19. Drory M.D., Glaeser A.M. (1985) J Am Ceram Soc 68:C14
20. Hickan S.H., Evans B. (1987) Phys Chem Miner 15:91
21. Huang P.Z., Li Z.H., Sun J., Gao H. (2003) Metal Mater Trans 34A:277
22. Huang P.Z., Li Z.H., Sun J. (2001) Model Simul Mater Sci Eng 9:193
23. Huang P.Z., Li Z.H., Sun J. (2002) Metall Mater Trans A 33:1117
24. Sun B., Suo Z., Evans A.G. (1994) J Appl Phys Solids 42:1653
25. Suo Z. (1997) Adv Appl Mech 33:193
26. Cocks A.C.F., Gill S.P.A. (1996) Acta Mater 44:4765
27. Sun B., Suo Z., Yang W. (1997) Acta Mater 45:1907
28. Huang J.M., Yang W. (1999) Model Simul Mater Sci Eng 7:87
29. Herring C. (1951) In: Kingston (ed) The physics of powder metallurgy. McGraw-Hill, New York, p 143
30. Yang W., Suo Z. (1996) Acta Mechanics Sinica 12:144
31. Pan J., Cocks A.C.F. (1995) Acta Mater 43:1395
32. Yu Y.H., Suo Z. (1999) J Mech Phys Solids 47:1131
33. Sun B., Suo Z. (1997) Acta Mater 45:4953
34. Okabayashi H. (1993) Mater Sci Eng 11:191
35. Hsueh C.H., Evans A.G. (1983)Acta Metall 31:189
36. Nichols F.A. (1976) J Mater Sci 11:1077
37. Mullins W.W. (1957) J Appl Phys 26:1205
38. Gitting A. (1988) Acta Metall 16:517
39. Walker G.K., Evans H.E. (1970) Met Sci J 4:155
40. Beere W.B., Greenwood G.W. (1971) Met Sci J 5:107
41. Shewmon P.G. (1964) Trans Metall Soc AIME 230:1134
42. Spears M.A., Evans A.G. (1982)Acta Metall 30:1281
43. Kraft O., Bader S., Sanchez J.E., Arzt E. (1993) Mater Soc Symp Proc 309:351
44. Sullivan T.D., Miller L.A. (1993) Mater Res Soc Symp Proc 309:169
45. Eshelby J.D. (1957) Proc R Soc Lond A241:376
46. Wang H., Li Z. (2003) J Mech Phy Solids 51:961
47. Wang H., Li Z. (2003) Metall Mater Trans A 34:1493
48. Wang H., Li Z. (2004) J Mater Sci 39:3425
49. Wang H., Li Z. (2004) J Appl Phys 95:6025
50. Hancock J.W. (1976) Met Sci 10:319
51. Takahashi T., Takahashi K., Nishiguchi K. (1991) Acta Metall 39:3199
52. Speight M.V., Beere W. (1975) Met Sci 9:190
53. Martin J.W., Doberty R.D. (1976) Stability of microstructure in metallic system. Cambridge University Press, p 178
54. Beere W., Speight M.V. (1978) Met Sci 12:172
55. Betekhtin V.I., Petrov A.I., Burenkov Y.A., Kartashov A.M., Azhimuratov U.N., Kadomtsev A.G., SKlenichka V. (1989) Fiz Met Metalloved 67:564
56. White S.R., Sottos N.R., Moore J., Geubelle P., Kessler M., Brown E., Suresh S., Viswanathan S. (2001) Nature 409:794
57. Kessler M.R., Sottos N.R., White S.R. (2003) Composites 34:743

Advances in Transmission Electron Microscopy: Self Healing or is Prevention better than Cure?

Jeff Th.M. De Hosson and Hiroyuki Y. Yasuda

Department of Applied Physics, the Netherlands Institute for Metals Research and Materials science Center, University of Groningen, Nijenborgh 4, the Netherlands
E-mail: j.t.m.de.hosson@rug.nl
Research center for Ultra-High Voltage Electron Microscopy, Osaka University, 7–1, Mihogaoka, Ibaraki, Osaka, 567-0047 Japan
E-mail: hyyasuda@mat.eng.osaka-u.ac.jp

Abstract In the field of transmission electron microscopy fundamental and practical reasons still remain that hamper a straightforward correlation between microscopic structural information and self healing mechanisms in materials. We argue that one should focus in particular on in situ rather than on postmortem observations of the microstructure. In this contribution this viewpoint has been exemplified with in situ TEM nanoindentation and in situ straining studies at elevated temperatures of metallic systems that are strengthened by either solid solution or antiphase boundaries in intermetallic compounds. It is concluded that recent advances in in situ transmission electron microscopy can provide new insights in the interaction between dislocations and interfaces that are relevant for self healing mechanisms in metallic systems.

1 Introduction

Undisputedly microscopy plays a predominant role in unraveling the underpinning mechanisms in autonomic healing phenomena. In general we may say that microscopy is devoted to linking microstructural observations to properties. Nevertheless, the actual coupling between the microstructure studied by microscopy on the one hand and the self healing property of a material on the other is almost elusive. The reason is that these properties are determined by the collective dynamic behavior of defects rather than by the behavior of an individual static defect. However, the situation is not hopeless and in this contribution we argue that for a more quantitative evaluation of the structure–property relationship of self healing material systems extra emphasis on in situ measurements is necessary.

There are at least two reasons that hamper a straightforward correlation between microscopic structural information and autonomic healing properties: one fundamental and one practical reason. Of course it has been realized for a long time that in the field of dislocations, disclinations, and interfaces we are facing nonlinear and nonequilibrium effects [1, 2]. The defects affected by self healing phenomena are in fact not in thermodynamic equilibrium and their behavior is very much nonlinear. This is a fundamental problem since adequate physical and mathematical bases for a sound analysis of these highly nonlinear effects do not exist. Another more practical reason why a quantitative evaluation of the structure–property relationship of self healing materials is rather difficult has to do with statistics. Metrological considerations of quantitative electron microscopy from crystalline materials put some relevant questions to the statistical significance of microscopy observations. In particular, situations where there is only a small volume fraction of defects present or a very inhomogeneous distribution statistical sampling may be a problem.

The importance of crystalline defects like dislocations lies in the fact that they are the carriers of plastic deformation in crystalline materials. The self healing properties of metallic systems may therefore be tailored by altering the extent to which dislocations can propagate or nucleate cracks. Since metals and alloys are most common in their polycrystalline form, i.e. they consist of many crystals separated by homophase (grain boundaries) or heterophase interfaces, the interaction between dislocations and planar defects is of particular interest. Grain boundaries act as obstacles to dislocation motion as conveyed through the classical Hall–Petch relation [3, 4] describing the increase in yield strength of polycrystalline metals with decreasing obstacle distance. Moreover, with the ongoing miniaturization of devices and materials, length scales have come within reach at which the deformation mechanisms may change drastically. A thorough understanding of such mechanisms is required to improve the mechanical properties of advanced materials and in particular to unravel more quantitatively autonomic healing mechanisms in materials.

As stated before, a major drawback of experimental and theoretical research in the field of crystalline defects is that most of the microscopy work has been concentrated on static structures. Obviously, the dynamics of moving dislocations are more relevant to the deformation mechanisms in metals. To this end we have developed nuclear spin relaxation methods in the past as a complementary tool to TEM for studying dislocation dynamics in metals [5]. A strong advantage of this technique is that it detects dislocation motion in the bulk of the material, as opposed to in situ transmission electron microscopy, where the behavior of dislocations may be affected by image forces due to the proximity of free surfaces. However, information about the local response of dislocations to an applied stress cannot be obtained by the nuclear spin relaxation technique and therefore in situ transmission electron microscopy remains a valuable tool in the study of dynamical properties of defects. Direct observation of dislocation behavior during indentation has recently become possible through in situ nanoindentation in a transmission electron microscope.

To make this contribution consistent and more attractive to study we have chosen to concentrate on the dynamic effects of pile-ups of dislocations interacting with planar defects, homophase as well as heterophase interfaces. The reason is that coplanar dislocation pile-ups may lead to cracks, which should be healed or even better should be prevented from nucleating. To exemplify the advantages and drawbacks of in situ TEM in relation to the interaction between dislocations and interfaces results are shown of in situ TEM nanoindentations and in situ TEM straining experiments at various temperatures.

The observations are discussed in relation to dynamical response of metallic systems and the influence of dislocations and interfaces on the formation of cracks. The objective of this contribution is not to address various autonomic healing mechanisms in metals but rather to discuss the various recent advances in in situ TEM techniques that can be helpful in attaining a more quantitative understanding of the dynamics of dislocations, interfaces, and cracks. In case of intermetallic compounds some basic ideas about reversible motion of dislocations are presented and discussed. With respect to self healing phenomena, we asked ourselves: is prevention not better than cure?

2   In situ TEM Nanoindentation

The observation of plastic deformation introduced by conventional nanoindentation has been restricted for a long time to postmortem studies of the deformed material, mostly by atomic force microscopy or scanning or transmission electron microscopy. This postmortem approach entails some significant limitations to the analysis of the deformation mechanisms. Most importantly, it does not allow for direct observation of the microstructure during indentation and thus lacks the possibility to monitor deformation events and the evolution of dislocation structures as the indentation proceeds. Moreover, the deformed microstructure observed after indentation is generally different from that of the material under load, due to recovery during and after unloading. In the case of postmortem analysis by transmission electron microscopy, the preparation of the indented surface in the form of a thin foil often leads to mechanical damage to the specimen or relaxation of the stored deformation due to the proximity of free surfaces, thereby further obscuring the indentation-induced deformation.

The recently developed technique of in situ nanoindentation in a transmission electron microscope [6–11], does not suffer from these limitations and allows for direct observation of indentation phenomena. Furthermore, as the indenter can be positioned on the specimen accurately by guidance of the TEM, regions of interest such as particular crystal orientations or grain boundaries can be specifically selected for indentation. In situ nanoindentation measurements [11] on polycrystalline aluminum films have provided experimental evidence that grain boundary motion is an important deformation mechanism when indenting thin films with a grain size of several hundreds of nanometers. This is a remarkable observation, since stress-induced grain boundary motion is not commonly observed at room temperature in this range of grain sizes.

Grain boundary motion in metals typically occurs at elevated temperatures driven by a free energy gradient across the boundary, which may be presented by the curvature of the boundary or stored deformation energy on either side of the boundary [12].

In the presence of an externally applied shear stress, it was found [13] that migration of both low-angle and high-angle grain boundaries in pure Al occurs at temperatures above 200°C. This type of stress-induced grain boundary motion (known as dynamic grain growth) is considered by many researchers to be the mechanism responsible for the extended elongations obtained in superplastic deformation of fine-grained materials. The occurrence of grain boundary motion in room temperature deformation of nanocrystalline fcc metals was anticipated recently by molecular dynamics simulations [14] and a simple bubble raft model [15]. Experimental observations of such grain boundary motion have subsequently been provided by in situ straining experiments of nanocrystalline Ni thin films [16] and in situ nanoindentation of nanocrystalline Al thin films [17]. In both the simulations and the experiments, grain boundary motion was observed for grain sizes below 20 nm. The dislocation mobility is greatly restricted at such grain sizes and other deformation mechanisms become more relevant. In contrast, the grain size for which grain boundary motion was found by in situ nanoindentation [11] was of the order of 200 nm.

In simple deformation modes such as uniform tension or compression, dislocation-based plasticity is still predominant and grain boundary motion generally does not occur. In the case of nanoindentation, however, the stress field is highly inhomogeneous and consequently involves large stress gradients [18]. These stress gradients are thought to be the primary factor responsible for the observed grain boundary motion at room temperature. Since the properties of high purity metals such as pure Al are less relevant for the design of advanced materials, here we focus on the indentation behavior of Al–Mg films and the effect of Mg on the deformation mechanisms described above. To this end, in situ nanoindentation experiments have been conducted on ultrafine-grained Al and Al–Mg films with varying Mg contents [19–21]. The classification "ultrafine-grained" in this respect is used for materials having a grain size of the order of several hundreds of nanometers.

2.1 Stage Design

In situ nanoindentation inside a TEM requires a special specimen stage designed to move an indenter towards an electron-transparent specimen on the optic axis of the microscope. The first indentation holder was developed in the late 1990s by Wall and Dahmen [6, 7] for a high-voltage microscope at the National Center for Electron Microscopy (NCEM) in Berkeley, California. In the following years, several other stages were constructed at NCEM with improvements made to the control of the indenter movement and the ability to measure load and displacement. In the work described in this contribution, two of these stages were used: a homemade holder for a JEOL 200CX microscope [8], and a prototype holder for a JEOL 3010 microscope with dedicated load and displacement sensors, developed in collaboration with Hysitron (Hysitron, Minneapolis, MN, USA).

The principal design of both holders is roughly the same. The indenter tip is mounted on a piezoceramic tube as illustrated in [8]. This type of actuator allows high-precision movement of the tip in three dimensions, the indentation direction being perpendicular to the electron beam. Coarse positioning is provided by manual screw drives that move the indenter assembly against the vacuum bellows. The indenter itself is a Berkovich-type diamond tip, which is boron-doped in order to be electrically conductive in the TEM. The goniometer of the TEM provides a single tilt axis, so that suitable diffraction conditions can be set up prior to indentation.

The motion of the indenter into the specimen during indentation is controlled by the piezoceramic tube. In the holder for the JEOL 200CX, the voltage applied to the tube is controlled manually and recorded together with the TEM image. Since the compliance of the load frame is relatively high, the actual displacement of the indenter into the material depends not only on the applied voltage, but also to a certain extent on the response of the material. Consequently, this indentation mode is neither load- nor displacement controlled. In the prototype holder for the JEOL 3010 and JEOL2010F, a capacitive sensor monitors the load and displacement during indentation. The displacement signal is used as input for a feedback system that controls the voltage on the piezoceramic tube based on a proportional-integral-derivative (PID) algorithm [22]. The indentation is therefore displacement controlled and can be programmed to follow a predefined displacement profile as a function of time.

The need for a separate load and displacement sensor as implemented in the prototype holder is mainly due to the complex response of the piezo tube. If the response were fully known, the load could be calculated at any time during indentation from the displacement (which can be determined directly from the TEM image) and the characteristics of the load frame [23]. Ideally, the correlation between the applied voltage and the displacement of the piezo element is linear. However, hysteresis and saturation effects lead to significant nonlinearities. Moreover, as lateral motion is achieved by bending the tube, the state of deflection strongly affects the response in the indentation direction as well. Calibration measurements of the piezo response in vacuo at 12 points across the lateral range showed an average proportionality constant of $0.12\,\mu m/V$ with a standard deviation as large as $0.04\,\mu m/V$. Although during indentation, the deflection of the tube is approximately constant and the response becomes more reproducible, the above-mentioned hysteresis and saturation effects still complicate the measurement of the load. The implementation of a dedicated load sensor, as in the new prototype holder, is therefore essential for obtaining reliable quantitative indentation data. Section 2.3 highlights the differences between the in situ indentation load displacement and displacement-controlled holders.

2.2 Specimen Geometry and Specimen Preparation

The geometry of the specimens used for in situ nanoindentation has to comply with two basic requirements: (i) an electron-transparent area of the specimen must be accessible to the indenter in a direction perpendicular to the electron beam; and (ii) this area of the specimen must be rigid enough to support indentation without bending or breaking. A geometry that fulfills both these requirements is a wedge

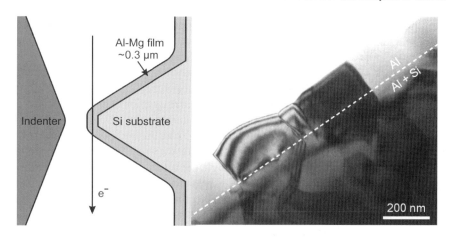

Fig. 1 (**a**) Schematic of in situ TEM indentation setup. The deposited Al–Mg film is electron-transparent and accessible to the indenter at the tip of the Si wedge. (**b**) Typical bright-field image of a deposited film. The dashed line shows the top of the Si ridge

that is truncated to a cap width large enough to provide the necessary rigidity while still allowing the electron beam to pass through. For the present investigation, wedge specimens were used as prepared by bulk silicon micromachining. Using this technique, wedge-shaped protrusions are routinely prepared on Si (001) substrates with a resolution of the order of 1 μm. The side planes of the ridge are aligned with {111} planes of the silicon crystal, so that repeated annealing and oxide removal subsequently leads to sharpening of the wedge driven by a reduction of the surface energy. The silicon ridge specimen geometry provides a means to investigate any material that can be deposited as a thin film onto the silicon substrate. Metals with a low atomic number such as aluminum are particularly suitable for this purpose, since films of these metals can be made to several hundreds of nanometers thickness and still be transparent at the cap of the wedge to electrons with typical energies of 200–300 kV, as schematically depicted in Figure 1a. An example of a resulting TEM image is shown in Figure 1b.

The Al and Al–Mg films that will be discussed were deposited by thermal evaporation. The substrate was kept at 300°C to establish a grain size of the order of the layer thickness, which was 200–300 nm for all specimens. After evaporation, the substrate heating was switched off, allowing the specimen to cool down to room temperature in approximately one hour. One pure Al film was prepared by evaporating a high purity (5N) aluminum source. Deposition of the Al–Mg alloy films was achieved by evaporating alloys with varying Mg contents. Since Al and Mg have different melting temperatures and vapor pressures, the Mg content of the deposited film is not necessarily equal to that of the evaporated material. Moreover, the actual evaporation rates depend on the quality of the vacuum and the time profile of the crucible temperature. The composition of the deposited alloy films was therefore determined by energy dispersive spectrometry (EDS) in a scanning electron microscope at 5 kV.

The measured Mg concentrations of the four Al–Mg films prepared were 1.1wt%, 1.8wt%, 2.6wt%, and 5.0wt%.

Since the solubility level of Mg in Al is 1.9wt% at room temperature [24], β′ and β precipitates were formed in the 2.6wt% and 5.0wt% Mg specimens due to the relatively long cooling time. The attainable image resolution in the indentation setup was not high enough to resolve these precipitates, being compromised by the thickness of the specimen and possibly by the fact that the electron beam travels very closely to the substrate over a large distance. Nevertheless, the presence of precipitates both in the matrix and at the grain boundaries could be confirmed by strain contrast and distorted grain boundary fringes, respectively, which were not observed in the 1.1wt% and 1.8wt% Mg specimens [20]. Furthermore, the presence of the brittle β phase on the grain boundaries leads to the appearance of intergranular cracks in the 2.6wt% and 5.0wt% Mg specimens. While Al deposited on a clean Si (001) surface may give rise to a characteristic mazed bicrystal structure due to two heteroepitaxial relationships [25], the Si substrates used in the present experiments were invariably covered with a native oxide film. Therefore, the orientations of the Al and Al–Mg grains of the film show no relation to that of the Si surface. An EBSD scan on the evaporated Al film showed a significant ⟨111⟩ texture, which can be explained by the fact that the surface energy of fcc materials has a minimum for this orientation. Furthermore, the EBSD measurements provided the distribution of the grain boundary misorientations, which showed that the grains are mostly separated by random high-angle grain boundaries with no significant preference for particular CSL orientations.

On each of the evaporated films, three to four in situ experiments were carried out with maximum depths ranging from 50 to 150 nm, using the indentation stage for the JEOL 200CX. The indentation rate, being controlled manually through the piezo voltage, was on the order of 5 nm/s. In addition, several quantitative in situ indentation experiments were conducted with the prototype holder for the JEOL 3010 microscope on the Al and Al–2.6% Mg films. These displacement-controlled indentations were made to a depth of approximately 150 nm with a loading time of 20 s. In order to be able to resolve grain boundary phenomena during each in situ indentation, the specimen was tilted to such an orientation that two adjacent grains were both in (different) two-beam conditions.

2.3   Load Control Versus Displacement Control

The onset of macroscopic plastic deformation during indentation is thought to correspond to the first deviation from elastic response in the load vs. displacement curve. For load-controlled indentation of crystalline materials, this deviation commonly has the form of a displacement burst at constant indentation load [33]. The elastic shear stress sustained prior to this excursion is often much higher than predicted by conventional yield criteria and can even attain values close to the theoretical shear strength, as was initially observed by Gane and Bowden [34]. The physical origin of the enhanced elastic loading and the subsequent displacement burst has been the subject of extensive discussions in literature [35]. While various mechanisms may be relevant in particular situations, many researchers [36–39], agree that the onset of macroscopic plastic deformation is primarily controlled by dislocation nucleation and/or multiplication, although the presence of an oxide film may significantly affect the value of the yield point.

The initial yield behavior of metals is in some cases characterized by a series of discontinuous yield events rather than a single one [33, 40, 41]. Because of the characteristic steps that result from these yield events during load-controlled indentation, this phenomenon is commonly referred to as "staircase yielding." The excursions are separated by loading portions that are predominantly elastic, and the plasticity is thus confined to the yield excursions at this stage of deformation. Staircase yielding may be explained in terms of the balance between the applied stress and the back stress on the indenter exerted by piled-up dislocations that are generated during a yield event at constant load. When the forces sum to zero, the source that generated the dislocations stops operating, and loading continues until another source is activated. This process repeats until fully plastic loading is established. Bahr et al. [39] suggested that staircase yielding occurs if the shear stress prior to the yield point is only slightly higher than the yield stress, so that upon yielding, the shear stress drops below the nucleation shear stress and further elastic loading is needed to activate the same or another dislocation source. From these viewpoints, the load at which each excursion occurs depends on the availability of dislocation sources under the indenter and on the shear stress required to nucleate dislocations from them. This accounts for the variation that is commonly observed between indentations in the number of excursions and their size.

Whereas extensive staircase yielding occurs during load-controlled indentation of pure Al thin films, it was recently found that Al–Mg thin films show essentially continuous loading behavior under otherwise identical conditions [20]. The apparent attenuation of yield excursions was attributed to solute drag on dislocations. In this contribution the influence of solute Mg on the plastic instabilities in indentation of Al–Mg is reported here in detail. It is shown that the effect of solute drag on the resolvability of discrete yield behavior depends strongly on the indentation parameters, in particular on the indentation mode, being either load- or displacement-controlled.

Conventional ex situ nanoindentation measurements were conducted both under load control and displacement control using a TriboIndenter (Hysitron, Minneapolis, MN) system equipped with a Berkovich indenter with an end radius of curvature of approximately 120 nm. Scanning probe microscopy (SPM) was used to image the surface prior to each indentation to select a target location on a smooth flat area of the specimen away from the wedge. The displacement-controlled experiments were performed at a displacement rate of 10 nm/s; the loading rate in the load-controlled experiments was 10 µN/s, which under the present circumstances corresponds to about 10 nm/s during the first tens of nanometers of loading.

In situ nanoindentation experiments under displacement control were performed in a TEM using a quantitative indentation stage, which has recently been developed and is described in [40]. The stage was equipped with a Berkovich indenter with an end radius of approximately 150 nm as measured by direct imaging in the TEM. The in situ indentations were carried out on the Al and Al–Mg films at the cap of the wedge, where the surface has a lateral width of the order of 300 nm. The displacement rate during indentation was 7.5 nm/s.

Given the significant rounding of the indenter in both types of experiments, the initial loading is well described by spherical contact up to a depth of the order of 10 nm. In Tabor's approximation, the elastoplastic strain due to spherical loading is proportional to $\sqrt{\delta/R}$, where δ is the indentation depth and R the indenter radius; the equivalent strain rate is therefore proportional to $1/\sqrt{(4\delta R)}d\delta/dt$. Using the above-mentioned values it is easily seen that at a depth of 10 nm the initial strain rates in both types of experiments compare reasonably well to one another, with values of $0.14\,\mathrm{s}^{-1}$ and $0.10\,\mathrm{s}^{-1}$ for the ex situ and in situ experiments, respectively.

The load-controlled indentation measurements show displacement bursts during loading on both the Al and the Al–Mg films, as illustrated in Figure 2. The curvature

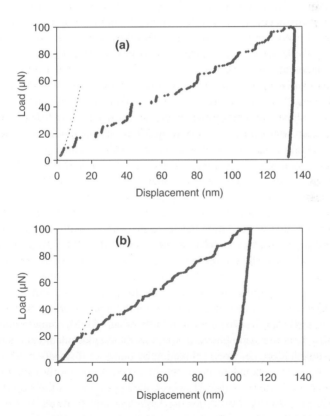

Fig. 2 Ex situ load-controlled indentation response of: (**a**) pure Al and (**b**) Al–2.6% Mg. The dashed lines represent elastic indentation by a spherical indenter with a radius of 120 nm, with the respective elastic moduli calculated from the slope of the unloading curve

of the loading portion prior to the first excursion is well described by elastic Hertzian contact, as indicated by the dashed curves. With the tip radii used, the depth over which the tip is rounded is larger than the depth over which the initial elastic behavior is expected; therefore, the expression for a spherical indenter is used. The subsequent yield behavior is classified as staircase yielding due the aforementioned dislocation-based mechanisms. Staircase yielding has been reported for indentation of both single crystal and ultrafine-grained polycrystalline Al thin films [9, 21] and therefore its occurrence is not expected to depend strongly on the presence of grain boundaries in our case.

The displacement bursts encountered in Al–Mg have a magnitude of up to 7 nm, which is substantially smaller than those observed in pure Al, being up to 15 nm in size. In fact, earlier load-controlled indentation measurements on Al–Mg [20] did not show clearly identifiable discrete yield events whatsoever, due to factors such as lower accuracy of the displacement measurement and more sluggish instrument dynamics with a resonance frequency of only 12 Hz as compared to about 125 Hz in the present instrument. In those experiments, it was observed that the attenuation of displacement bursts occurs for Mg concentrations both below and above the solubility limit in Al, from which it was inferred that the effect is due to solute Mg, which impedes the propagation of dislocation bursts through the crystal. Consequently, at a constant indentation load and for a given amount of stored elastic energy, fewer dislocations can be pushed through the solute atmosphere of Al–Mg than through a pure Al crystal, which accounts for the observed difference in size of the yield excursions. Comparison of Figures 2a and 2b further reveals that the loading portions in between consecutive yield events in Al–Mg show significant plastic behavior, whereas in Al they are well described by elastic loading, which at these higher indentation depths is attested by a slope that is intermediate between spherical [42] and Berkovich [43] elastic contact. The plasticity observed in Al–Mg can be explained in terms of the solute pinning of dislocations that were already nucleated during the preceding yield excursion. As the load increases further, some of the available dislocations are able to overcome the force associated with solute pinning, thereby allowing plastic relaxation to proceed more smoothly. Since dislocation motion is less collective than in pure Al, the measured loading response has a more continuous appearance.

When carried out under displacement control, the ex situ indentations show a much more evident effect of the solute drag on the initial yielding behavior, as illustrated in Figure 3. The loading curves of both Al and Al–Mg show pronounced load drops, which have the same physical origin as the displacement excursions in load-controlled indentation, i.e. stress relaxation by bursts of dislocation activity. Also in this case, the loading behavior up to the first load drop closely follows the elastic response under spherical contact. However, the appearance of the load drops is very different: in pure Al, the load drops are large and mostly result in loss of contact, while in Al–Mg they are smaller and more frequent, and contact is maintained during the entire loading segment. The forward surges occurring with each load drop are a result of the finite bandwidth of the feedback system. In the case of pure Al,

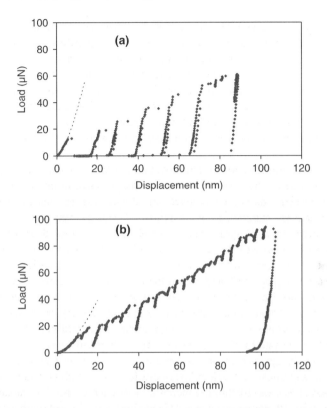

Fig. 3 Ex situ displacement-controlled indentation response of: (**a**) pure Al and (**b**) Al–2.6% Mg

the observations of complete load relaxation and loss of contact are indicative of the stored elastic energy being fully released in the forward surge before the feedback system is able to reduce the load. In Al–Mg however, solute pinning strongly reduces the dislocation velocity, which enables the feedback system to respond fast enough to maintain elastic contact. Thus, not all of the stored elastic energy is inputted back into the specimen.

The comparison between the load-controlled and displacement-controlled experiments shows that discrete yield events are far more resolvable under displacement control. This is particularly the case for solid solution strengthened alloys, in which the somewhat ragged appearance of the loading portions in between clearly identifiable strain bursts in the load-controlled data is clarified as a series of small but easily distinguishable discrete yield events in the displacement-controlled data. This enhanced sensitivity may be rationalized as follows. When the critical shear stress for a dislocation source under the indenter is reached under load control, a discernable strain burst results only if the source is able to generate many dislocations at constant load, i.e. the load-displacement curve must shift from a positive slope to an extended range of zero slope for the slope change to be readily detected. This again is possible only if the newly nucleated dislocations can freely propagate through the lattice, as in pure Al. Under displacement control however, provided that the feedback bandwidth is sufficiently high, the system may respond to the decrease in contact stiffness when only a few dislocations are nucleated, causing a distinct shift from a positive to a steeply negative slope in the load-displacement curve. Therefore, a detectable load drop can occur without collective propagation of many dislocations and as such may easily be observed even under solute drag conditions. This result cautions against using only load-controlled indentation to determine whether yielding proceeds continuously.

The in situ TEM indentations on both Al and Al–Mg show a considerable amount of dislocation activity prior to the first macroscopic yield point. This is a remarkable observation, as the initial contact would typically be interpreted as purely elastic from the measured loading response. The observations of incipient plasticity are illustrated in Figure 4 by the TEM images and load-displacement data recorded during an in situ displacement-controlled indentation on Al–Mg. While the indented grain is free of dislocations at the onset of loading (Figure 4a), the first dislocations are already nucleated within the first few nanometers of the indentation (Figure 4b), i.e. well before the apparent initial yield point that would be inferred from the load vs. displacement data only. At the inception of the first macroscopic yield event, dislocations are present throughout the entire grain (Figure 4c). The yield event itself is associated with a rearrangement of these dislocations, which significantly changes the appearance of the dislocation structure (Figure 4d). However, the number of newly nucleated dislocations between (c) and (d) is relatively small, as also becomes clear from the limited increase in indentation depth (3 nm corresponds to approximately 10 Burgers vectors). This supports our perception that only a small number of dislocations need to be nucleated in order for a yield event to be detected under displacement control, although the first dislocations nucleated between (a) and (b) do not provide an obvious signature in the load-displacement curve. In the case of in situ displacement-controlled indentation of pure Al [10], the onset of dislocation nucleation/propagation coincides with a barely detectable yet unambiguous load drop that occurs well before the initial macroscopic yield event, which is further evidence of more collective dislocation motion in Al in comparison to Al–Mg. The in situ observations of Al–Mg furthermore provide a self-consistency check for the dynamics of a yield event. With solute drag preventing full load relaxation, the size of a forward surge Δh is essentially determined by the dislocation velocity v and the mechanical bandwidth of the transducer f. Therefore, ignoring the drag exerted by the feedback system, the dislocation velocity may to a first approximation be estimated as $v \sim \Delta h \cdot f$, which, using $\Delta h = 7\,\mathrm{nm}$ and $f = 125\,\mathrm{Hz}$, yields a velocity of the order of $1\,\mu\mathrm{m/s}$. This is of the same order as observed in situ for the initial dislocations in Figure 4b, which traversed the 300 nm film thickness in about 130 ms (four video frames at a frame rate of 30 frames per second, see video at http://www.dehosson.fmns.rug.nl/).

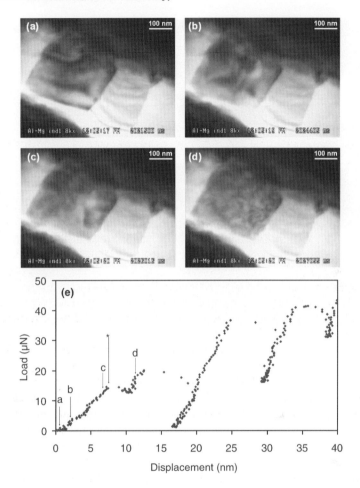

Fig. 4 TEM bright-field Image sequence (**a–d**) from the initial loading portion; (**e**) of the indentation on Al–2.6% Mg depicted in Figure 9b. The first dislocations are nucleated between (**a**) and (**b**), i.e. prior to the apparent yield point. The nucleation is evidenced by an abrupt change in image contrast: before nucleation, only thickness fringes can be seen, whereas more complex contrast features become visible at the instant of nucleation. (See http://www.dehosson.fmns.rug.nl/.)

2.4   *Dislocation and Grain Boundary Motion*

The effect of Mg on the propagation of dislocations is particularly visible during the early stages of loading. While in pure Al the dislocations instantly spread across the entire grain (i.e. faster than the 30 frames per second video sampling rate), they advance more slowly and in a jerky type fashion in all observed Al–Mg alloys. Figure 5 shows a sequence of images from an indentation in Al–2.6% Mg. The arrows mark the consecutive positions where the leading dislocation line is pinned by solutes. From these images, the mean jump distance between obstacles is estimated

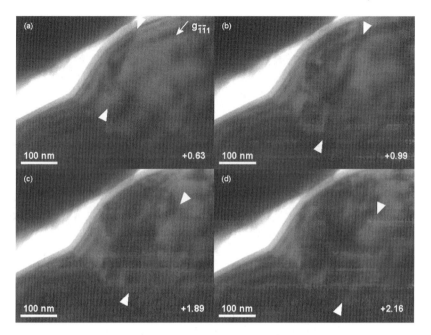

Fig. 5 Series of bright-field images showing jerky motion of dislocations during indentation of Al–2.6wt% Mg. The time from the start of the indentation is given in seconds. Note the presence of a native oxide layer on the surface

to be of the order of 50 nm. Due to the single-tilt axis limitation of the indentation stage, the orientation of the slip plane relative to the electron beam is unknown; therefore, the measured jump distance is a projection and a lower bound of the actual jump distance.

At the low strains for which jerky-type dislocation motion is observed, solute atoms are the predominant barriers to mobile dislocations, as has been shown in earlier in situ pulsed nuclear magnetic resonance (NMR) experiments [5, 26–28]. Consequently, the mean jump distance can be predicted by Mott–Nabarro's model of weakly interacting diffuse forces between Mg solutes and dislocations in Al [27]. A calculation of the effective obstacle spacing, assuming that the maximum internal stress around a solute atom has a logarithmic concentration dependence, yields a value of 30 nm in Al–2.6% Mg. This is in fair agreement with our experimental observation of a mean jump distance of the order of 50 nm.

Besides solute atoms, (semi-) coherent β'/β precipitates in Al–Mg alloys can also provide significant barriers to dislocation motion. As aforementioned, the mean spacing of these precipitates could not be measured very accurately due to the limited resolution of the microscope combined with the specific indentation stage. However, we can make an estimate based on the solid solubility of Mg in Al at room temperature of 1.9wt%. The calculated volume fraction f_V is 2.4% for the β phase at 300K. The mean planar separation, which is a relevant measure for the interaction of a gliding dislocation with a random array of obstacles in its slip plane, is given by [30, 31]

$$\lambda \cong \frac{2\sqrt{2}\pi r}{3 f_V} \tag{1}$$

provided that the size of the particles r is negligible in comparison with their center-to-center separation, i.e. if $\lambda >> r$. It is reasonable to assume that the minimum size of the semicoherent precipitates is at least 10 nm to produce sufficient strain contrast. As a result, the mean planar separation of the precipitates is calculated to be at least 92 nm, i.e. larger than the mean separation between the solutes. In this approach, the obstacles are assumed to be spherical and consequently, we ignore the effect that the precipitation in Al may become discontinuous or continuous depending on the temperature. However, even in the case of a Widmanstätten structure, the effective separation between the needle-shaped precipitates is larger than the effective solute obstacle spacing [32]. Therefore, based on the experimental observations in the alloys below and above the solid solubility of Mg, the strain contrast and the abovementioned theoretical considerations, solute atoms are assigned as the main obstacles to dislocation motion.

To confirm the occurrence of grain boundary movement in aluminum as had been reported earlier [11], several in situ indentations were performed near grain boundaries in the pure Al film. Indeed, significant grain boundary movement was observed for both low- and high-angle boundaries. It should be emphasized that the observed grain boundary motion is not simply a displacement of the boundary together with the indented material as a whole; the boundary actually moves through the crystal lattice and the volume of the indented grain changes accordingly at the expense of the volume of neighboring grains. The trends observed throughout the indentations suggest that grain boundary motion becomes more pronounced with decreasing grain size and decreasing distance from the indenter to the boundary. Moreover, grain boundary motion occurs less frequently as the end radius of the indenter increases due to tip blunting or contamination. Both these observations are consistent with the view that the motion of grain boundaries is promoted by high local stress gradients as put forward in the introduction of this paper. The direction of grain boundary movement can be both away from and towards the indenter, and small grains may even completely disappear under indentation [17]. Presumably, the grain boundary parameters play an important role in the mobility of an individual boundary, since the coupling of the indenter-induced stress with the grain boundary strain field depends strongly on the particular structure of the boundary.

The quantitative in situ indentation technique offers the possibility to directly relate the observed grain boundary motion to features in the load-displacement curve. While this relation has not been thoroughly studied in the present investigation, preliminary results suggest that the grain boundary motion is associated with softening in the loading response. Softening can physically be accounted for by the stress relaxation that occurs upon grain boundary motion. However, the quantification of overall mechanical behavior is complicated by the frequent load drops at this stage of indentation, and further in situ indentation experiments are needed to investigate

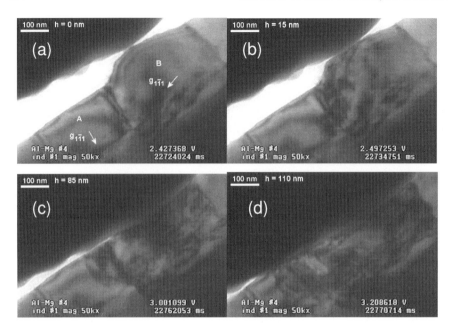

Fig. 6 Series of bright-field images from an indentation on Al–1.8% Mg. No movement of the high-angle grain boundaries is observed

this phenomenon more systematically and quantitatively. The movement of grain boundaries as observed in Al was never found for high-angle boundaries in any of the Al–Mg specimens, even when indented to a depth greater than half of the film thickness (see Figure 6). Our observations as such indicate a significant pinning effect of Mg on high-angle grain boundaries in these alloys. In contrast to high-angle grain boundaries, the mobility of low-angle boundaries in Al–Mg was found to be less affected by the presence of Mg.

The Al–Mg films investigated include compositions both below and above the solubility limit of Mg in Al. However, no differences in indentation behavior between the solid solution and the precipitated microstructures were observed. Consequently, the observed pinning of high-angle boundaries in Al–Mg is attributed to solute Mg. The pinning is presumably due to a change in grain boundary structure or strain fields caused by solute Mg atoms on the grain boundaries.

Relatively few direct experimental observations have been reported of this type of interaction. Sass and coworkers observed that the addition of Au and Sb impurities to bcc Fe changes the dislocation structure of $\langle 100 \rangle$ twist boundaries of both low-angle [44] and high-angle [45] misorientation. Rittner and Seidman [46] calculated solute distributions at $\langle 110 \rangle$ symmetric tilt boundaries with different boundary structures in an fcc binary alloy using atomistic simulations. However, the influence of solutes on the structure of such boundaries has not been experimentally identified.

The fact that low-angle grain boundaries were found to be mobile regardless of the Mg content can be explained by their different boundary structure. Up to a misorientation of 10–15°, low-angle boundaries can be described as a periodic array of edge and screw dislocations by Frank's rule [47]. In such an arrangement, the strain fields of the dislocations are approximated well by individual isolated dislocations and their interaction with an external stress field can be calculated accordingly. Since there is no significant interaction between the individual grain boundary dislocations, the stress required to move a low-angle boundary is much lower than for a high-angle boundary. Low-angle pure tilt boundaries consisting entirely of parallel edge dislocations are fully glissile and therefore particularly mobile. In general, a combination of glide and climb is required to move a low-angle boundary [48]. As a corollary, the structural difference between low and high-angle boundaries also affects the extent of solute segregation. Because solutes generally segregate more strongly to high-angle boundaries [49], the observed difference in mobility may partly be a compositional effect. As aforementioned grain boundary motion is regarded as an essential mechanism in superplastic behavior of metals and that will be the topic of the following section, focusing on in-situ TEM straining-heating experiments.

3   In situ TEM Straining and Heating: Superplasticity

Although initial experimental observations of superplasticity in metals date back to the 1920s, for a long time the phenomenon was mainly regarded as a laboratory curiosity [49]. However, research interests in superplasticity greatly increased in the 1960s [51, 52], when it was demonstrated that in this regime metal sheets could easily be formed to complex shapes. Research efforts are increasingly being directed towards new classes of superplastic materials, some of which exhibit superplastic behavior at considerably higher forming rates [53, 54]. This so called high-strain-rate superplasticity is expected to receive broad industrial interest and may replace existing forming techniques if such materials can efficiently be produced on a large scale [55].

The hallmark of superplastic deformation is a low flow stress σ that shows a high strain rate sensitivity m. A high-strain-rate sensitivity is necessary to stabilize the plastic flow so as to avoid necking during tensile deformation. The incipient formation of a neck leads to a local increase of the strain rate, which, in the case of a positive strain rate sensitivity, leads to an increase of the flow stress in the necked region. If the strain rate sensitivity is sufficiently high, the local flow stress increases to such an extent that further development of the neck is inhibited. Most common metals show a strain-rate sensitivity exponent lower than 0.2, whereas values around 0.3 or higher are needed to delay necking long enough to produce the strains characteristic of superplasticity. Besides a high-strain-rate sensitivity, a low rate of damage accumulation (e.g. cavitation) is required to allow large plastic strains to be reached.

For coarse grained Al–Mg alloys deformed in the viscous glide regime, values for the maximum strain in excess of 300% can be obtained [56, 57, 58]. Such elongations are close to those found in conventional superplasticity of fine-grained Al–Mg alloys and are sufficient for many practical applications. Moreover, forming by viscous-glide controlled creep has two important advantages over conventional superplastic forming: (i) the rate of viscous glide is not restricted by dislocation climb and consequently higher strain rates can be achieved, and (ii) since viscous glide is independent of grain size, the preparation of the materials is less complex. It should be noted that since the deformation under viscous-glide control does not follow the original definition of superplasticity in the strictest sense, the deformation behavior has also been referred to as "enhanced ductility" or "quasi-superplasticity" by some researchers [56, 57, 59]. There are two competing mechanisms in the viscous-glide regime, dislocation glide and climb; the slower of the two is rate controlling. A physical interpretation of the empirically found three-power-law relation for viscous-glide creep is readily based on the Orowan equation, relating the macroscopic strain rate $\dot{\varepsilon}$ to the mobile dislocation density ρ_m and the average dislocation velocity $\bar{v}_d$. Although no direct measurements of the relation between applied stress and dislocation velocity under solute-drag conditions are available, most models suggest that $\bar{v} \propto \sigma$ in this regime, i.e. the stress exponent $n_v = 1$ [2, 60]. Furthermore, experimental observations have shown that the stress exponent n_d of the mobile dislocation density $\rho_m \propto \sigma^{n_d}$ lies between 1.6 and 1.8 for Al–Mg alloys [2, 60, 62]. This is in reasonable agreement with theoretical predictions [63] suggesting that $\rho_m \propto \sigma^2$ [64]. The strain rate sensitivity index depends critically on the stress dependence of the product $\rho(\sigma)\bar{v}_d(\sigma)$. Assuming $n_v \approx 1$ and $n_d \approx 2$, it follows from the Orowan equation that $\dot{\varepsilon} \propto \sigma^3$, as also is found in the model by Weertman [63]. It follows that the stress exponent $n \approx 3$ and hence, the strain rate sensitivity $m = 1/n \approx 0.33$. From the abovementioned considerations it is clear that the three-power law ($m = 0.33$) is no more than an approximate relationship arising from the stress dependence of the dislocation density and the drag stress.

The alloys presented here are two coarse grained Al–4.4% Mg and Al–4.4% Mg–0.4% Cu alloys with minor amounts of Ti, Mn, and Cr ($<0.1\%$) and an average initial grain size of 70 µm, and a fine grained Al–4.7% Mg–0.7% Mn alloy (AA5083) with an average grain size of 10 µm. (see also [65, 66]). In order to directly observe the evolution of dislocation structures during superplastic deformation, the coarse-grained Al–Mg alloys were subjected to in–situ TEM tensile experiments at elevated temperature in a TEM. Such experiments require a specimen stage that is capable of straining TEM specimens while maintaining a controllable temperature of the order of 400°C. At present, only one type of stage with combined heating and straining capability is commercially available (Gatan, Pleasanton, CA). The design of this stage relies on direct physical contact between a heating element and the specimen to control the specimen temperature. The temperature of the specimen is tacitly assumed to be equal to the furnace temperature as measured by a thermocouple. This is approximately valid at high temperatures ($\sim 1,000°C$) when the specimen is mostly heated by radiation. However, at the intermediate temperatures used in this study, radiation is negligible, and the specimen temperature can only reach the furnace temperature if the thermal contact between the two is very good. The requirement that the specimen be movable for tensile testing results in poor thermal contact; moreover, the degree of contact fluctuates during the course of a tensile experiment.

This was confirmed by calibration measurements in low vacuum on TEM tensile specimens with a thermocouple spot-welded close to the electron-transparent area. Applying a thermally conductive paste between the heating element and the specimen greatly improved the performance of the holder. However, such viscous agents are not suitable for high vacuum systems such as TEMs.

A few homemade heating–straining stages have been developed over the last two decades [67–70], most of which use a filament to heat the specimen by radiation [67, 68]. The temperature is measured by a thermocouple that is positioned as close as possible to the observed area of the specimen. Since the specimen is heated exclusively by radiation, the thermocouple attains approximately the same temperature as the specimen and therefore provides a very accurate temperature measurement. The high temperature in situ experiments reported in this paper were partially conducted at the Institut National Polytechnique in Grenoble, France, using the double tilt heating straining holder described in reference [68]. Calibration experiments have shown the measured temperature of this holder to be accurate to within 10°C [71]. In the intermediate temperature range ($\sim$150°C), in situ tensile experiments were also performed using the Gatan heating straining holder described above.

Figure 7 shows three micrographs representative of the microstructural evolution observed during superplastic forming of the coarse grained Al–Mg alloy. At a strain of a few percent, just beyond the yield point, random configurations of dislocations are visible (Figure 7a). This stage of deformation is characterized by a drop of the flow stress [72], which indicates dislocation multiplication from an initially low dislocation density pinned by Mg solutes [55]. During further straining, subgrain formation occurs primarily along the original grain boundaries, as in Figure 7b showing subgrain boundaries near a high-angle boundary triple junction. At this stage, the substructure shows many incomplete subgrain boundaries, i.e. boundaries with a very low misorientation ($<1°$) that do not fully enclose a subgrain. Only when a strain of the order of 1 is attained, the subgrains completely fill the grain interior. Figure 7c shows the refined subgrain structure at a strain of 170% and an average subgrain size of approximately 5 μm. Note that the size distribution is fairly broad, with observed subgrain sizes ranging from 1 to 10 μm. [72–75]

The effect of the Mg content on the tensile ductility is twofold. On the one hand, a higher Mg content increases the extent of solute drag, thereby stabilizing the plastic flow. However, beyond a few percent Mg, the effect on the strain-rate sensitivity becomes fairly marginal [57]. On the other hand, the presence of Mg significantly reduces dynamic recovery as evidenced by the slow formation of subgrains. As a result, Mg concentrations above 5% can easily give rise to dynamic recrystallization within a certain domain of temperature and strain rate, which in the absence of grain refining second phase particles leads to rapid coarsening of the microstructure. The currently used composition with 4.4% Mg appears to be a good balance between solute drag and dynamic recovery, leading to enhanced tensile ductility in excess of 300%. In torsional deformation, where a high strain rate sensitivity to avoid necking is less important, the ductility benefits most from dynamic recovery (leading to geometric dynamic recrystallization at high strains) and is consequently higher for pure Al than for Al–Mg alloys [76].

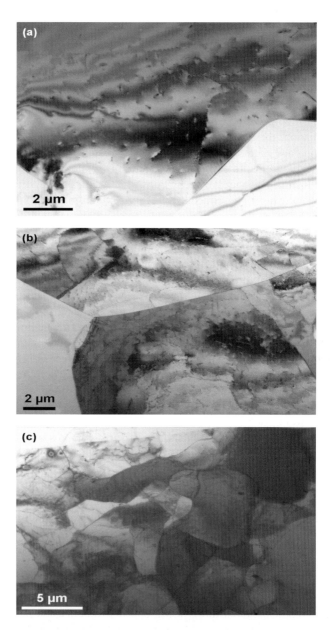

Fig. 7 Dislocation substructure in Al–Mg deformed at 440°C and $5 \cdot 10^{-3}$ s^{-1} to (**a**) 4%; (**b**) 20%; and (**c**) 170%

The initially inhomogeneous formation of subgrains gives rise to a "core and mantle" microstructure, in which most deformation is concentrated along the grain boundaries. In fine-grained superplasticity, this type of microstructure has been associated with grain mantle deformation processes as an accommodating mechanism for grain boundary sliding [73, 74, 77]. In the present case, dynamic recovery is initially confined to the mantle region, but extends throughout the microstructure at higher strains. The evolution of the dislocation substructure during superplastic deformation can be directly observed by in situ tensile experiments in a TEM. A difficulty inherently associated with this technique is presented by the image forces resulting from the proximity of free surfaces, which may significantly influence the dislocation dynamics compared to bulk behavior (e.g. [69]). In the present investigation we have attempted to minimize such effects by preparing tensile TEM specimens from macroscopically prestrained alloys and studying only the initial motion of dislocations from their starting configuration. However, for the present case of Al–Mg alloys, it turns out that surface diffusion of Mg severely limits the temperature range at which the in–situ experiments can be conducted.

At temperatures in excess of 200°C, the tensile TEM specimens were consistently found to fracture intergranularly at very low loads (typically ~30 gf). This is evidently not representative of the bulk behavior at high temperature showing very high tensile ductility. Below 200°C, the specimens showed ductile transgranular failure at loads of the order of 300 gf. By EDS and electron diffraction analysis it was found that the intergranular fracture areas of the TEM specimens deformed at high temperature contained large amounts of Mg and MgO. Presumably, surface diffusion of Mg becomes appreciable at high temperature and leads to segregation of Mg to the grain boundaries and consequently to grain boundary embrittlement.

In the temperature range below 200°C, dislocation climb is not activated and therefore extensive dynamic recovery is not to be expected. However, even at low temperature, some rearrangements of the substructure were observed that may be illustrative of those occurring during dynamic recovery. The absorption of dislocations by a subgrain boundary is shown in Figure 8; this process contributes to the increase in grain boundary misorientation that is associated with dynamic recovery. In other

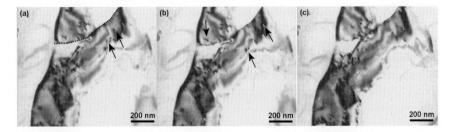

Fig. 8 Absorption of dislocations into a subgrain boundary in Al–Mg during in situ straining at ~150°C. The subgrain boundary is marked by a dotted line in (**a**). The arrows in (**a**) and (**b**) indicate the dislocations that are absorbed by the boundary in the next image respectively

words: although dislocation motion is solely due to glide at low temperatures, the observed rearrangements resemble the processes that contribute to dynamic recovery at high temperature.

4   In situ TEM Straining: Pseudo-Elasticity

Since dislocations are the basic carriers of plasticity in crystalline materials and fracture is due to crack propagation, one may also expect a relationship between cracks and dislocations. Since dislocations represent discontinuities in displacement, they can also be used, at least in a mathematical sense, to describe a macroscopic static crack and its dynamic behavior. At the same time dislocations facilitate crack opening and relaxation effects and altogether dislocations can be considered as the basic building block of a crack. The idea that a crack can be thought of as an array of discrete coplanar and parallel dislocations was worked out mathematically by Eshelby et al. [78]. Leibfried [79] suggested a continuum approximation and instead of dealing with a coplanar row of discrete dislocations, it was proposed that the crack plane contains a dislocation density smeared out over the crack plane [80]. The stress field of a crack running from $-a$ to a can then simply be related to the stress fields of individual dislocations and via the dislocation distribution $B(x_1)$ on the crack plane $(x_2 = 0)$ by integration

$$\sigma_{ij}^{\text{crack}}(x_1, x_2) = \int_{-a}^{a} B(x_1')\sigma_{ij}^{\text{dis}}(x_1 - x_1', x_2)\mathrm{d}x_1' \tag{2}$$

However, close to the crack tip these stresses are above the yield stress of the material. As a consequence plastic flow due to dislocations around the crack-tip region continues until the stress singularity is removed, either by the stresses of the dislocations created during plastic flow, or by blunting of the crack tip due to the emission of dislocations. Mathematically, the pile-up of dislocations produces an infinite stress at the crack tip that can be avoided by abandoning linear continuum mechanics in the tip region [81]. In terms of dislocation theory, the nonlinearity can be dealt with quite simply by allowing some of the leading dislocations in the pile-up to leak away forward into the material ahead of the crack [82].

  Besides the conversion of a pile-up of climb and glide edge dislocations into a crack (mode I and mode II, respectively) another feature that can occur is that a grain boundary represented by a wall of stacked edge dislocations is forced apart by an effective shear stress resulting in a crack (see Fig. 9 [83]). This situation of crack nucleation by dislocation pile-ups was analyzed theoretically by Frank and Stroh [84, 85, 86] both for isotropic and anisotropic media. With respect to the topic of autonomic healing, i.e. healing of existing cracks that may propagate, one may ask what should one do for preventing crack formation in the first place. Would not

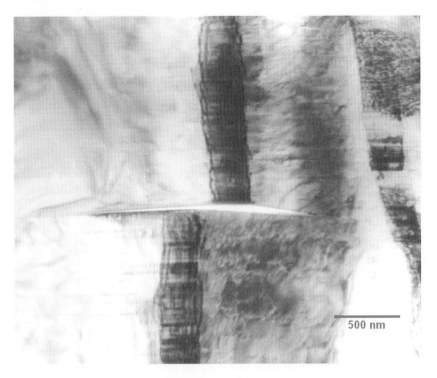

Fig. 9 Stroh crack: a tilt grain boundary is forced apart resulting in a crack. TEM image of a kink band viewed along the $\langle 11\bar{2}0 \rangle$ direction in Ti_3SiC_2

prevention better than cure? The answer for ordered intermetallic compounds under cyclic loading is rather simple. Based on the above considerations about similarities between cracks and dislocations and with reference to sections 2 and 3, an interesting strategy is to avoid extensive dislocation pile-up formation and to make sure that the dislocation slip path is fully reversible so that cracks will not nucleate. In a sense the macroscopic mechanical behavior will look fully elastic although it is based on plasticity. Therefore the phenomenon can be coined pseudo-elasticity, although in literature pseudo-elasticity is linked to a thermoelastic transformation. Here we mean "plasticity induced pseudo-elasticity" [87–90].

Clearly, what is needed for this pseudo-elasticity phenomenon is a low friction stress and a high back stress on dislocations. In pure metals and solid solution strengthened metals pseudo-elasticity (see sections 2 and 3) will not occur because the friction stress τ_0 is rather high in bcc metals (about 80% of the flow stress) and a considerable back stress is not available. In ordered compounds the situation is completely different. Because of the translational invariance of the Burgers vector the dislocations move as superlattice dislocations, i.e. as superpartials that are bound by an antiphase boundary energy. The latter provides a back stress τ_{APB} on the

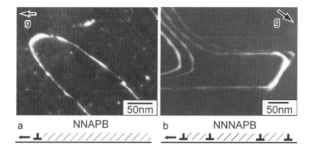

Fig. 10 Dislocation structure in Fe–Al-ordered single crystals (**a**) Fe–23 atom % Al deformed to $\varepsilon_p = 5.0\%$. Superpartials are pulled back by γ NN APB; (**b**) Fe–28 atom % Al deformed to $\varepsilon_p = 0.5\%$. No pseudo-elasticity because of absence of a back stress

leading superpartial dislocation. For the loading τ_l and unloading τ_u situation the balance of stresses reads:

$$\tau_\ell = \tau_0 + \tau_{APB} \tag{3}$$

$$\tau_u = \tau_{APB} - \tau_0 \tag{4}$$

If all the superpartials constituting the superlattice dislocations move, $\tau_\ell = \tau_0$ and no pseudo-elasticity will be observed (see Figure 10). Therefore, pinning of the leading partial is necessary by an obstacle that is strong enough to prevent shearing by the full superlattice dislocation. Ordered compounds provide these obstacles via ordered domains. For example, D0$_3$ ordered compounds like Fe$_3$Al contains B2 domains. The unit superdislocation in the case of D0$_3$ consists of four $1/4\langle 111 \rangle$ partial dislocations. In the D0$_3$ structure this type of superlattice dislocation creates two types of shear APBs. The domains are bound by the inner two superpartial dislocations that are flanked by two $1/4\langle 111 \rangle$ APBs on both sides. In the latter case the corresponding APB energy or rather the APB drag tension is formulated as:

$$\gamma_{NN} = \frac{2\sqrt{2}}{a_0^2}\left\{ 4V_1 S_1^2 + V_2\left(S_2^2 - 4S_1^2\right)\right\} \tag{5}$$

whereas the former APB tension on the superpartial is written as:

$$\gamma_{NNN} = \frac{2\sqrt{2}}{a_0^2}\left\{ 2V_2 S_2^2\right\} \tag{6}$$

Here: S_1 is the first-neighboring order parameter, S_2 is the second-neighboring order parameter; V_1 represents the first-neighboring energy; V_2 is the second-neighboring energy, a_0 is the lattice parameter and b represents the magnitude of the Burgers vector. Obviously S_i depends on the chemical composition and temperature but one may infer that the back stress τ_{APB} on the leading partial is larger than on the second partial dislocation, i.e. causing a better recovery ($\gamma_{NN} > \gamma_{NNN}$). If the leading $1/4\langle 111 \rangle$ crosses a B2 domain boundary a step is created with a high energy that effectively

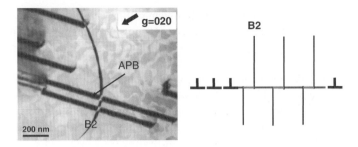

Fig. 11 TEM observation and schematic representation of the intersection between a leading $1/4\langle111\rangle$ superpartial dislocation in $D0_3$ Fe–Al system and a B2-type domain creating a high-APB energy step

will block the passage of the trailing three superpartials in its wake (see Figure 11). Upon unloading the superpartials are pulled back by the high APB drag tension and a full recovery will be achieved. Indeed pseudo-elasticity was observed in Fe$_3$Al single crystals with the $D0_3$ crystallographic structure, but interestingly without a martensitic transformation [87–90]. Also Fe$_3$Ga single crystals with the $D0_3$ structure showed pseudo-elasticity without a martensitic transformation [88]. Recently we have observed emitting dislocations from a crack tip with in situ TEM straining and indeed the fully reverse motion upon unloading provides evidence for the theoretical explanation of pseudo-elasticity, i.e. not based on martensitic transformations but on the motion of partial dislocations comprising superlattice dislocations.

Nevertheless, the details of this mechanism of "prevention is better than cure" approach are not clear. There are several interesting questions about the proposed mechanism. Many dislocations ($\sim10^{13}$ m^{-2}) have to move to achieve a considerable plastic strain. How do they interact and what is the mean jump distance of the individual or grouped superlattice dislocations? If the dislocations cross-slip, can they move backwards during unloading or are low-energy dislocation structures generated? Pseudo-elastic behavior depends strongly on loading axis and grain boundary engineering is needed for polycrystalline material. Further, it is important whether shear deformation occurred in the twinning or anti-twinning sense, since the $D0_3$ structure is based on bcc.

5   Conclusions and Outlook

This contribution highlights recent advances in transmission electron microscopy, in particular concentrating on in situ TEM experiments. The objective of this contribution is not to address various self healing mechanisms in metals but rather to discuss the various recent advances in in situ TEM techniques that can be helpful in attaining a more quantitative understanding of the dynamics of dislocations, interfaces, and cracks. We have reviewed the various possibilities of in situ TEM indentation, straining, and heating. The recently developed techniques of in situ TEM allows for direct observation of indentation-induced dynamical processes, reversible motion of dislocations in ordered intermetallic compounds and crack blunting phenomena. The in situ TEM indentation, in situ TEM straining, and heating experiments are summarized as follows:

The appearance of discrete yield events during nanoindentation of metallic systems depends on the ability of dislocations to propagate into the crystal, and is therefore substantially affected by solute pinning. Under load control, the characteristic yield excursions commonly observed in pure metals are strongly attenuated by solute pinning, leading to a more continuous loading response. Under displacement control, pure metals mostly exhibit full load relaxation during discrete yielding, but in alloys, solutes impede dislocation motion and thereby prevent the load from relaxing completely upon reaching a plastic instability. Yield events are resolved more clearly under displacement control, particularly in the presence of solute drag, since displacement-controlled indentation does not require collective dislocation motion to the extent required by load-controlled indentation in order to resolve a yield event. This perception is confirmed by in situ TEM displacement-controlled indentations, which show that many dislocations are nucleated *prior* to the initial macroscopic yield point and that the macroscopic yield event is associated with the *rearrangements* of the dislocations.

During indentation of the ultrafine-grained Al film, extensive movement of both low- and high-angle grain boundaries is observed. The occurrence of this deformation mechanism, which under uniform stress conditions is restricted to nanocrystalline materials, is attributed to the high-stress gradients involved in sharp indentation. In contrast to the observations in pure Al, no such movement of high-angle grain boundaries is found in any of the Al alloy films.

In coarse-grained Al–Mg alloys, superplastic deformation is accomplished by viscous glide of dislocations. The viscous glide is accompanied by dynamic reconstruction of the microstructure, the appearance of which depends on the deformation parameters and can be both detrimental and beneficial to the ductility. During dynamic recovery, grain refinement occurs by the formation of subgrain boundaries and low-angle grain boundaries. An advantage of plasticity based on viscous glide is that this mechanism has virtually no grain size dependence and therefore the preparation of such materials is less complex compared with the traditional fine-grained superplastic materials.

Intermetallic DO_3-based compounds show fully reversible dislocation motion in in-situ TEM straining experiments, i.e. plasticity induced pseudo-elasticity without a martensitic transformation. Emitting dislocations were observed from a crack tip providing support for the theoretical explanation that is based the motion of partial dislocations comprising superlattice dislocations.

In summary, new insights in the interaction between dislocations and planar defects such as grain boundaries in solution strengthened alloys and heterophase interfaces in intermetallic compounds have been achieved. It is concluded that with recent advances in in situ transmission electron microscopy the nucleation and propagation of cracks due to dislocation pile-ups can be scrutinized in self healing metallic materials.

Acknowledgements The authors are grateful to the support and collaboration with Wouter Soer (Philips Research), Yukichi Umakoshi (Osaka), Andy M. Minor, Eric A Stach, all from LBL Berkeley USA, with Steven Shan, SA Syed Asif, Oden L Warren, all from Hysitron, USA, and with Béatrice Doisneau-Cottignies, Grenoble, France.

References

1. Nabarro FRN (1967) Theory of crystal dislocations. Oxford University Press, Oxford
2. Hirth JP, Lothe J (1968) Theory of dislocations. McGraw-Hill, New York
3. Hall EO (1951) Proc Phys Soc Lond B 64:747
4. Petch NJ (1953) J Iron Steel Inst 174:25
5. De Hosson JTM, Kanert O, Sleeswyk AW (1983) In: Nabarro FRN (ed) Dislocations in solids. North-Holland, Amsterdam, pp 441–534
6. Wall MA, Dahmen U (1997) Microsc Microanal 3:593
7. Wall MA, Dahmen U (1998) Microsc Res Tech 42:248
8. Stach EA, Freeman T, Minor AM, Owen DK, Cumings J, Wall MA, Chraska T, Hull R, Morris Jr JW, Zettl A, Dahmen U (2001) Microsc Microanal 7:507
9. Minor AM, Morris Jr JW, Stach EA (2001) Appl Phys Lett 79:1625
10. Minor AM, Lilleodden ET, Stach EA, Morris Jr JW (2002) J Electron Mater 31:958
11. Minor AM, Lilleodden ET, Stach EA, Morris Jr JW (2004) J Mater Res 19:176
12. Doherty RD, Hughes DA, Humphreys FJ, Jonas JJ, Juul Jensen D, Kassner ME, King WE, McNelley R, McQueen HJ, Rollett AD (1997) Mater Sci Eng A 238:219
13. Winning M, Gottstein G, Shvindlerman LS (2001) Mater Sci Eng A 317:17
14. Van Swygenhoven H, Caro A, Farkas D (2001) Mater Sci Eng A 309–310:440
15. Van Vliet KJ, Tsikata S, Suresh S (2003) Appl Phys Lett 83:1441
16. Shan Z, Stach EA, Wiezorek JMK, Knapp JA, Follstaedt DM, Mao SX (2004) Science 305:654
17. Jin M, Minor AM, Stach EA, Morris Jr JW (2004) Acta Mater 52:5381
18. Larsson PL, Giannakopoulos AE, Söderlund E, Rowcliffe DJ, Vestergaard R (1996) Int J Solids Struct 33:221
19. Soer WA, De Hosson JTM, Minor AM, Stach EA, Morris Jr JW (2004) Mater Res Soc 795: U9 3 1
20. Soer WA, De Hosson JTM, Minor AM, Morris Jr JW, Stach EA (2004) Acta Mater 52:5783
21. De Hosson JTM, Soer WA, Minor AM, Shan Z, Stach EA, Syed Asif SA, Warren OL (2006) J Mat Sci 41:7704
22. Warren OL, Downs SA, Wyrobek TJ (2004) Z Metall 95:287
23. Minor AM (2002) Ph.D. thesis, University of California, Berkeley
24. Mondolfo LF (1979) Aluminum alloys: structure and properties. Butterworth, London, p 313
25. Dahmen U, Westmacott KH (1988) Scripta Metall 22:1673
26. Schlagowski U, Kanert O, De Hosson JTM, Boom G (1988) Acta Metall 36:865
27. De Hosson JTM, Kanert O, Schlagowski U, Boom G (1988) J Mater Res 3:645
28. Detemple K, Kanert O, De Hosson JTM, Murty KL (1995) Phys Review 52 B:125
29. Nabarro FRN (1975) in: Hirsch PB (ed) The Physics of metals, vol 2. Cambridge University Press, Cambridge, p 152
30. Foreman AJE, Makin MJ (1966) Philos Mag 14:911
31. Kocks UF (1966) Philos Mag 14:1629
32. De Hosson JTM, Alsem WHM, Tamler H, Kanert O (1983) In: Sih GC, Provan JW (eds) Defects, fracture and fatigue. Martinus Nijhoff, The Hague, p 23
33. De Hosson JTM, Soer WA, Minor AM, Shan Z, Syed Asif SA, Warren OL (2006) Microsc Microanal 12(S02):890
34. Gane N, Bowden FP (1968) J Appl Phys 39:1432

35. Kramer DE, Yoder KB, Gerberich WW (2001) Philos Mag A 81:2033
36. Gouldstone A, Koh HJ, Zeng KY, Giannakopoulos AE, Suresh S (2000)Acta Mater 48:2277
37. Gerberich WW, Venkataraman SK, Huang H, Harvey SE, Kohlstedt DL (1995) Acta Metall Mater 43:1569
38. Gerberich WW, Nelson JC, Lilleodden ET, Anderson P, Wyrobek JT (1996) Acta Mater 44:3585
39. Bahr DF, Kramer DE, Gerberich WW (1998) Acta Mater 46:3605
40. Minor AM, Shan Z, Stach EA, Syed Asif SA, Cyrankowski E, Wyrobek T, Warren OL (2006) Nat Mater 5:697
41. Soer WA, De Hosson JTM, Minor AM, Shan Z, Syed Asif SA, Warren OL (2007) Applied Phys Lett (submitted)
42. Johnson KL (1985) Contact mechanics. Cambridge University Press, Cambridge
43. Larsson PL, Giannakopoulos AE, Söderlund E, Rowcliffe FJ, Vestergaard R (1996) Int J Solids Struct 33:221
44. Sickafus K, Sass SL Scripta Metall 18:165
45. Lin CH, Sass SL Scripta Metall 22:735
46. Rittner JD, Seidman DN (1997) Acta Mater 45:3191
47. Frank FC (1950) Plastic deformation of crystalline solids office of naval research. Washington, DC, p 151
48. Read WT (1953) Dislocations in crystals. McGraw-Hill, New York
49. Sutton AP, Balluffi RW (1995) Interfaces in crystalline solids. Clarendon Press, Oxford
50. Grimes R (2003) Mater Sci Technol 19:3
51. Underwood EE (1962) J Metals 14:914
52. Backofen WA, Turner IR, Avery DH (1964) Trans Am Soc Metals 57:980
53. Chokshi AH, Mukherjee AK, Langdon TG (1993) Mater Sci Eng R 10:237
54. Higashi K, Mabuchi M, Langdon TG (1996) ISIJ Int 36:1423
55. E Usui, T Inaba, N Shinano (1986) Z Metall 77:179
56. Taleff EM, Lesuer DR, Wadsworth J (1996) Metall Mater Trans A 27:343
57. Taleff EM, Henshall GA, Nieh TG, Lesuer DR, Wadsworth J (1998) Metall Mater Trans A 29:1081
58. Yoshida H, Tanaka H, Takiguchi K, European Patent No 0846781
59. Woo SS, Kim YR, Shin DH, Kim WJ (1997) Scripta Mater 37:1351
60. Friedel J (1967) Dislocations. Pergamon press, Oxford.
61. Horiuchi R, Otsuka M (1972) Trans Jap Inst Metals 13:284
62. Oikawa H, Matsuno N, Karashima S (1975) Met Sci J 9:209
63. Weertman J (1975) J Appl Phys 28:1185
64. Mills MJ, Gibeling JC, Nix WD (1985) Acta Metall 33:1503
65. Henshall GA, Kassner ME, McQueen HJ (1992) Metall Trans A 23:881
66. Drury MR, Humphreys FJ (1986) Acta Metall 34:2259
67. Kubin LP, Veyssière P(1982) Electron Microsc 531
68. Pelissier J, Debrenne P (1993) Microsc Microanal Microstruct 4:111
69. Couret A, Crestou J, Farenc S, Molenat G, Clement N, Coujou A, Caillard D (1993) Microsc Microanal Microstruct 4:153
70. Messerschmidt U, Bartsch M (1994) Ultramicroscopy 56:163
71. Doisneau-Cottignies B (2005) private communication
72. Chezan AR, De Hosson JTM (2005) Mater Sci Forum 405–497:883
73. Chezan AR, De Hosson JTM (2006) Mater Sci Eng A410–411:120
74. Soer WA, Chezan AR, De Hosson JTM (2006) Acta Mater 54:3827
75. McQueen HJ, Knustad O, Ryum N, Solberg JK (1985)Scripta Metall 19:73
76. Gourdet S, Montheillet F (2000) Mater Sci Eng A 283:274
77. Gifkins RC (1976) Metall Trans A 7:1225
78. Eshelby JD, Frank FC, Nabarro FRN (1951) Phil Mag 42:351
79. Leibfried G (1951) Z Phys 130:214

80. Weertman J (1996) Dislocation based fracture mechanics. World Scientific, Singapore
81. Barenblatt GI (1962) Adv Appl Mech 7:55
82. Bilby BA, Cottrell AH, Swinden K (1963) Proc Roy Soc A272:304
83. Kooi BJ, Poppen RJ, Carvalho NJM, De Hosson JTM, Barsoum MW (2003) Acta Materialia 51:2859
84. Frank FC and Stroh AN (1952) Proc Phys Soc 65:811
85. Stroh AN (1954) Proc Roy Soc 223A:404
86. Stroh AN (1958) Philos Mag 3:597
87. Yasuda HY, Nakano K, Nakajima T, Ueda M, Umakoshi Y (2003) Acta Materialia 51:5101
88. Yasuda HY, Aoki M, Takaoka A, Umakoshi Y (2005) Scripta Materialia 53:253
89. Yasuda HY, Nakajima T, Nakano K, Yamaoka K, Ueda M, Umakoshi Y (2005) Acta Materialia 53:5343
90. Yasuda HY, Nakajima T, Murakami S, Ueda M, Umakoshi Y (2006) Intermetallics 14:1221

Self Healing in Coatings at High Temperatures

Wim G. Sloof

Delft University of Technology, Department of Materials Science and Engineering, Mekelweg 2,
The Netherlands
E-mail: w.g.sloof@tudelft.nl

1   Introduction

Alloys for high temperature applications in an oxidizing environment depend on the formation of a protective and slow growing oxide scale. The composition of these alloys is such that a continuous layer of a thermodynamically stable oxide is formed through selective oxidation of one of the constituting elements. Then, the oxide layer forms a barrier between the environment and the underlying alloy. The alloys for high temperature applications can be divided into alumina (Al_2O_3), silica (SiO_2), or chromia (Cr_2O_3) formers, such as stainless steels, superalloys (Reed 2006), and intermetallics (MX, where M is Ti, Fe, Co or Ni, and X denotes Al, Si, or Cr). These materials are successfully applied in for example gas turbine engines (aero, marine, and industrial), heating equipment and automotive converters etc. In this chapter, the focus will be on alumina forming alloys encountered as coating material for blades and vanes in gas turbine engines. However, the principles addressed also apply to the other mentioned classes of high temperature alloys.

The increasing demand for higher efficiency of turbine engines, thereby saving fuel and reducing CO_2 emission, and simultaneously enhancing the lifetime of components, has been strong incentives for the development of protective coatings on high-temperature alloys (Nicholls 2003). Improvement of the turbine engine efficiency by ever-higher operating temperatures has been achieved by novel material design, improved cooling technologies and better manufacturing methods. The development of nickel-based superalloys, where the turbine blades and vanes are made of, has similarly progressed from wrought alloys, to cast alloys, to single-crystal alloys (Reed 2006). The improvement of high temperature strength and creep resistance of the superalloys has been obtained at the expense of its high-temperature oxidation resistance. As a result, diffusion or overlay coatings are applied to protect the load bearing components against high temperature oxidation and hot-corrosion, i.e. attack of sulphur from molten salt deposits in the range of 650–950°C (Saunders and Nicholls 1996).

The diffusion aluminide coatings are produced by various techniques, such as pack cementation and chemical vapour deposition (CVD) (Tamarin 2002). Through diffusion of Al into the component surface and subsequent heat treatment a coating of β-NiAl (bcc) is formed. Under severe hot-corrosion conditions, or at temperatures above 1,050°C, aluminide coatings offer only limited protection. To overcome this problem, platinum-modified aluminide diffusion coatings were developed. Prior to alumizing the component, a few microns thick Pt layer is electrodeposited onto the surface. The Pt in the coating promotes the selective oxidation of aluminium and this results in a purer, slower-growing alumina scale. Diffusion coatings, by nature of their formation, imply a strong interdependence on substrate composition in determining both their corrosion resistance and mechanical properties. Hence, deposition of a coating with a good balance between oxidation, corrosion resistance, and mechanical properties has led to the development MCrAlY overlay coatings (Gleeson 2000), where M denotes Ni and/or Co. These coatings are deposited on the superalloy components by vacuum plasma spraying, electroplating or electron beam physical vapour deposition (EB-PVD) (Tamarin 2002). The MCrAlY coatings contain typically 18–22 wt% Cr, Al 8–12 wt% Al, and about 0.5 wt% Y as a reactive element to promote oxide-scale adhesion. This coating alloy consists of an Al-rich β-NiAl type (bcc) phase and an Al-poor γ-Ni type (fcc) solid solution phase with NiY-rich precipitates.

At the surface of both the aluminide diffusion and MCrAlY overlay coatings, a protective alumina scale will be formed in an oxidizing environment at high temperatures; see Figure 1. An alumina oxide scale compared with other oxide scales is stable at high temperatures. For example, chromia and silica become volatile at very high temperatures, above 1,000°C and 1,500°C respectively. Furthermore, an alumina scale is slow growing due to the very low diffusion of both aluminium and oxygen (Schütze 2000).

High compressive stresses (up to 5 GPa) develop in the oxide scale upon cooling from the operation temperature to ambient temperature due to the mismatch between the coefficients of thermal expansion of the aluminium oxide and the underlying coating and substrate (Evans et al. 2001). These stresses are the driving force for delamination of the oxide scale. Depending on the oxide thickness and adhesion, parts of the oxide-scale flake off; see Figure 2. Upon reheating to the operation temperature the oxide-scale repairs itself by selective oxidation. The coating serves as a reservoir of aluminium. The lifetime of the coating under thermal cyclic oxidation is set by the fracture and delamination behaviour of the oxide scale and the depletion of aluminium from the coating. Ultimately, the coating does not offer protection to

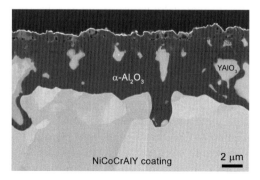

Fig. 1 A protective α-Al$_2$O$_3$ oxide layer thermally grown at 1,100°C on a NiCoCrAlY overlay coating. An aluminium-depleted zone occurs in the overlay coating due to oxidation

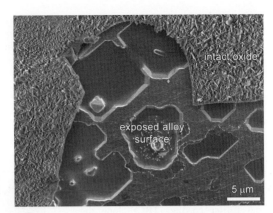

Fig. 2 An α-Al₂O₃ oxide layer on top of a β-NiCrAl doped with Hf damaged due to spallation

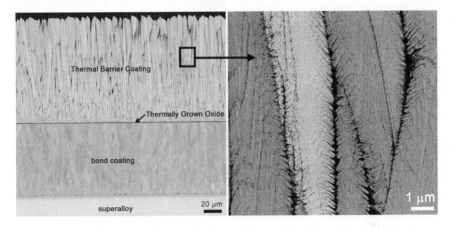

Fig. 3 A thermal barrier high-temperature coating system. A thin thermally grown oxide (TGO) layer is present between the strain tolerant EB-PVD thermal barrier coating (TBC) (ZrO₂ with 7 wt% Y₂O₃) and a NiCoCrAlY bond coating (BC)

the underlying component when the Al reservoir in the coating is exhausted. Then, self-repair by forming alumina cannot occur anymore.

In modern high-temperature coatings systems, a thermal barrier coating (TBC) is applied as a top coating on a diffusion or overlay coating to lower metal surface temperatures. In combination with internal cooling of the component, a temperature gradient of 100–150°C can be realized across the thickness of the TBC (Peters et al. 2001). The current TBC's are made of yttria stabilized zirconia (ZrO₂ with 6–8 wt% Y₂O₃) and deposited by plasma spraying or EB-PVD (Clarke and Levi 2003). The advantage of an EB-PVD over a plasma sprayed TBC is its high-strain compliance due to the columnar microstructure; see Figure 3. However, EB-PVD coatings have a higher thermal conductivity (about 1.5 W/m K) compared with plasma sprayed coatings (about 1 W/m K). In the high-temperature TBC coating system, the diffusion

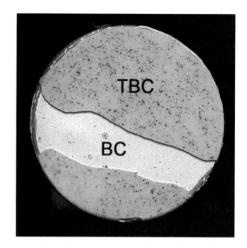

Fig. 4 Spallation of an EB-PVD thermal barrier coating (TBC) along the bond coating (BC)/thermally grown oxide (TGO) interface (button diameter 10 mm)

or overlay coating, often referred to as bond coating (BC), provides the protection against high-temperature oxidation. In this case, the oxide layer that forms between the BC and the TBC is referred to as the thermally grown oxide (TGO) layer.

Failure of the high-temperature TBC coating systems is confined mainly to the surrounding of the TGO layer, although cracks initiated at the TGO layer can run vertically through the TBC and reach its free surface (Evans et al. 2001). The brittle TBC experiences cracks that run predominantly parallel to the TGO layer. Coalescence of these cracks and buckling of the TGO layer due to the large compressive stresses that develop upon cooling eventually generate spallation. The lifespan of the coating system critically depends on the BC oxidation, i.e. the ultimate TGO thickness. Other factors that affecting the lifetime include the strength of the bond coat, changes in bond-coat microstructures and chemistry, the surface roughness of the component, and possible sintering of the ceramic that can modify the overall system compliance. EB-PVD coatings tend to fail at the TGO/BC interface (see Figure 4), unless substantial defects like voids in the TGO (Nijdam et al. 2006) or rumpling of the BC surface occur (Evans et al. 2001). In contrast with the aluminide diffusion or MCrAlY overlay coatings, the TBC however, does not show any self healing capacity despite being one of the most critical parts of the coating system for the lifetime of a coated component.

2   Current Design Routes to Get Better Properties

The protection offered against high-temperature oxidation and hot corrosion by diffusion or overlay coatings relies on forming a thermodynamically stable oxide scale at the surface. The developed oxide must have a low growth rate, be uniform and closed, and experience a good adhesion with the underlying alloy (see section 1).

This can be realized by exclusive formation and growth of an α-Al_2O_3 layer at high temperatures (say above 900°C).

In order to understand the design strategies for coatings at high temperatures, lets consider the strain energy associated with the high thermal mismatch stresses developed in the oxide layer, which are the driving force for delamination at the oxide/coating interface (see section 1). The amount of strain energy per unit of interface area G for a biaxial state of stress equals:

$$G = \frac{d_{ox}\sigma_{ox}^2}{E_{ox}(1 - \nu_{ox}^2)} \tag{1}$$

where d_{ox} is the thickness, E_{ox} the Young's modules, and ν_{ox} the Poisson's ratio of the oxide scale, and σ_{ox} denotes the stress in the oxide layer. Hence, the driving force for delamination is proportional to the oxide-scale thickness. Generally, the growth of the oxide scale at high temperatures follows a parabolic growth rate law:

$$d_{ox}^2 = k_p t \tag{2}$$

where k_p is the parabolic growth rate constant and t the oxidation time. Since the parabolic growth rate constant is related to the diffusion of the metal and oxygen, a low value of their diffusion coefficients or a small number of fast diffusion paths (e.g. grain boundaries) will result in a slow growing oxide scale.

The buckle-driven delamination of the oxide scale is resisted by the fracture toughness of the interface Γ_i. Failure may occur if the strain energy released is larger than the interface toughness, i.e. if $G > \Gamma_i$. However, only some of the strain energy in the oxide sale is transmitted into the interface (Wright and Evans 1999). Nevertheless, the interface toughness is a key factor for the lifetime of the high-temperature protective coating. Besides the chemical bonding between the oxide and metallic coating also mechanical keying, bifurcation of cracks, and possible plastic deformation contribute to the interface toughness.

Although not all the underlying mechanisms that determine the properties and behaviour of the coatings for high temperatures are yet fully understood, the main strategies of alloy design and treatments to improve and optimize these coatings are summarized in the following (details can be found in the references).

To promote the exclusive formation of an α-Al_2O_3 surface layer usually Cr is added to the alumina forming alloy, as e.g. is the case with the MCrAlY coating alloy. The addition of Cr reduces the oxygen solubility of the alloy and thereby preventing internal oxidation of Al. This has the advantage that less Al is required in the alloy to form a protective to α-Al_2O_3 scale, which is beneficial in view of the mechanical properties and melting point of the alloy. A similar effect can be expected from third or higher order alloying elements that reduce the oxygen solubility such as Si and Ti (Brady et al. 2000).

Addition of a small amount of reactive element (up to 0.5 wt%) like Yttrium improves the scale adhesion enormously (Pint 2003). It is generally recognized that the main effect of the reactive elements is scavenging of impurities like sulphur that are inevitably present in the coating after deposition (Bennett et al. 2005). Segregation of S to the coating/oxide interface causes a significant decrease in scale adhesion. Experiments with alloys where S was removed by a hydrogen heat treatment displayed a similar effect as adding a reactive element (Smialek 2000). Thus, elements with a strong interaction with S are candidate reactive elements; examples of such elements are: Sr, Y, Zr, La, and Hf (Bennett et al. 2005). However, for a reactive element to be effective, it also must be homogenously distributed and have some solubility in the alloy. Further, a reactive element may not reduce the adhesion when segregated to the coating oxide interface (e.g. K, Na, and Ca, which have also a high affinity to S). Besides scavenging of impurities, reactive elements can significantly promote the formation of α-Al_2O_3, reduce the oxide growth rate, suppress formation of voids at the interface, and create pegs.

Precious metal additions to the alumina forming alloy, such as Pt, Pd, Ru, and Re are also known to improve the oxidation resistance of high-temperature coatings (Leyens et al. 2003), as demonstrated with the Pt-modified diffusion coatings. The Pt in the coating promotes the selective oxidation of aluminium and this results in a purer, slower growing alumina scale. Further, Pt improves scale adhesion due to reduced growth of interfacial voids, interact with indigenous sulphur, and limit the outward diffusion of minor alloying elements from the superalloy substrate.

Refinement of the coating alloy microstructure by surface melting using a powerful laser or electron beam can improve significantly the oxidation resistance at high temperatures (Nijdam et al. 2006). The surface melting and subsequent fast solidification of a multiphase MCrAlY alloy results in a fine-grained microstructure. In addition, the reactive element is more homogeneously distributed. Hence, the diffusion of Al towards the interface with the oxide scale is faster and the efficiency of the reactive element is enhanced.

A pre-annealing and pre-oxidation treatment of a BC can be very beneficial to enhance the lifetime of a high-temperature coating system with a TBC (Nijdam and Sloof 2006). A short annealing treatment in a vacuum furnace after deposition of the coating causes segregation of the reactive element, which during subsequent oxidation may lead to the formation of pegs. These pegs may hinder crack growth and so increase the resistance for fracture along the oxide layer/coating interface. A controlled oxidation treatment with low oxygen partial can promote the exclusive formation of α-Al_2O_3 during the initial, relatively fast, stage of oxidation. Then, the formation of less protective oxides (e.g. NiO, Cr_2O_3, or spinel oxides) will be suppressed. In addition, a course grained α-Al_2O_3 oxide layer develops which eventually leads to a lower the growth rate of the oxide layer during high-temperature oxidation.

3 Self Healing Concept

Oxygen from the environment reacts with an element (in our case aluminium) from a high-temperature coating alloy to form a protective oxide scale. This is a self healing process, as is manifested most clearly when the metal component experiences

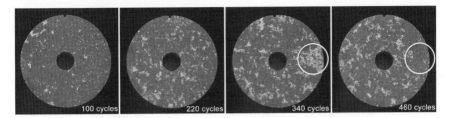

Fig. 5 Damage and healing of protective oxide scale on a β-NiCrAl doped with Y high-temperature coating alloy in a thermal cycle test. The bright spots are damaged areas. By comparing the macrographs for different numbers of cycles, it is evident that new oxide has formed at the damaged areas. The thermal cycle test was carried out with buttons of the alloy with a diameter of 10 mm in air with temperature cycles of 9 min heating to 1, 163°C, a high-temperature oxidation dwell time of 45 min, and subsequently 10 min cooling in ambient air

a series of thermal cycles; see Figure 5. Then, due to thermal mismatch stresses damage by fragmentarily spallation of the oxide layer occurs upon cooling. Upon subsequent high-temperature exposure, the damaged areas of the oxide layer on the coating will be repaired. However, the continuous repairing of the oxide layer slowly consumes the reservoir of the preferred oxide-forming element in the coating. Consequently, the self healing capacity of the coating is reduced. Once the self healing effect ends, less-protective oxides form internally in the coating and underlying substrate, thereby deteriorating the properties of the entire component. Therefore, it is vital to find strategies that optimize the self healing capacity of the coating and thereby prolong the lifespan of the metal component.

A useful method to quantify the self healing behaviour of a coating alloy is by simply monitoring the weight change in a thermal cycle test; see Figure 6. In a thermal cycle test, a specimen of the coating alloy experiences a series of temperature cycles in an oxidizing environment. Each cycle comprises heating to the oxidation temperature, a period at the oxidation temperature and cooling to room temperature.

The experimentally determined net mass change curves as function of cycle number are the sum of the mass *gain* due to healing at the high temperature dwell of a thermal cycle and the mass *loss* due to damage upon cooling; see Figure 7.

To extract the amount of healing and damage during each cycle from the experimental data, a thermal cycling oxidation and spallation model can be used. From the various models available, the model developed by 2004 was adopted. In this model, the sample surface of the alloy is envisioned to be divided into n_0 segments of equal area $F_A = 1/n_0$. For each cycle j, it is assumed that during the high temperature dwell time Δt, each segment in the oxide scale grows according to parabolic growth kinetics (a valid assumption if the scale consists of only α-Al$_2$O$_3$). Furthermore, it is assumed that during each cycle, spalling of one of the thickest segments of the scale occurs at the oxide/metal interface upon cooling. The model is schematically represented in Figure 8 and displays both oxide growth at high temperature and spallation

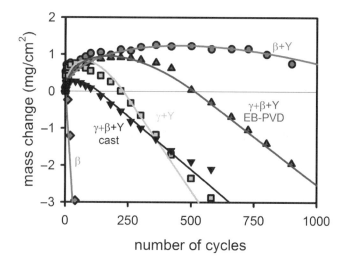

Fig. 6 Weight change per unit of area as a function of the cycle number in a thermal cycle test of different alloys for high temperature coatings (cf. Table 1); one undoped β-phase alloy, the other alloys were doped with 0.2 atom % Y. The thermal cycle test was carried out with buttons of the alloys with a diameter of 10 mm in air with temperature cycles of 9 min heating to 1, 163°C, a high-temperature oxidation dwell time of 45 min, and subsequently 10 min cooling in ambient air. The solid lines represent the fitting of the oxidation and spallation model (Equation (5)) to the data points

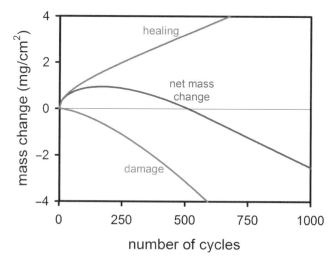

Fig. 7 Illustration of the course of the net mass change (blue line) as function of number of cycles upon thermal cycle oxidation testing calculated using a thermal cycle oxidation and interface spalling model (from Smialek 2004). The net mass change is the sum of the mass gain due to oxide healing (red line) and the mass loss due to oxide spallation (magenta line)

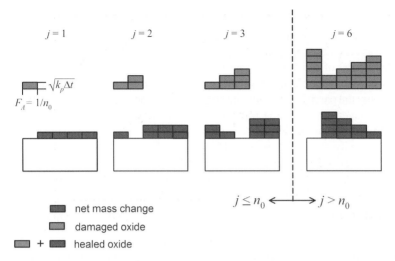

Fig. 8 Illustration of the thermal cycle oxidation and interface spalling model (from Smialek 2004). The mass gain due to oxide healing (blue and magenta), the mass loss due to oxide damage (magenta), and the net mass gain on the alloy surface (blue) for cycle numbers $j = 1, 2, 3$, and 6 are displayed. For simplicity, only five oxide segments ($n_0 = 5$) are presented, with k_p the parabolic oxide growth rate constant and F_A the area fraction of spallation

upon cooling for $n_0 = 5$ and $j = 1, 2, 3$, and 6. Following these model assumptions, the total amount of healing H as function of the cycle number can be described as:

$$H = \begin{cases} F_A\sqrt{k_p t}\left(n_0 j^{1/2} + \frac{1}{3}j^{3/2}\right) & j \le n_0 \\ F_A\sqrt{k_p t}\left(j n_0^{1/2} + \frac{1}{3}n_0^{3/2}\right) & j > n_0 \end{cases} \tag{3}$$

where the parabolic growth rate constant k_p is used to characterize the oxide layer growth (cf. Equation 2).

The corresponding total amount of damage D occurring upon cooling as function of cycle number can be expressed as:

$$D = \begin{cases} -S_c F_A \sqrt{k_p t}\left(\frac{1}{2}j^{1/2} + \frac{2}{3}j^{3/2}\right) & j \le n_0 \\ -S_c F_A \sqrt{k_p t}\left(\left(j + \frac{1}{2}\right)n_0^{1/2} - \frac{1}{3}n_0^{3/2}\right) & j > n_0 \end{cases} \tag{4}$$

where S_c is the molecular weight ratio of oxide to oxygen in a given oxide ($S_c = 2.12$ for α-Al_2O_3). Finally, the net mass change per unit area ($\Delta W / A$) as function of cycle number, given as the sum of H and D, can be written as:

$$\frac{\Delta W}{A} = \begin{cases} F_A\sqrt{k_p t}\left[\frac{1}{2}(2n_0 - S_c)j^{1/2} + \frac{1}{3}(1 - 2S_c)j^{3/2}\right] & j \le n_0 \\ F_A\sqrt{k_p t}\left[\left((1 - S_c)j - \frac{1}{2}S_c\right)(n_0)^{1/2} + \frac{1}{3}(1 + S_c)(n_0)^{3/2}\right] & j > n_0 \end{cases} \tag{5}$$

In Equations (3)–(5) a different expression is used for cycle numbers j larger than n_0, since there is an additional weight loss due to spallation that occurred previously up to n_0 cycles; see also Figure 8.

Examination of Equation (5) reveals that the data in a thermal cycle experiment, i.e. net mass change versus the number of cycles, can be described with only two parameters: the parabolic growth rate constant k_p (a measure for the capacity to repair the oxide layer) and the area fraction of spallation F_A (a measure for the damage susceptibility of the oxide layer). Together, k_p and F_A are a measure for the self healing capacity of the coating alloy. Optimum self healing capacity is achieved when k_p and F_A have small values.

4 Examples

As an example, it will be demonstrated how guidelines can be found for the design of MCrAlY coatings to optimize their self healing capacity from a combination of experiments and modelling. A series of high-temperature coating alloys of different design (i.e. composition and microstructure; see Table 1) were subjected to thermal cycle testing. First, three baseline compositions in the Ni-rich portion of the Ni–Cr–Al phase diagram were selected: (i) γ Ni–27Cr–9 Al (fcc); (ii) β Ni–4.5Cr–38 Al (bcc); and (iii) $\gamma + \beta$ Ni–20Co–20Cr–24 Al (compositions in atom %). All three alloys have different Cr/Al ratio, but have sufficient Cr and Al to produce a protective α-Al$_2$O$_3$ layer on the coating surface at high temperatures. Each baseline composition was tested with and without addition of 0.2 atom % Y as reactive element. Next, for the β-phase alloy coating, the type and amount of minor element additions were optimized. Besides Y, also Zr and Hf as reactive element addition were tested, and their amount was varied between 0.05 and 0.2 atom % (Bennett 2006). Finally, for the two-phase $\gamma + \beta$ alloy coatings, the size of the β precipitates was varied by using three different alloy processing routes (Nijdam et al. 2006): (i) casting, resulting in to a coarse microstructure (β size $\sim$20 µm); (ii) laser surface melting (LSM), producing a fine microstructure (β size $\sim$3 µm); and (iii) EB-PVD, resulting in a very fine microstructure (β size $\sim$1 µm).

Table 1 Alloy compositions and microstructures used for optimizing self healing capacity of high-temperature coatings (compositions in atom %)

Base Composition	Type of Reactive-Element Addition	Quantity of Reactive-Element Addition	Processing/Microstructure
γ Ni–27Cr–9Al	–/Y	0.2%	Cast
$\gamma + \beta$ Ni–20Co–20Cr–24Al	–/Y	0.2%	Cast/LSM/EB-PVD
β Ni–4.5Cr–38Al	–/Y/Zr/Hf	0.05%/0.1%/0.2%	Cast

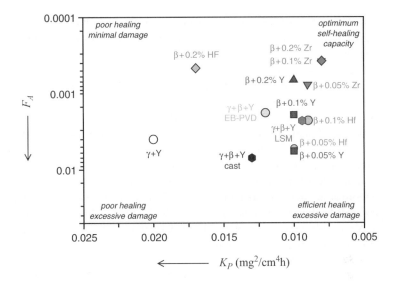

Fig. 9 Self healing capacity map of alloys for high-temperature coatings (cf. Table 1). The parabolic oxide growth rate constant k_p expresses the capacity to repair the oxide layer and the area fraction of spallation F_A expresses the damage susceptibility of the oxide layer for adhesive failure. The poor self healing undoped alloys are outside the range of this graph

The two parameters (k_p and F_A) expressing the self healing capacity of the coating alloys were determined for each coating alloy tested, by fitting the weight change curves recorded in the thermal cycle tests to Equation (5) using a linear least-square method. The results are displayed as a map in Figure 9, clearly show that the best self healing capacity is obtained with the β-phase alloy doped with Zr. The self healing capacity of the γ-phase alloy is significantly lower than that of the more brittle β-phase alloys. Comparing the γ + β phase alloys, then those with the fine microstructure outperform the alloy with the coarse microstructure. In this way, directions are found to improve the self healing capacity of high-temperature coating alloys.

5   Future Perspectives and Conclusions

The oxide-scale formation on high-temperature diffusion and overlay coatings is a classical example of self healing behaviour. In this case, the self healing is manifest, because at high temperatures the mobility of the elements involved in the self healing process are sufficiently high that no additional stimuli are necessary. There are still various opportunities to further improve the properties of these coatings or tailor them for other applications than addressed here. For example, a similar self healing concept may be utilized for metal components or coated surfaces exposed to wear or erosion. Then, the surface that is generally covered with a thin oxide film will be damaged. By re-oxidation of a suitable element added to the material and acting as a healing agent, the damage may be restored. It may be necessary to stimulate such a process with radiation or heat.

The trend to apply TBCs more and more in the hot section of a turbine engine, in order to further increase the operation temperature of gas turbine engines, diminishes the importance of the self healing aspect of the BC, although this coating remains vital for the protection of the component against high-temperature oxidation. The current TBCs, however, do not exhibit any self healing capacity to repair cracks and thereby reducing its ultimate failure by spallation. This is in particular relevant for the less strain tolerant plasma sprayed TBCs that provides better thermal isolation than the more strain tolerant EB-PVD TBCs. It is a great challenge to develop TBCs with some self healing properties, to enhance the lifetime of high-temperature TBC systems. A promising route may be to add particles that oxidize when cracks develop in the TBC and in doing so as a healing agent.

References

Bennett I.J., Kranenburg J.M., Sloof W.G. (2005) Modeling the influence of reactive elements on the work of adhesion between oxides and metals. J Am Cerem Soc 88:2209–2216

Bennett I.J., Sloof W.G. (2006) Modeling the influence of reactive elements on the work of adhesion between a thermally grown oxide and a bond coat alloy. Mater Corros 57:223–229

Brady M.P., Gleeson B., Wright I.G. (2000) Alloy design strategies for promoting protective oxide scale formation. JOM 52:16–21

Clarke D.R., Levi C.G. (2003) Materials design for the next generation thermal barrier coatings. Annu Rev Mater Res 33:383

Evans A.G., Mumm D.R., Hutchinson J.W., Meier G.H., Pettit F.S. (2001) Mechanisms controlling the durability of thermal barrier coatings. Prog Mater Sci 46:505–553

Gleeson B. (2000) High-temperature corrosion of metallic alloys and coatings. In: Cahn RW, Haasen P, Kramer EJ (eds) Materials science and technology, vol 19. Wiley-VCH, New York, pp 67–130

Leyens C., Pint B.A., and Wright I.G. (2003) Effect of composition on the oxidation and hot corrosion resistance of NiAl doped with precious metals. Surf Coat Technol 133–134:5–22

Nicholls J.R. (2003) Advances in coating design for high performance gas turbines. MRS Bull 28:659–670

Nijdam T.J., Sloof W.G. (2006) Combined pre-annealing and pre-oxidation treatment for processing of thermal barrier cotings on NiCoCrAlY bond coatings. Surf Coat Technol 201:3894–3900

Nijdam T.J., Kwakernaak C., Sloof W.G. (2006) The effects of alloy microstructure refinement on short-term thermal oxidation of NiCoCrAlY alloys. Metall Mater Trans A 37A:683–693

Nijdam T.J., Marijnissen G.H., Vergeldt E., Kloosterman A.B., Sloof W.G. (2006) Development of a pre-oxidation treatment to improve the adhesion between thermal barrier coatings and NiCoCrAlY bond coatings. Oxid Met 66:269–294

Peters M., Leyens C., Schulz U., Kayser W.A. (2001) EB-PVD thermal barrier coatings for aeroengines and gas turbines. Adv Eng Mater 3:193–203

Pint B.A. (2003) Optimization of reactive-element additions to improve oxidation resistance of alumina-forming alloys. J Am Ceram Soc 86:686–695.

Reed R.C. (2006) The superalloys fundamentals and applications. Cambridge University Press, Cambridge

Saunders S.R.J., Nicholls J.R. (1996) Oxidation, hot corrosion and protection of metallic materials. In: Cahn R.W., Haasen P. (eds) Physical metallurgy 4th edn. Elsevier, Amsterdam, pp 1291–1361

Schütze M. (2000) Fundamentals of high temperature corrosion. In: Cahn R.W., Haasen P., Kramer E.J. (eds) Materials science and technology, vol 19. Wiley-VCH, New York, pp 67–130

Smialek J.L. (2000) Maintaining adhesion of protective Al_2O_3 scales. JOM 52:22–25

Smialek J.L. (2004) Universal characteristics of an interface spalling cycle oxidation model. Acta Materialia 52:2111–2121

Tamarin Y. (2002) Protective coatings for turbine blades. ASM International, Metals Park, OH

Wright P.K., Evans A.G. (1999) Mechanisms governing the performance of thermal barrier coatings. Curr Opin Solid State Mater Sci 4:255–265

Hierarchical Structure and Repair of Bone: Deformation, Remodelling, Healing

Peter Fratzl and Richard Weinkamer

Max Planck Institute of Colloids and Interfaces, Research Campus Golm, Potsdam, Germany
E-mail: fratzl@mpikg.mpg.de

1 Introduction

The design of natural materials follows a radically different paradigm as compared to engineering materials: organs are growing rather than being fabricated. As a main consequence, adaptation to changing conditions remains possible during the whole lifetime of a biological material. As a typical example of such a biological material, bone is constantly laid down by bone forming cells, osteoblasts, and removed by bone resorbing cells, osteoclasts. With this remodelling cycle of bone resorption and formation, the skeleton is able to adapt to changing needs at all levels of structural hierarchy. The hierarchical structure of bone is summarized in the second part of this chapter.

Inspiration for the design of artificial materials with self-repair properties may be gained at three levels. First, bone is a composite of mineral particles and organic matrix which are arranged in such a way, that the stiff mineral carries most of the load while the majority of the deformation takes place in the organic phase. Reversible bonds, which are breaking and reforming under load, provide a means of damage repair at the molecular level. This is discussed in the third part of this chapter as a first approach towards self-repair at the (supra-)molecular level.

Second, bone remodels permanently, removing damaged bone and replacing it by new material. The general assumption is that cells entrapped in the bone tissue (and called osteocytes) act as strain sensors and signal large deformations to the cells responsible for resorbing old and for forming new bone. In this way, material which has been damaged (and is less able to carry loads) will be removed and replaced by new material. This is a second level of self-repair where local damage is sensed by a dedicated sensor (the osteocyte) and material modification occurs by actuators (osteoblasts and osteoclasts) in response to this. Bone remodelling is discussed in the fourth part of this chapter.

Finally, bone healing is as third phenomenon (discussed in the last part of this chapter) and quite different from remodelling. Healing occurs in natural systems when a serious damage cannot be recovered anymore by the normal adaptive feedback loop. It is a complex repair mechanism, which involves a sequence of events

S. van der Zwaag (ed.), *Self Healing Materials. An Alternative Approach to 20 Centuries of Materials Science*, 323–335.
© 2007 *Springer*.

and the transient deposition of tissues, such as the callus, which are totally different from normal bone. However, in contrast to most other tissues in our body where a scar tissue remains after healing, bone has the ability to regenerate completely by remodelling the fracture callus back into bone.

2   Hierarchical Bone Structure

Bones [7] have diverse shapes depending on their function. Long bones, such as the femur or the tibia, are found in our extremities and provide stability against bending and buckling. Others, such as the vertebra are mostly loaded in compression. In some cases, the bone shell is filled with a "spongy" material called cancellous or trabecular bone (see Figure 1). The cortical bone shell constitutes the outer walls of

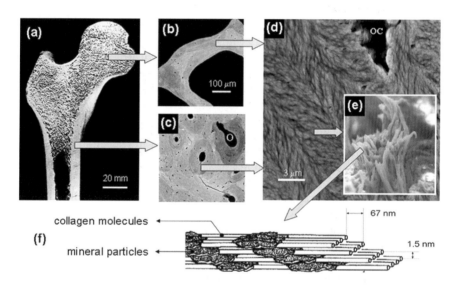

Fig. 1 Hierarchical structure of bone in the human femur [10]. A section across the femur (**a**) reveals its tube-like structure with the walls made of cortical (or compact) bone, shown enlarged in (**c**). The femoral head is filled with cancellous (or trabecular) bone (**b**). The grey-level in these images taken with back-scattered electrons is a measure for the local content of calcium phosphate mineral. Cortical bone (**c**) is fairly dense with a porosity in the order of 6%, mainly due to the presence of osteons (labelled "O" in (**c**)), which are blood vessels surrounded by concentric layers of material. Cancellous bone has a porosity in the order of 80% and can be considered a foam-like network of bone trabeculae, one of which is visible in (**d**). Both cortical and cancellous bone is made of a laminate of collagen fibrils (**e**) reinforced with calcium phosphate particles of nanometric dimensions (**f**). The bone tissue contains isolated living cells (osteocytes, labelled "OC" in (**d**)) connected by thin capillaries which might play a role in sensing mechanical strains in the tissue. (**f**) Mineral crystals arranged parallel to each other and parallel to the collagen fibrils in a regularly repeating, staggered arrangement. [25, 26]

the bones and can reach a thickness between several tenths of a millimetre (in vertebrae) to several millimetres or even centimetres (in the mid-shaft of long bones). The thickness of the struts in the "spongy" cancellous bone (Figure 1) is typically between 100 and 300 μm. The bone tissue contains a number of osteocytes (visible by their lacunae in Figure 1d) which are cells held responsible for sensing and signalling strains. The struts (or trabeculae) of cancellous bone (Figure 1c) also show some osteocyte lacunae. The trabeculae are fully surrounded by bone marrow which contains blood and, therefore, the nutrients needed by the osteocytes inside the bone material as well as by the bone cells sitting on the surface of trabeculae. Figure 1d reveals a lamellar structure which is a very common motif in bone material. Indeed, bone is a composite of collagen fibres reinforced with calcium phosphate particles. Based on scanning and transmission electron microscopy, it has been proposed that the arrangement in lamellar bone corresponds to a rotated plywood structure, where the fibres are parallel within a thin sub-layer and where the fibre direction rotates around an axis perpendicular to the layers [43, 44]. The origin of the rotated plywood structure could be a twisted-nematic (or cholesteric) liquid crystalline arrangement of collagen [15, 23, 28].

At the lower levels of hierarchy, bone is a composite of collagen and mineral nanoparticles. Structure and properties have been reviewed recently [10]. The organic matrix of bone consists of collagen and a series of non-collageneous proteins and lipids. About 85–90% of the total bone protein consists of collagen fibrils [37]. The mineralized collagen fibril of about 100 nm in diameter is the basic building block of the bone material (the insert in Figure 1d clearly reveals the fibrillar nature of the tissue in a fracture surface). The fibrils consist of an assembly of 300 nm long and 1.5 nm thick collagen molecules, which are deposited by the osteoblasts (bone forming cells) into the extracellular space and then self-assemble into fibrils. Adjacent collagen molecules within the fibrils are staggered along the axial direction by $D \approx 67$ nm, generating a characteristic pattern of gap zones with 35 nm length and overlap zones with 32 nm length [21, 25, 45]. Collagen fibrils are filled and coated by tiny mineral crystals. Crystals occur at regular intervals along the fibrils, with an approximate repeat distance of 67 nm [20], which corresponds to the distance by which adjacent collagen molecules are staggered. Crystal formation is triggered by collagen or – more likely – by other non-collageneous proteins acting as nucleation centres [36]. After nucleation, the crystals are elongated, typically plate-like [25, 26, 39], but extremely thin and they grow in thickness later [11, 33]. In bone tissue from several different mammalian and non-mammalian species, bone mineral crystals have a thickness of 1.5–4.5 nm [10, 34, 41]. The basic hydroxyapatite mineral of bone – $Ca_5(PO_4)_3OH$ – often contains other elements that replace either the calcium ions or the phosphate or hydroxyl groups, one of the most common occurrences being the replacement of the phosphate group by a carbonate group [10, 42]. In addition to crystals embedded in fibrils, there is also extrafibrillar mineral [34], which probably coats the 50–200 nm thick collagen fibrils [9].

3   Reversible Bonds

The recurring structural motif at the supramolecular level of bone is an anisotropic stiff inorganic component reinforcing the soft organic matrix. The high toughness and defect tolerance of natural biomineralized composites is believed to arise from these nanometre scale structural motifs. In recent work [17], it was shown that both mineral nanoparticles and the mineralized fibrils deform at first elastically, but to different degrees. Using in situ tensile testing with combined high brilliance synchrotron x-ray diffraction and scattering, it was found that tissue, fibrils, and mineral particles take up successively lower levels of strain, in a ratio of 12:5:2. The maximum strain seen in mineral nanoparticles (0.15–0.20%) can reach up to twice the fracture strain calculated for bulk apatite. The results are consistent with a staggered model of load transfer in the bone matrix, exemplifying the hierarchical nature of bone deformation (see Figure 2).

The long (>5–10 μm [3, 5, 9]) and thin (100–200 nm diameter) mineralized fibrils lie parallel to each other and are separated by a thin layer (1–2 nm thick) of extrafibrillar matrix. When external tensile load is applied to the tissue, it is resolved

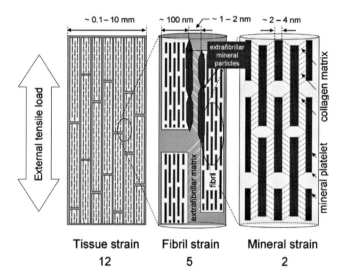

Fig. 2 Schematic model for bone deformation in response to external tensile load at three levels in the structural hierarchy (from Gupta et al. [17]): at the tissue level (left), fibril array level (center), and mineralized collagen fibrils (right). (Center) The stiff mineralized fibrils deform in tension and transfer the stress between adjacent fibrils by shearing in the thin layers of extrafibrillar matrix (white dotted lines show direction of shear in the extrafibrillar matrix). The fibrils are covered with extrafibrillar mineral particles, shown only over a selected part of the fibrils (red hexagons) so as not to obscure the internal structure of the mineralized fibril. (Right) Within each mineralized fibril, the stiff mineral platelets deform in tension and transfer the stress between adjacent platelets by shearing in the interparticle collagen matrix (dashed lines indicate shearing qualitatively and do not imply homogeneous deformation)

into a tensile deformation of the mineralized fibrils and a shearing deformation in
the extrafibrillar matrix [19]. While we do not have precise data on its mechanical
behaviour or its composition, it is likely that it is comprised of non-collagenous proteins like osteopontin and proteoglycans like decorin. Single molecule spectroscopy
of fractured bone surfaces showed that the extrafibrillar matrix has properties similar to a glue layer between the fibrils – specifically, it is relatively weak but ductile
and deforms by the successive breaking of a series of "sacrificial bonds" [9, 38]. The
matrix may also be partially calcified [2], which would increase its shear stiffness
and reduce its deformability.

These results point towards a deformation mechanism where the matrix–fibril
interface is disrupted beyond the yield point, and the matrix moves past the fibrils,
forming and reforming the bonds with the fibrils (see Figure 3). Such a situation is
analogous to a viscous flow of a liquid past a solid substrate, and likewise, for a constant velocity of flow (constant strain rate) a constant shear stress is transmitted to the
substrate (mineralized fibril) which thus holds its strain at a non-zero value. Mechanical tests established a high sensitivity of the macroscopic plastic deformation of bone
to the strain rate and temperature. These results suggest that the elementary process-
controlling bone plasticity at the molecular level is localized to within $1 \, nm^3$, and has
an activation energy in the order of $1 \, eV$ [18]. Most likely, this process is localized
in a small fraction of the bone tissue – the extrafibrillar matrix – and corresponds
to the disruption of calcium-mediated ionic bonds between the long and irregular

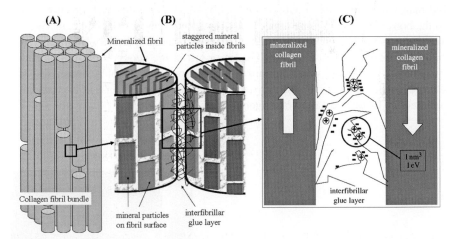

Fig. 3 Sketch of the putative function of the thin glue layer between collagen fibrils in bone (according to Gupta et al. [18, 19]): (**A**) fibril bundle; (**B**) fibrils with intra- and interfibrillar mineral joined
by a thin glue layer. (The thickness of this layer is very likely just a few nanometers and appears
highly exaggerated in the figure); (**C**) putative structure of the glue layer formed by chains of molecules interacting by charges, probably with the help of multivalent cations, such as calcium (circles).
A number of charges located on a given molecular segment have to be broken simultaneously (as
also observed in other polyelectrolyte systems), giving rise to the experimentally observed [18] activation enthalpy of $\sim 1 \, eV$ within a typical volume of $1 \, nm^3$

chains of molecules constituting this matrix. The picture which emerges is that plastic deformation is controlled by an elementary process where segments of molecules in the interfibrillar layer are connected by charge interactions (see Figure 3). This is in excellent agreement with other works showing that the deformation in bone might be associated with (calcium-dependent) sacrificial bonds [9, 38] and with the observation that plastic deformation occurs in a thin "glue" layer between fibrils [17, 19]. Self-repair at this structural level therefore consists in the breaking and reforming of molecular bonds.

4   Adaptation and Remodelling

Biological structures and materials must not be thought as statically determined by the genetic blueprint, but systems actively responding to the biophysical stimuli of their environment. During growth they are able to adapt their architecture to improve their functionality according to external constraints. Also later on, biological systems are able for an adaptive response to occurring changes in the environment and to reestablish the "agreement" with the outside world. And even in the case without external changes, repair mechanisms can be active to reduce the damage present in the system and thereby improving the performance.

Figure 4 shows the trabecular structure inside the vertebral body of an 85-year-old man with trabeculae running mainly in vertical and horizontal directions. Due to the bony projection on the upper left, called bone spur or osteophyte, in vivo the vertebra was not loaded via the vertebral disks as usual, but directly from one vertebra to the next via the osteophyte. The lower left of the vertebral body is pervaded by diagonal trabeculae, which are unknown from patients without osteophytes. The change in loading due to the osteophyte caused an adaptive response of the trabecular architecture by the formation of diagonal stress-bridges [30].

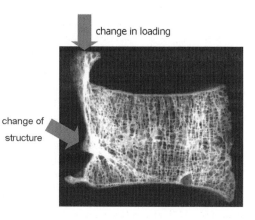

change in loading

change of structure

Fig. 4 The trabecular architecture responded with a change of structure to a change of loading. Vertebral body from an 85-year-old man with an osteophyte on the upper left and a diagonal stress-bridge. (Adapted from Mosekilde et al. [30])

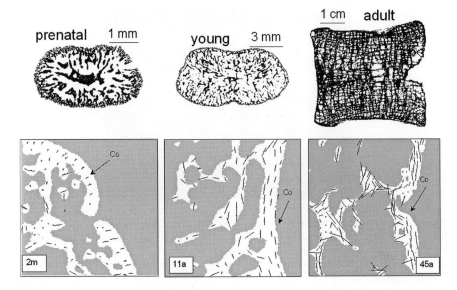

Fig. 5 Evolution of the trabecular architecture (top row) and the orientation of the collagen–mineral composite (bottom row) in human vertebra from the prenatal to the adult state. Only a small fraction of the outer part of the vertebra is shown at the bottom row, where the rim labelled Co marks the outer cortical shell; white areas correspond to bone; grey areas to marrow space. The age of the individuals was 2 months, 11 and 45 years. The direction of the bars indicates the predominant orientation of the elongated (plate-like) mineral nanoparticles. The length of the bars indicates the degree of alignment

As a second example, the top row of Figure 5 shows the evolution of the trabecular architecture inside a vertebra from the embryonal state to adulthood. The pattern of the trabeculae changes from a preferred radial orientation of the mineralized cartilage before birth to a pattern in which the trabeculae are oriented predominantly along vertical and horizontal directions in bone [33]. This change in structure can be interpreted as a response to a change in the loading from a more isotropic loading before birth to a vertical loading in compression along the spine for an adult human. These changes on the architectural level of the trabeculae are accompanied by changes taking place at the nanostructural level [33]. As visible in Figure 5, the orientation of mineral particles changes with age.

The ability of bone to adapt to a changing mechanical environment is based on a permanent remodelling of the material. Historically the two processes of bone modelling and bone remodelling are distinguished, where bone modelling specifically refers to the adaptation to new mechanical requirements including adaptations during growth. The term bone remodelling is then reserved for the continuous renewal and the maintenance of bone. Since this distinction is rather vague and, above all, recent investigations suggest that a singular regulation mechanism is enough to explain both processes [22], we prefer to refer with bone remodelling to all processes involving resorption and desposition of bone [40].

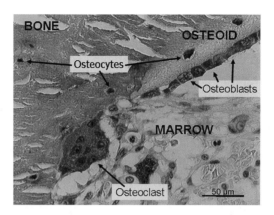

Fig. 6 Activity on bone cells in the remodelling cycle: multinucleated osteoclasts are resorbing bone. Osteoblasts on the bone surface synthesize osteoid which is later mineralized to become new bone. Some of these bone-forming cells get entrapped inside the matrix and become osteocytes. (Light micrograph with courtesy from Dr. Paul Roschger, Ludwig Boltzmann Institute of Osteology, Vienna, Austria)

The renewal of bone occurs in an interplay between different bone cells (see Figure 6). In trabecular bone osteoclasts resorb a bone packet and leave behind a cavity with a depth of about 60 μm [24]. This hole is then filled with new bone matrix by the osteoblasts. The newly deposited matrix, initially unmineralized, increases its mechanical stiffness by increasing the mineral content. While the resorption phase takes 1–3 weeks, the time for the formation of new material is about several months. The mineralization process starts with an initial surge to reach a level of more than half of its final mineralization after about 10 days, but then slows down and proceeds over years [10, 35]. In cortical bone the outer surface accessible to the osteoclasts is strongly reduced compared to trabecular bone. Cortical bone is, therefore, remodelled by forming new osteons. Osteoclasts resorb a tunnel with a speed of about 20–40 μm/day [24]. Osteoblasts then fill the tunnel again, leaving in the middle a channel for a blood vessel. As a consequence of the different remodelling geometry, only about 5% of cortical bone is renewed each year compared to 25% of trabecuar bone, but these percentages are site dependent with a reduced remodelling rate in the peripheral skeleton [31]. In general, bone remodelling occurs at about 1–2 million microscopic sites in the adult skeleton.

Dating back already to the end of the nineteenth century the study of the trabecular structure in the femoral head led to the conclusion that bone remodelling is not a random exchange of old bone matrix with new one, but a mechanically controlled process. The so-called Wolff–Roux law states that bone is deposited wherever mechanically needed and is resorbed wherever there is no mechanical need. This idea was further developed by Frost who proposed that a mechanically controlled feedback loop is active in bone and regulating bone mass and architecture. In analogy with a thermostat, which switches the heater on or off according to its set points, the term mechanostat was coined [12, 13]. Once there is a local mechanical overloading, e.g. due to a new intensive sport activity or due to a locally reduced bone mass caused by osteoclast resorption, the mechanostat gives the signal for bone deposition. New bone will be added as long as normal strains in this local region are regained. The same principle, but with opposite signs occurs in the case of mechanical disuse, e.g. due to prolonged bedrest. The mechanostat makes sure that the mechanically dispensable bone is removed.

In conclusion, the bone cells described in Figure 6 create a feedback loop, where positive actuation (bone formation) is due to osteoblasts, negative actuation (bone resorption) to osteoclasts, and the mechanical sensing most likely to osteocytes. The modelling and permanent remodelling has at least three functions:

- Functional adaptation during growth
- Adjustment of the structure to modified external conditions
- Removal of damaged material and replacement by new (undamaged) bone

This represents a second level of self-repair whereby micro-damage is continuously removed and replaced by new material.

5   Healing

The aim of materials science to design and produce self-healing materials causes increasing interest how Nature realizes and controls actual healing responses. Most striking examples are the complete regrowth of lost limbs in salamanders [1]. Although such limb regeneration does not occur in humans, bone healing is an example of how after significant damage the human biological system cannot only return to functionality, but achieve a complete restoring by itself. In contrast to other tissues, which are repaired by producing scar tissue, bone has the capability to regenerate itself, thereby returning basically to the prefracture state including its mechanical performance. Since the underlying biology and gene expression resembles strikingly what happens during the bone development in the embryonic phase, it was hypothesized that in bone regeneration basic steps of skeletal development are recapitulated [14]. Beside the importance of biological factors [8] – here only the importance of restoring the vascularisation to avoid cell death due to lack of oxygen and the overwhelming presence of signalling molecules should be mentioned – the mechanical loading within the fracture site plays a crucial role for the healing [4, 32].

Already the fundamental distinction between primary and secondary fracture healing is based on how the fracture is stabilized. Primary (or direct) fracture healing occurs when the fracture fragments are immobilized and ideally compressed against each other. The remaining gap is refilled by bone remodelling creating osteons that bridge the two bone fragments. Although such a direct healing seems the ideal healing response, it has the disadvantage that, beside surgical problems, the broken bone

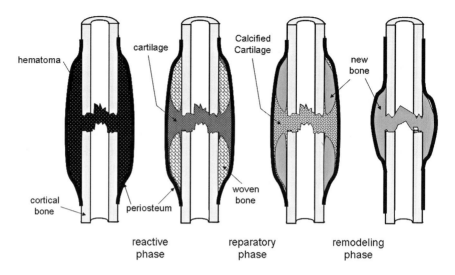

Fig. 7 Different phases in secondary fracture healing: the formation of a hematoma in the reactive (inflammatory) phase (left); new bone formation at the outside of the bone fragments by intramembranous ossification (middle); the cartilage within the fracture gap mineralizes and a bone bridge is formed at the outside of the fracture fragments which eventually leads to full restoration in the remodelling phase (right). (Adapted from Li et al. [27])

remains unstable for months and even years till the process is completed [16]. In this time reloading should be avoided.

In case small movement of the fracture fragments is allowed, healing proceeds via a stabilisation of the bone fragments by forming an external callus [29]. The different stages of secondary bone healing in a long bone like a femur or a tibia can be summarized as follows: the starting point of a bone fracture is typically an incident in which the bone is loaded beyond its maximal strength. Beside the disruption of the bone material, the blood supply, most important in the outer "skin" layer of the bone called the periosteum, is interrupted and cells die. The incipient inflammation has the task to clean the fracture region form dead material and to initiate processes to restore the blood supply and to congregate mesenchymal stem cells (Figure 7). A key process, which is thought to be mechanically regulated, is the differentiation of these precursor cells into cells that can produce different tissues: cartilage, fibro-cartilage, fibrous tissue, and bone. Interestingly, in secondary healing, not only the direct formation of bone called intramembranous ossification occurs, but also an ossification process, which involves a "detour" via the formation of cartilage, is active (endochondral ossification) (Figure 7). Instead of constructing an osseous bridge between the fracture ends, a hard bone shell is formed around the fracture site, while the fracture gap itself is filled with soft tissue (cartilage), which mineralizes and is finally substituted by bone. In the final phase, bone remodelling and resorption take over and lead to a removal of the bone around the original fracture and therefore complete the restoration of the broken bone [4, 32]. The essence of secondary fracture healing is, consequently, a fast reaction which aims at stabilizing the fracture gap by the formation of a transient overdimensionized biological splint, which is later removed. This intricate way of bone healing allows a fast return to prefracture values of bone strength and stiffness even before the process is complete.

As a general conclusion, we see from the example of bone that self-repair operates at very different scales and in a more or less permanent way. At the lowest level, broken bonds are reforming, which leads to the possibility of some plastic deformation. Defects and damage of intermediate size are removed and replaced by new bone in the permanent remodelling process. Finally, a catastrophic failure is repaired in a multi-step process (known as healing) where the integrity is restored first by joining the ends by soft tissue and a complete restoration takes much longer time.

References

1. Alvarado A.S. (2000) Regeneration in the metazoans: why does it happen? *Bioessays* 22: 578–590
2. Bonar L.C., Lees S., Mook H.A. (1985) Neutron-diffraction studies of collagen in fully mineralized bone. J Mol Biol 181:265–270
3. Braidotti P., Bemporad E., D'Alessio T., Sciuto S.A., Stagni L. (2000) Tensile experiments and SEM fractography on bovine subchondral bone. J Biomech 33:1153–1157
4. Carter D.R., Beaupré G.S. (2001) Skeletal function and form, mechanobiology of skeletal development, aging, and regeneration. Cambridge University Press, Cambridge
5. Craig A.S., Birtles M.J., Conway J.F., Parry D.A.D. (1989) An estimate of the mean length of collagen fibrils in rat tail-tendon as a function of age. Connect Tissue Res 19:51–62
6. Currey J.D. (1999) The design of mineralised hard tissues for their mechanical functions. J Exp Biology 202:3285–3294
7. Currey J.D. (2002) Bones – structure and mechanics. Princeton University Press, Princeton, p 436
8. Einhorn T.A. (1998) The cell and molecular biology of fracture healing. Clin Orthop Relat Res S7–S21
9. Fantner G., Hassenkam T., Kindt J.H., Weaver J.C., Birkedal H., Pechenik L., Cutroni J.A., Cidade G.A.G., Stucky G.D., Morse D.E., Hansma P.K. (2005) Sacrificial bonds and hidden length dissipate energy as mineralized fibrils separate during bone fracture. Nat Mater 4:612–616
10. Fratzl P., Gupta H.S., Paschalis E.P., Roschger P. (2004) Structure and mechanical quality of the collagen-mineral nano-composite in bone. J Mater Chem 14:2115–2123
11. Fratzl P., Fratzl-Zelman N., Klaushofer K., Vogl G., Koller K. (1991) Nucleation and growth of mineral crystals in bone studied by small-angle x-ray-scattering. Calcif Tissue Int 48:407–413
12. Frost H.M. (1987) Bone mass and the mechanostat – a proposal. Anat Rec 219:1–9
13. Frost H.M. (2003) Bone's mechanostat: a 2003 update. Anat Rec Part A – Discov Mol Cell Evol Biol 275A:1081–1101
14. Gerstenfeld L.C., Cullinane D.M., Barnes G.L., Graves D.T., Einhorn T.A. (2003) Fracture healing as a post-natal developmental process: molecular, spatial, and temporal aspects of its regulation. J Cell Biochem 88:873–884
15. Giraud-Guille M.M. (1988) Twisted plywood architecture of collagen fibrils in human compact-bone osteons. Calcif Tissue Int 42:167–180

16. Goodship A.E., Cunningham J.L. (2001) Pathophysiology of functional adaptation of bone in remodeling and repair in vivo. In: Cowin S.C. (ed) Bone mechanics handbook. CRC Press, Boca Raton, FL

17. Gupta H.S., Seto J., Wagermaier W., Zaslansky P., Boesecke P., Fratzl P. (2006) Cooperative deformation of mineral and collagen in bone at the nanoscale. PNAS USA 103:17741–17746

18. Gupta H.S., Fratzl P., Kerschnitzki M., Benecke G., Wagermaier W., Kirchner H.O.K. (2007) Evidence for an elementary process in bone plasticity with an activation enthalpy of 1eV. J Roy Soc Interf 4:277–282

19. Gupta H.S., Wagermaier W., Zickler G.A., Aroush D.R.B., Funari S.S., Roschger P., Wagner H.D., Fratzl P. (2005) Nanoscale deformation mechanisms in bone. Nano Lett 5:2108–2111

20. Hassenkam T., Fantner G.E., Cutroni J.A., Weaver J.C., Morse D.E., Hansma P.K. (2004) High-resolution AFM imaging of intact and fractured trabecular bone. Bone 35:4–10

21. Hodge A.J., Petruska J.A. (1963) In: Ramachandran GN (ed) Aspects of protein structure. Academic Press, New York, pp 289–300

22. Huiskes R., Ruimerman R., van Lenthe G.H., Janssen J.D. (2000) Effects of mechanical forces on maintenance and adaptation of form in trabecular bone. Nature 405:704–706

23. Hulmes D.J.S. (2002) Building collagen molecules, fibrils, and suprafibrillar structures. J Struct Biol 137:2–10

24. Jee W.S.S. (2001) Integrated bone tissue physiology: anatomy and physiology. In: Cowin S.C. (ed) Bone mechanics handbook. CRC Press, Boca Raton, FL

25. Landis W.J. (1996) Mineral characterization in calcifying tissues: atomic, molecular and macromolecular perspectives. Connect Tissue Res 35:1–8

26. Landis W.J., Hodgens K.J., Arena J., Song M.J., McEwen B.F. (1996) Structural relations between collagen and mineral in bone as determined by high voltage electron microscopic tomography. Microsc Res Tech 33:192–202

27. Li R., Ahmad T., Spetea M., Ahmed M., Kreicbergs A. (2001) Bone reinnervation after fracture: a study in the rat. J Bone Miner Res 16:1505–1510

28. Martin R., Farjanel J., Eichenberger D., Colige A., Kessler E., Hulmes D.J.S., Giraud-Guille M.M. (2000) Liquid crystalline ordering of procollagen as a determinant of three-dimensional extracellular matrix architecture. J Mol Biol 301:11–17

29. McKibbin B. (1978) Biology of fracture healing in long bones. J Bone Joint Surg (Br) 60: 150–162

30. Mosekilde L., Ebbesen E.N., Tornvig L., Thomsen J.S. (2000) Trabecular bone structure and strength – remodelling and repair. J Musculoskel Neuron Interact 1:25–30

31. Parfitt A.M. (2002) Misconceptions (2): turnover is always higher in cancellous than in cortical bone. Bone 30:807–809

32. Prendergast P.J., Van der Meulen M.C.H. (2001) Mechanics of bone regeneration. In: Cowin S.C. (ed) Bone mechanics handbook. CRC Press, Boca Raton, FL

33. Roschger P., Grabner B.M., Rinnerthaler S., Tesch W., Kneissel M., Berzlanovich A., Klaushofer K., Fratzl P. (2001) Structural development of the mineralized tissue in the human L4 vertebral body. J Struct Biol 136:126–136

34. Rubin M.A., Rubin J., Jasiuk W. (2004) SEM and TEM study of the hierarchical structure of C57BL/6J and C3H/HeJ mice trabecular bone. Bone 35:11–20

35. Ruffoni D., Fratzl P., Roschger P., Klaushofer K., Weinkamer R. (2007) The bone mineralization density distribution as a fingerprint of the mineralization process. Bone 40:1308–1319

36. Sodek J., Ganss B., McKee M.D. (2000) Osteopontin. Crit Rev Oral Biol Med 11:279–303

37. Termine J.D., Robey P.G. (1996) Bone matrix proteins and the mineralization process. In: Favus M.J. (ed) Primer on the metabolic bone diseases and disorders of mineral metabolism, 3rd edn. An official publication of the American Society for Bone and Mineral Research. Lippincott-Raven Publishers, Philadelphia, PA

38. Thompson J.B., Kindt J.H., Drake B., Hansma H.G., Morse D.E., Hansma P.K. (2001) Bone indentation recovery time correlates with bond reforming time. Nature 414:773–776

39. Traub W., Arad T., Weiner S. (1992) Origin of mineral crystal-growth in collagen fibrils. Matrix 12:251–255
40. Weinans H., Odgaard A. (1995) Bone structure and remodeling: an introduction. In: Odgaard A., Weinans H. (eds) Bone structure and remodeling. World Scientific, Singapore
41. Weiner S., Traub W. (1992) Bone-structure – from Angstroms to microns. Faseb J 6:879–885
42. Weiner S., Wagner H.D. (1998) The material bone: structure mechanical function relations. Annu Rev Mater Sci 28:271–298
43. Weiner S., Traub W., Wagner H.D. (1999) Lamellar bone: structure-function relations. J Struct Biol 126:241–255
44. Weiner S., Arad T., Sabanay I., Traub W. (1997) Rotated plywood structure of primary lamellar bone in the rat: orientations of the collagen fibril arrays. Bone 20:509–514
45. White S.W., Hulmes D.J.S., Miller A., Timmins P.A. (1977) Collagen-mineral axial relationship in calcified turkey leg tendon by x-ray and neutron-diffraction. Nature 266:421–425

Modeling of Self Healing of Skin Tissue

FJ Vermolen, WG van Rossum, E Javierre[1], and JA Adam[2]

[1] *Delft University of Technology, Department of Applied Mathematics, Mekelweg 4, The Netherlands*
[2] *Old Dominion University, Department of Mathematics and Statistics, Norfolk, USA*
E-mail: F.J. Vermolen@tudelft.nl

Abstract A suite of mathematical models for epidermal wound healing is presented. The models deal with the sequential steps of angiogenesis (neovascularization) and wound contraction (the actual healing of a wound). An innovation is the combination of the two processes which do not take place in a complete sequential manner but overlap partially. The models consist of nonlinearly coupled diffusion–reaction equations, in which transport of oxygen, growth factors, and cells, and mitosis are taken into account. Further, Adam's alternative model, which is based on the assumption of the presence of an active layer at the wound edge, is described and some implications are presented. An important feature of the model due to Adam is that the wound edge is tracked explicitly as a part of the solution. In this work several numerical methods to solve the moving boundary problem are described.

Keywords: finite elements, wound contraction, neovascularization

1   Introduction

Models for bone regeneration and wound healing often rely on experiments on animals. When a wound occurs, blood vessels are cut and blood enters the wound. Due to coagulation of blood inside the wound, the wound is temporarily closed and as a result the blood vessels adjacent to the wound are also closed. In due course contaminants will be removed from the wounded area and the blood vessel network will be restored, but initially due to insufficient blood supply, there will be a low concentration of nutrients which are necessary for cell division and wound healing. Wound healing, if it occurs, proceeds by a combination of several processes: chemotaxis (movement of cells induced by a concentration gradient), neovascularization, synthesis of extracellular matrix proteins, and scar modeling. Previous models incorporate cell mitosis, cell proliferation, cell death, capillary formation, oxygen supply, and growth factor generation, including studies by Sherratt et al. [1], Fillion et al. [2],

Maggelakis [3] and Gaffney et al. [4], to mention just a few. A recent work devoted to mathematical biology has been written by Murray [5], in which the issue of wound healing is also treated. The wound healing process can roughly be divided into the following partly overlapping consecutive stages:

1. Formation of a blood clot on the wound to prevent undesired chemicals from entering the tissue of the organism (blood clotting/inflammatory phase)
2. Formation of a network of tiny arteries (capillaries) for blood flow to supply the necessary nutrients for wound healing
3. Division and growth of epidermal cells (mitosis), taking place during the actual healing of the wound (remodeling phase)

A good supply of nutrients and constituents is necessary for the process of cell division and cellular growth. For this purpose tiny capillaries are formed during the process of angiogenesis. Some models for capillary network formation have been proposed by Gaffney et al. [4] and Maggelakis [12].

Wound contraction is modeled by Sherratt and Murray [1] who consider cell division and growth factor generation simultaneously for healing of epidermal wounds. Their model consists of a system of reaction–diffusion equations and an active layer at the wound edge is not taken into account explicitly, but this layer is merely a result of an increased cell division near the wound edge as predicted by their model. Among many others a recently developed model for wound contraction was proposed by Adam et al. [8, 9, 14, 15–17]. Until now, the conditions for wound healing were only analyzed for geometries where only one spatial coordinate could be used. Hence, the key innovations in the present study are the following: finite element solutions for arbitrary wound geometries; and a combination of models for angiogenesis and wound contraction as partly overlapping consecutive processes.

The present paper contains a compilation of recent work and some ideas for future studies, such as the use of the level set method for wound contraction and it is organized as follows. First, an existing model for angiogenesis (neovascularization) with an extension is described. This is followed by a treatment of a model for wound contraction and a proposition of how to combine these models. We continue with a description of the numerical method to solve the resulting nonlinearly coupled system of partial differential equations and present some numerical experiments, followed by some concluding remarks.

2   The Mathematical Models

In this section several models for epidermal wound healing is presented. The model consists of neovascularization and wound contraction and the construction of the model relies on a combination of the ideas developed by Maggelakis [12], Gaffney et al. [4] and Sherratt and Murray [1].

2.1 Neovascularization

Maggelakis constructed this model on angiogenesis in 2004 [12]. It is assumed that
the tips of the capillaries act as the only sources for oxygen supply. The domain of
computation is assumed to be divided into the wound region Ω_w and the undamaged
region Ω_u, that is $\Omega = \Omega_w \cup \Omega_u \cup (\overline{\Omega}_w \cap \overline{\Omega}_u)$ and Ω_w is embedded within Ω_u.
The closures of Ω_w and Ω_u respectively are denoted by $\overline{\Omega}_w$ and $\overline{\Omega}_u$. Let n and c_o
respectively denote the capillary density and oxygen concentration and let them be
functions of time t and space within the domain of computation Ω; then a mass
balance results into the following partial differential equation (PDE):

$$\frac{\partial c_o}{\partial t} = D_o \operatorname{div} \operatorname{grad} c_o + \lambda_n n - \lambda_o c_o, \text{ for } \mathbf{x} \in \Omega_w,$$
$$\text{subject to } c_o(\mathbf{x}, t) = c_i \text{ for } \mathbf{x} \in \overline{\Omega}_w \cap \overline{\Omega}_u. \tag{1}$$

The initial oxygen concentration is assumed to be zero in Ω_w. The above equation
is based on the assumption that the oxygen supply and oxygen consumption depend
linearly on the capillary density and oxygen concentration respectively. When the
oxygen level is low, macrophages appear at the wound site. The macrophages release
MDGFs (macrophage derived growth factors) which are hormones that enhance the
growth of blood vessels and collagen deposition and hence help to restore the capillaries that provide the skin with the necessary nutrients and oxygen for cell division
needed for wound contraction. An assumption in the model is that macrophages are
produced if the oxygen level is below a threshold value, say c_θ. The production rate,
Q, is assumed to depend linearly on the lack of oxygen, that is

$$Q = Q(c_o) = \begin{cases} 1 - \dfrac{c_o}{c_\theta}, & \text{if } c_o < c_\theta, \\ 0, & \text{if } c_o \geq c_\theta. \end{cases} \tag{2}$$

The mass balance of MDGFs results into the following PDEs in the wound region
Ω_w and out of the wound region Ω_u:

$$\frac{\partial c_m}{\partial t} = D_m \operatorname{div} \operatorname{grad} c_m + \lambda_m Q(c_o) - \lambda c_m, \quad \text{for } \mathbf{x} \in \Omega_w,$$

$$\frac{\partial c_m}{\partial t} = D_m \operatorname{div} \operatorname{grad} c_m - \lambda_c c_m, \quad \text{for } \mathbf{x} \in \Omega_u. \tag{3}$$

The initial MDGF concentration, denoted by c_m, is assumed to be zero in the entire
domain of computation Ω and a homogeneous Neumann boundary condition is used.
The capillary density is assumed to grow as a result of the MDGFs in a logistic
manner, that is

$$\frac{\partial n}{\partial t} = D_n \operatorname{div} \operatorname{grad} n + \mu c_m n \left(1 - \frac{n}{n_m}\right), \quad \text{for } \mathbf{x} \in \Omega, \tag{4}$$

where n_m denotes the maximum capillary density. The capillary density is assumed to satify the following initial condition

$$n(\mathbf{x}, 0) = \begin{cases} 0, & \text{for } \mathbf{x} \in \Omega_w, \\ n_m, & \text{for } \mathbf{x} \in \Omega_u. \end{cases} \tag{5}$$

A homogeneous Neumann boundary condition is used for n. We assume the capillary tips to migrate via a random walk. But it is reasonable since away from the wound the capillary network is assumed to be undamaged. This random walk was not incorporated into Maggelakis' model. The random walk was biased by Gaffney et al. [4]. The bias is neglected in this paper but it will be dealt with in future work. Further, Maggelakis sets in a nonzero artificial starting value for the capillary density to have the capillary density to increase up to the equilibrium value. Analytic solutions are considered in [12] for the above equations under the assumptions that the diffusion of oxygen and growth factors is instantaneous and that $D_n = 0$. The induced growth of MDGFs due to a lack of oxygen can be classified as a negative feedback mechanism. An extension of this model with model hypotheses due to Gaffney et al. [4], who incorporate the interaction of the capillary tip density with the endothelial cell density and use a biased random walk resulting into nonlinear cross diffusion, is omitted in this text, but it will be a topic in future studies.

2.2 Wound Contraction

In this section we describe the extension of two models for wound contraction with neovascularization. The first model is due to Sherratt and Murray [1] whose model is based on a system of nonlinearly coupled diffusion-reaction equations for a growth factor and epidermal cell density. The second model is due to Adam who assumes that the wound contraction rate is proportional to the local curvature of the wound. The wound edge is followed explicitly as a part of the solution.

2.2.1 The Extension of the Model due to Sherratt and Murray

The mechanism for wound contraction is cell division and growth (mitosis). This mechanism is triggered by a complicated system of growth factors. In the present model we use the simplification that wound contraction is influenced by one generic growth factor only. The growth factor concentration influences the production of epidermal cells. Following Sherratt and Murray [1], we assume that the main source of growth factors is the epidermal cells. Growth factors diffuse through the tissue region and the concentration decays due to reactions with other chemicals present in the tissue. Further, the epidermal cells need the supply of nutrients by the capillaries in order to be able to grow and divide. A constant blood flow through the capillaries is assumed, hence, the mitotic rate can be related to capillary density n. The mitotic rate is assumed to depend on the capillary density in a nonlinear fashion by a function $\phi_p(y)$, given by

$$\phi_p(y) = \frac{y^p}{(1-y)^p + y^p}, \tag{6}$$

where p is assumed to be positive. Herewith, the expression of Sherratt and Murray [1] where the accumulation of the epidermal cells is determined by proliferation (diffusive transport), mitosis and cell death, is adjusted to

$$\frac{\partial m}{\partial t} = D_1 \operatorname{div} \operatorname{grad} m + \phi_p\left(\frac{n}{n_m}\right) s(c)m\left(2 - \frac{m}{m_0}\right) - km, \qquad \mathbf{x} \in \Omega_w, \tag{7}$$

subject to $m(\mathbf{x}, t) = \hat{m}$, for $\mathbf{x} \in \overline{\Omega}_w \cap \overline{\Omega}_u$.

in order to depend on the capillary density. In the above equation $m = m(\mathbf{x}, t)$ and $c = c(\mathbf{x}, t)$ respectively denote the epidermal cell density and growth factor concentration. The unwounded cell density is denoted by m_0. The function $s = s(c)$ is a nonlinear function of the growth concentration describing the mitotic rate, see [5, 1]. For the growth factor accumulation a similar relationship due to diffusive transport, production and decay is obtained with a similar adaptation for the dependence of the capillary density:

$$\frac{\partial c}{\partial t} = D_2 \operatorname{div} \operatorname{grad} c + \phi_p\left(\frac{n}{n_m}\right) f(m) - \lambda c, \qquad \mathbf{x} \in \Omega_w. \tag{8}$$

subject to $c(\mathbf{x}, t) = \hat{c}$, for $\mathbf{x} \in \overline{\Omega}_w \cap \overline{\Omega}_u$.

As initial conditions for m and c we have $m(\mathbf{x}, 0) = 0$ and $c(\mathbf{x}, 0) = 0$. In the above equation $f(m)$ denotes a nonlinear relation for the growth factor regeneration. In the model due to Sherratt and Murray [1], two different types of growth factors are considered: (1) activators; and (2) inhibitors, both with their characteristic functions for s and f.

A growth factor is referred to as an activator if a large concentration of it enhances mitosis. Whereas if a large concentration of growth factors slows down mitosis, then, the growth factor is called an inhibitor. This implies that $s'(c) > 0$ and $s'(c) < 0$ correspond to an activator and an inhibitor respectively for $m \in (0, 2m_0)$. The function $f(m)$ must reflect the appropriate cellular response to injury depending on whether the growth factor activates or inhibits mitosis. Hence, if the growth factor is an activator, then, to restore the skin, a large concentration of growth factors is needed. Hence, if the cell density is low, the cells should be stimulated to produce the growth factor. Hence, f should be large if the cell density is low. Except whenever $m = 0$, then, there is no source for the production of growth factors, and hence $f(0) = 0$. Further, on the contrary, if the growth factor inhibits mitosis, then, the production of growth factors should be small in the wounded area, that is, if the cell density is small. More on this complicated biological issue can be found in Murray

[5], Chapter 9. For the activator, we have

$$s(c) = k \cdot \left(\beta + \frac{2c_m(h-\beta)c}{c_m^2 + c^2} \right),$$

$$\text{where } \beta = \frac{c_o^2 + c_m^2 - 2hc_oc_m}{(c_o - c_m)^2}. \tag{9}$$

$$f(m) = \lambda c_o \frac{m}{m_0} \cdot \frac{m_0^2 + a^2}{m^2 + a^2},$$

For an inhibitor, we have

$$s(c) = k \cdot \frac{(h-1)c + hc_o}{2(h-1)c + c_o},$$

$$f(m) = \frac{\lambda c_o}{m_0} \cdot m. \tag{10}$$

Here β, a, h, and c_m are constants. We refer to Morray [5] for their meaning and values used in the calculations. We assume that within the wound initially $m = c = 0$ and at the wound edge, $m = m_0$ and $c = c_o$. Hence, we want the initial state to be unstable and the boundary conditions to be stable, so that the functions m and c converge to the values of the boundary conditions m_0 and c_o as $t \to \infty$. In other words, the unwounded state is stable with respect to small perturbations, whereas the wounded state is unstable. Note that the capillary density from the neovascularization model described in the previous subsection is needed. Furthermore, $\lim_{p \to \infty} \phi_p(y) = u_{1/2}(y)$, which is in the class of Heaviside functions with the jump at $y = 1/2$. The behavior is motivated by the intuitively obvious notion that the mitotic rate and production rate of the growth factors increase with the capillary density. The rest of the equations remains the same. The present coupling between the two models is an innovation. Further, the model will be extended to include the interaction of the capillary density with the endothelial cell density, which is the building block of the capillaries. A further future implementation is the feedback from the contraction model to the neovascularization model.

2.2.2 The Extension of the Model due to Adam

Due to the damage of tiny blood vessels around the wound, there is an increased activity of cellular growth, cell division, and production of the growth factor that enhances wound healing. We assume that this layer, commonly referred to as the *active layer*, has a constant thickness of d. The situation is as sketched in Figure 1. The model due to Adam is based on the production, decay, and transport of a generic growth factor. The wound moves if and only if the growth factor concentration at the wound edge exceeds a threshold value. The wound edge moves with a rate proportional with the local wound edge curvature. In Figure 1 we use Ω_w, Ω_L, and Ω_u to denote the wound itself, the active layer and the outer tissue respectively. Far away

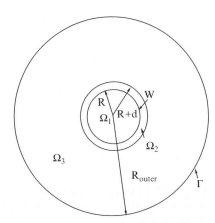

Fig. 1 The geometry of a circular wound, Ω_w, Ω_L, and Ω_u respectively denote the wound, active layer and the outer tissue. W denotes the wound edge

from the wound, that is at the boundary of the domain of computation Γ we assume that the concentration of the growth factor is zero. The wound edge, the interface between the wound (Ω_w) and the active layer (Ω_L), is indicated by W.

Let the total domain of computation be given by $\Omega = \Omega_w \cup \Omega_L \cup \Omega_u$, then, following Adam [8], we state the fundamental equations for the transport, production, and decay of the growth factor concentration, c, which are:

$$\frac{\partial c}{\partial t} - D \operatorname{div} \operatorname{grad} c + \lambda c = P f(x, y), \quad \text{for } (x, y) \in \Omega, \tag{11}$$

$$c(x, y, t) = 0, \quad \text{for } (x, y) \in \Gamma, \tag{12}$$

$$c(x, y, 0) = 0, \quad \text{for } (x, y) \in \Omega, \tag{13}$$

$$\text{further } f(x, y) = \begin{cases} \phi_p\left(\frac{n}{n_m}\right), & \text{for } (x, y) \in \Omega_L \\ 0, & \text{for } (x, y) \in \Omega_w \cup \Omega_u \end{cases}. \tag{14}$$

In the equations D, P, and λ denote the constant diffusion coefficient, production rate constant and the decay coefficient of the growth factor respectively. The same function for the capillary density ϕ_p as in the previous section on the model due to Sherratt and Murray [1] is used here. A new feature is that the production of the growth factors is assumed to depend on the capillary tip density. This is motivated by the need of a supply of nutrients to the cells to produce growth factors, to grow and to divide. The growth factor concentration, c, is to be determined. Further, the second and third term in Equation (11) respectively account for growth factor transport and growth factor loss. The right-hand side of Equation (11) accounts for the production of the growth factor. Equation (12) represents the boundary condition and the step-function f accounts for the growth factor production taking place in the active layer only. Adam [8] points out that for the derivation of a *critical size defect*, which is the smallest wound that does not heal, the time derivative in the diffusion reaction equation does not have to be taken into account. Healing at a certain location of the wound edge implies that the inward normal component of the velocity, v_n, of the wound edge W, between wound and active layer, is positive. In the present paper we use the assumption from Adam [8] that the wound heals if and only if the growth factor concentration exceeds a threshold concentration $\hat{c}$, hence

$$v_n > 0 \text{ if and only if } c(x, y, t) \geq \hat{c} \text{ for } (x, y) \in W,$$
$$\text{else } v_n = 0. \tag{15}$$

This implies that in order to determine whether the wound heals at a certain location at W at a certain time t, one needs to know the growth factor concentration there, obviously requiring the concentration equation to be solved.

Adam considers analytic expressions for the time independent case for several geometries: planar (linear) geometry [8], a circular wound on a spherical surface [14], a circular wound on a planar surface [16]. A wound with spherical symmetry is considered in terms of analytic expressions by Arnold [17]. Since the finite element method is used for the growth factor concentration, the model is easily extended to any other geometry.

Far away from the wound, we assume that the growth factor concentration is zero. The results will be presented by colored contour plots and graphs of the concentration over the wound edge in section 3. In the following section the fundamental differential equation is discussed, based on the derivation in the appendix.

2.2.3 Wound Contraction as a Function of Growth Factor Concentration

Consistent with the general discussion in the appendix, we assume that the healing rate S is a linear function of the local curvature of the wound, that is

$$S(n, R(t)) = \phi_p(\frac{n}{n_m}) \left(\alpha + \frac{\beta}{R(t)} \right), \tag{16}$$

where R is the local radius of curvature and $\alpha, \beta > 0$ are considered as constants, prohibiting growth of the wound. Since the inflow of nutrients and building blocks for the epidermal cells is crucially important for the cell division to take place, the healing rate S is assumed to depend on the capillary tip density which takes care of the nutrient inflow. If S is Lipschitz continuous in R in a neighborhood of R_0, then, the solution of Equation (41), which relates the wound edge velocity to the healing rate (see the appendix), is

$$\int_{R(t)}^{R_0} \frac{ds}{S(s)} = \frac{t}{2}. \tag{17}$$

First, we take $\phi_p = 1$ to get insight into the healing rate using analytic expressions. For the case of Equation (16) and a circular wound, this yields

$$t = \frac{2}{\alpha}\left\{R_0 - R - \frac{\beta}{\alpha}\ln\left(\frac{\alpha R_0 + \beta}{\alpha R + \beta}\right)\right\}, \quad \text{provided } \alpha \neq 0, \tag{18}$$

and for $\alpha = 0$, one obtains

$$t = \frac{1}{\beta}\left(R_0^2 - R^2\right). \tag{19}$$

This implies that there exists a healing time, t_h, which is given by

$$t_h = 2\int_0^{R_0}\frac{ds}{S(s)}, \tag{20}$$

provided that the above integral exists. For the case of Equation (16), this becomes for a circular wound:

$$t_h = \frac{2}{\alpha}\left\{R_0 - \frac{\beta}{\alpha}\ln\left(\frac{\alpha R_0 + \beta}{\beta}\right)\right\}, \quad \text{provided } \alpha \neq 0, \tag{21}$$

and for $\alpha = 0$

$$t_h = \frac{R_0^2}{\beta}. \tag{22}$$

2.2.4 Threshold of Growth Factor Concentration to Wound Contraction

Here we consider an extension of the above model to account for the likelihood, as in many physiological systems, of threshold behavior. Equation (41) is solved subject to the initiation of healing once the growth factor concentration is high enough [6, 11]. This is done by redefining the rate function $S(n, R(t))$ as

$$S(n, R(t)) := S(n, R(t))H(c(R(t), t) - \hat{c}), \tag{23}$$

where $H(s)$ represents a heaviside function, defined as

$$H(s) = \begin{cases} 0, & s < 0, \\ 1, & s \geq 0. \end{cases}$$

The threshold concentration of the growth factor for wound healing is denoted by $\hat{c}$. In the remainder of this paper, expression (23) is used for S. This requires knowledge of the growth factor concentration at the wound edge at all times in the simulation of the healing process, since the wound heals further if and only if $c(R(t), t) \geq \hat{c}$ (consistent with the introduction of the Heaviside function in Equation (16)). This implies that once the wound edge is inwardly displaced, the equation for the growth factor concentration has to be solved. Since the position of the wound edge moves in time, the position of the interface conditions at the wound edge, and hence, the location where the conditions for continuity of the growth factor concentration and its normal derivative are imposed, also change in time. Thus we are faced with a moving boundary problem where the velocity of the wound edge is determined by Equation (41).

Furthermore, the thickness of the active layer d is permitted to change with the wound radius during the healing process, according to the simple choice

$$d(R(t)) = c_1 R_0 + c_2 R(t). \tag{24}$$

This implies for the circular wound that the active layer thickness is proportional to the sum of the reciprocals of the initial curvature and the subsequent curvature at time t (see references in [8] for clinical intimations of this effect). Hence, during the healing process, if $c_1 > 0$ and $c_2 > 0$, then the thickness of the active layer decreases, and the amount of growth factor that is produced in this layer decreases. Note, however, that the wound area decreases as well. So for a given β in Equation (16) there is a trade-off as to whether this gives rise to a decrease or an increase of the growth factor concentration at the wound edge. If the growth factor concentration at the wound edge decreases, then this may cause the healing process to cease prematurely.

We remark here that if the distribution of the growth factor is assumed to be instantaneous, implying that the time derivative in the growth factor equation vanishes (this being effectively a steady-state environment) then, the growth factor concentration at the edge of the wound is well defined by the solution of the growth factor equation as a function of the wound radius $R(t)$. Further, the wound radius decreases continuously and since the wound edge concentration depends on the wound radius continuously, the wound edge concentration depends on time continuously as well. This implies that Equation (41) can be integrated up to $\hat{R}$ for which $c(\hat{R}) = \hat{c}$, hence, the formal solution is given by

$$2 \int_{\hat{R}(t)}^{R_0} \frac{ds}{S(s)} = t, \text{ provided that } R(t) \geq \hat{R}. \tag{25}$$

In general the determination of this $\hat{R}$ involves a zero-point iteration method. However, for the case that the time derivative is not dropped or that a numerical solution for the concentration is used, then there is no explicit relation between $c(R(t), t)$ and $R(t)$. Hence, the determination of $\hat{R}$ is less straightforward and a time-stepping method has to be used, where at each time step the growth factor concentration at the wound edge has to be monitored. As an example we show a computation with an active layer whose thickness varies with the wound radius in Figure 2. The concentration at the wound edge is determined by the use of the Bessel function solution of Arnold et al. [16]. Then, at a certain wound radius, the healing criterion is checked and then the wound radius is displaced accordingly. It can be seen in Figure 2 that the growth factor concentration at the wound edge decreases during the healing process down to the threshold concentration. This is a consequence of the decrease of the thickness of the active layer, which becomes too thin to support a sufficiently high concentration at the wound edge. Then, the healing process stops and hence the final wound radius is nonzero.

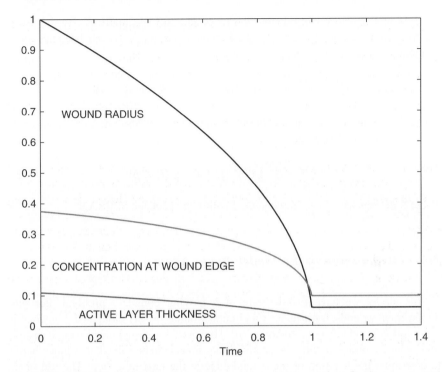

Fig. 2 The healing of a wound for $\alpha = 0$ and $\beta = 1$, $c1 = 0.01$, $c_2 = 0.1$, $D = 1/144$, and $\hat{c} = 0.1$. Note that the units for the wound radius and active layer thickness will differ from the units of the concentration. The concentration is expressed in arbitrary units

3 Numerical Method

3.1 The Solution of the Model due to Sherratt and Murray

The numerical method is explained for the wound contraction model since this model contains most of the difficulties. To solve the system of partial differential equations a finite element strategy is used. In this section we present the solution method for our preliminary results consisting of the solution of the equations of Sherratt and Murray. Piecewise linear basis functions are used for the triangular elements with a Newton–Cotes integration, which makes the mass matrix diagonal. For the time integration method, we use the following IMEX method:

$$
\begin{aligned}
M\frac{\underline{m}^{k+1}}{\Delta t} &= M\frac{\underline{m}^k}{\Delta t} + D_1 S\underline{m}^{k+1} + M_1(\underline{m}^k, \underline{c}^k)\underline{m}^{k+1}, \\
M\frac{\underline{c}^{k+1}}{\Delta t} &= M\frac{\underline{c}^k}{\Delta t} + D_2 S\underline{c}^{k+1} - \lambda M\underline{c}^{k+1} + \underline{b}^k.
\end{aligned}
\tag{26}
$$

Here M and S respectively denote the mass and stiffness matrices. By the use of the Newton–Cotes integration rule, the element mass matrix for M_1 becomes for an element Ω_ε surrounded by nodes i, j, and p:

$$M_1^{\Omega_\varepsilon} = \frac{|\Delta|}{6} \, \mathrm{diag}(s(c_\alpha^k)(2 - n_\alpha^k)), \text{ for } \alpha \in \{i, j, p\}, \tag{27}$$

where $\frac{|\Delta|}{2}$ represents the area of the triangular element. The αth component of the element vector for $\underline{b}$ becomes

$$b_\alpha^{\Omega_\varepsilon} = \frac{|\Delta|}{6}(f(n_\alpha^k)), \text{ for } \alpha \in \{i, j, p\}. \tag{28}$$

These functions have been implemented within the package SEPRAN.

3.2   The Solution of the Model due to Adam

In this model we have to take care of a moving boundary problem. For one-dimensional wounds, for instance for circular wounds or very elongated thin wounds, the solution is relatively straightforward. However, for complicated wound geometries, where a problem with two spatial dimensions is to be solved, it is much harder to determine the location of the wound edge at the next time step. The use of the moving grid method, though possible, will be very laborous here. Therefore, it is suggested here to use the level set method. The level set method is also capable of dealing with topological changes of the wound, such as splitting of a wound into two wounds, which possibly takes place if the wound geometry is non-convex. The level set method has been used successfully in various applications, such as Stefan problems and two phase problems, to mention a few of them. An interesting book on the level set method is due to Osher and Fedkiw [7]. Though, the level set method has not yet been implemented, the basic principle is sketched for the sake of completeness. We will certainly implement the level set method for the wound healing problems in future studies. Let ϕ be the level set function, which is a signed distance function, let further $\phi = 0$ correspond with the wound edge, then, we have

$$\frac{d}{dt}\phi(x(t), y(t), t) = 0 \Rightarrow \frac{\partial \phi}{\partial t} + \mathbf{v} \cdot \mathrm{grad}\phi = 0. \tag{29}$$

Here $\mathbf{v}$ denotes the velocity of the level curve $\phi = 0$, which corresponds to the wound edge. The wound edge velocity is obtained by Equation (41). Note that we should monitor whether the concentration of the growth factors exceeds the threshold concentration. Hence the wound edge will move at some locations, whereas at some other locations the wound edge does not move (temporarily). The above equation is extended to the entire domain of computation to track the moving interface emplicitly. Hence, it is necessary to extend $\mathbf{v}$ continuously over the entire domain of computation $\Omega = \Omega_w \cup \overline{\Omega}_L \cup \Omega_u$. At time $t = 0$ the level set function is initialized as a "signed distance function" by

$$
\phi(\mathbf{x}, 0) = \begin{cases} +\text{dist}(\mathbf{x}, S(0)) & \text{for } \mathbf{x} \in \Omega_L(0) \cup \Omega_u(0), \\ 0 & \text{for } \mathbf{x} \in W(0), \\ -\text{dist}(\mathbf{x}, S(0)) & \text{for } \mathbf{x} \in \Omega_w(0). \end{cases} \tag{30}
$$

Note that since the wound edge moves, the domains Ω_w, Ω_L, and Ω_u are functions of time. 'By this choice we have that at each point in Ω, the level set function ϕ is given the value of the distance between the point and the closest wound edge position, with a positive sign for locations outside the wound and a negative sign for locations inside the wound. In general, as time proceeds the level set function is no longer a distance function. To keep ϕ a distance function, a reinitialization step is taken by solving

$$
\frac{\partial \psi}{\partial \tau} = \text{Sign}(\phi(\mathbf{x}, t)) \left(1 - \|\text{grad } \psi\|\right), \tag{31}
$$

where τ is a fictitious time, with initial condition $\psi(\mathbf{x}, 0) = \phi(\mathbf{x}, t)$. The above Equation (31) guarantees that whenever $\phi = 0$, then, $\psi = 0$, *i.e.* the zero level set of ϕ and ψ are the same. The steady-state solution of the above equation is given by $\|\nabla \psi\| = 1$, which is a characteristic of a distance function. After solution of Equation (31) to obtain ψ as a signed distance function, we set $\phi = \psi$. There is an abundant variety of solution methods of equation (31), see for instance Osher and Fedkiw [7]. In this work we consider a time integration method for Equation (31), and the reinitialization procedure is done such that within a narrow bandwidth of several meshsizes around the interface, the level set function indeed approximates a signed distance function, as presented in [18]. It is only there where this is important. At locations far away from the interface, only the sign of the level set function matters.

The solution of the equation for the growth factor concentration is straightforward by the level set method since the level set function clearly indicates the thickness of the active layer. The velocity $\mathbf{v}$ is obtained for the wound edge from the Equation of motion (41). It is extended to the whole domain of computation by solving

$$
\frac{\partial v_q}{\partial \tau} + \text{Sign}(\phi \frac{\partial \phi}{\partial x_q}) \frac{\partial v_q}{\partial x_q} = 0, \tag{32}
$$

where the subscript q denotes the coordinate direction and τ represents a fictitious time. This procedure was introduced by Chen et al. [19] for the solution of a Stefan problem. Some preliminary results are shown in Figures 3 and 4 for an elliptic wound and for a butterfly-shaped wound, which has a non-convex shape. In the preliminary plots, the concentration at the wound edge is assumed to exceed the threshold value at all times. This assumption will be relaxed in a future study. It can be seen that the level set method is capable of dealing with complicated geometries and topological changes of the wound. This is clearly illustrated by the butterfly-shaped wound, which does not reflect a realistic clinical example.

To compare the healing kinetics of wounds of several geometry, the wound area has been plotted as a function of time for wounds with various initial geometry

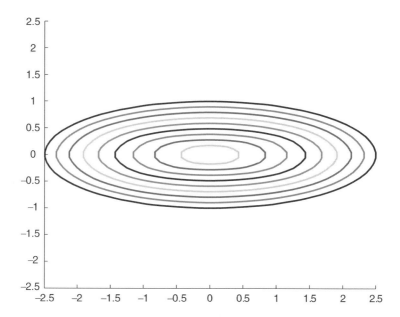

Fig. 3 The healing of an elliptic wound at several stages

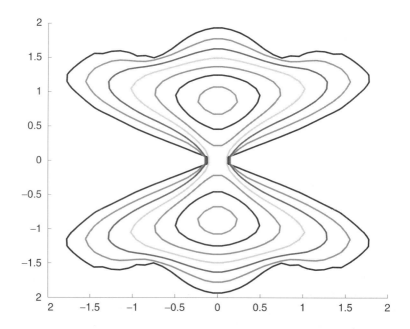

Fig. 4 The healing of a butterfly-shaped wound at several stages

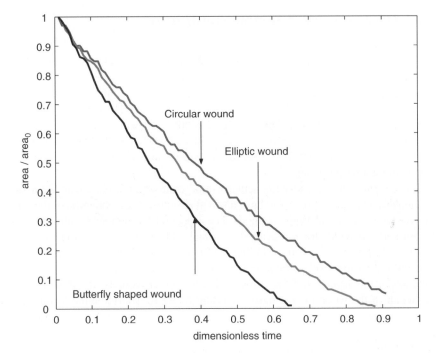

Fig. 5 The healing rate of wounds with various geometries

in Figure 5. All wounds have the same initial area. The circular wound has a radius of 1.6, the initial elliptic wound equation is given by $\left(\frac{x}{2.57}\right)^2 + (y)^2 = 1$. It can be seen that the butterfly-shaped wound heals with the largest rate. It appears from Figure 5 that the minimal curvature of the wound is crucially important. The wound becomes flatter and more like a cutaneous wound if the curvature is small. However, the healing kinetics is similar for all wounds. The small wiggles in the curve are caused by interpolation routine used to determine the zero level set of the level set function ϕ to identify the wound edge. Furthermore, once the wound domain has been determined, the area needs to be computed numerically. Hence, the wiggles are due to numerical methods in the postprocessing steps.

4 Numerical Experiments

Since the thickness of human skin is in the order of a millimeter, the model can be considered as two dimensional. In this section, we show some results using the combined models for angiogenesis and wound contraction due to Sherratt and Murray [1]. Subsequently, we show some results by the use of the model due to Adam which is based on the presence of an active layer adjacent to the wound edge.

4.1   The Model due to Sherratt and Murray

First, we show some finite element results for a circular wound with radius 1 of the
neovascularization model. Our simulations indicate that the oxygen and growth factor
concentrations reach the steady-state solution rapidly. This confirms the assumption
under which Maggelakis [12] found the analytic solutions. The capillary tip density
exhibits a transient behavior. Due to diffusion, the tip density decreases for a while
outside the wound region, whereas it increases at all stages within the wound region.
As time proceeds, it converges towards its equilibrium value. The results are not
shown in the present paper.

Subsequently, we consider the actual wound contraction time, where we assume
the neovascularization process to have taken place entirely, as a function of the aspect
ratio of rectangular and elliptic wounds. The aspect ratio is defined by the ratio of the
length and the width for a rectangular wound and by the ratio of the longer and
shorter axis for elliptic wounds. Following Sherratt and Murray [1] the wound edge
is identified by the level curve $n = 0.8$. In Figure 6 we show the healing time as a
function of the aspect ratio for ellipsoidal and circular wounds. It can be seen that
the width of the wound is the most important factor to determine the healing time.

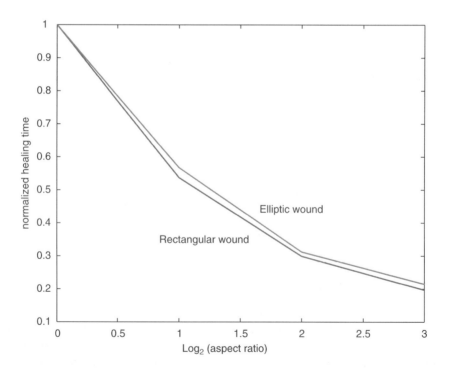

Fig. 6 Healing times as a function of the aspect ratio of rectangular and ellipsoidal wounds with the
same area

However, the relation is not linear, the decrease of the healing time as a function of the aspect ratio is much less than for the linear case.

Subsequently, when the models are combined, the full system of PDE is solved, and the influence of the oxygen production on the healing rate is shown in Figure 7. For larger values of λ_n wound healing slows down. This is caused by the decrease of the MDGF growth factor concentration as more oxygen is supplied. The decrease of MDGF growth factors, decreases the speed of capillary formation and hence the supply of nutrients needed for cell division in the wound contraction process deteriorates. This results into a delay of mitosis, and hence, the wound contraction slows down. Finally, the influence of the capillary density on the healing time is investigated. The influence is quantified by the power p in the function $\phi_p(y)$. For $0 < p < 1$ the influence is large for $n/n_m < 0.5$ and small elsewhere. For $p > 1$, the behavior is opposite. The calculations, not shown here, result into a strange behavior for $0 < p \leq 1$ that the wound heals from the edge at the early stages and starts healing from the wound center at the later stages. This behavior is not realistic. Note that $p = 1$ corresponds to a linear coupling. For $p > 1$ the wound only heals from the edge, which is more realistic from an intuitive point of view. Therefore, the expectation is that $p > 1$ reflects the right behavior. For the input values we used the same values as Sherratt and Murray [1] and Maggelakis [12].

4.2 The Model due to Adam

The growth factor distribution in a general-shaped wound is computed by the use of a standard Galerkin finite element method with piecewise linear basis functions. The time-dependent problem is solved by the use of the Euler backward method.

In this section we consider the growth factor distribution obtained by the use of the finite element method in the vicinity of an *elliptic* wound. In the calculations of this section, we take $\phi_p = 1$ (no coupling with neovascularization). In the first subsection the accumulation of the growth factor is taken into account. The second subsection deals with the influence of the wound geometry on the growth factor distribution. Finally, we consider the healing rate for a circular wound in relation to the concentration profile of the growth factor. In all the calculations that are presented in this section the outer boundary is taken far enough away such that its influence is negligible. On the outer boundary we apply the condition that the growth factor concentration is zero. The results of the finite element method have been validated by the analytical solution in [8] for a rectangular (one-dimensional) wound.

4.2.1 The Growth Factor Accumulation

For the sake of illustration we compute the growth factor profile for the following data following Adam [8]

$$P = \lambda = 5 \cdot 10^{-5} s^{-1}, D = 5 \cdot 10^{-5} cm^2/s, R = 1cm, \delta = 0.2cm.$$

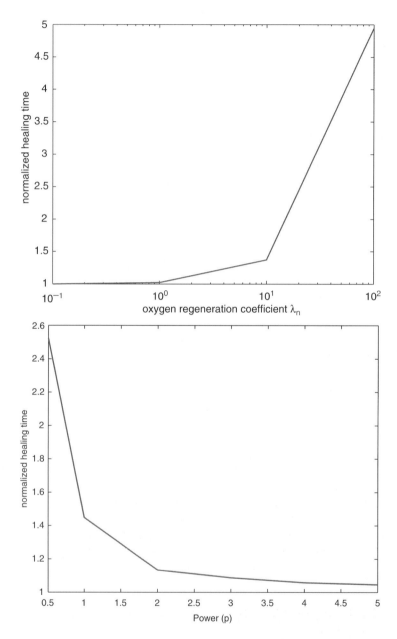

Fig. 7 Circular wound:left: The influence of oxygen production on the healing time ($p = 5$); right:
The influence of the power p in ϕ on the healing time ($\lambda_o = 0.1$)

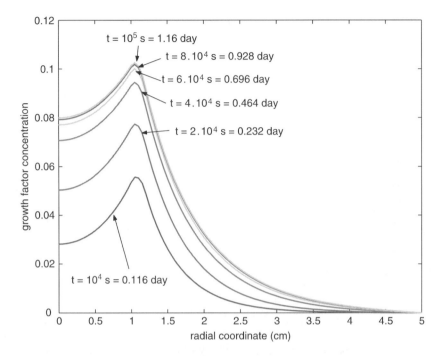

Fig. 8 The radial growth factor concentration profile at sequential instants of time for a circular wound. The concentration is expressed in arbitrary units

The results are shown in Figure 8.

In Figure 8 the radial profile of the growth factor concentration has been plotted at consecutive times up to $t = 10^5$ s. From the calculations it can be seen that the differences between the solutions at consecutive times become smaller and smaller and hence it follows that the profile at $t = 10^5$ s does not differ much from the profile for the time-independent solution, that is as $t \to \infty$. From this it can be seen that the growth factor accumulates at the wound edge up to the threshold concentration. This time is referred to as an incubation time for the healing process to start. In Figure 9 the growth factor concentration at the wound edge is plotted for the same circular wound as a function of time. It can be seen that at approximately a day the growth factor concentration is about to converge to its maximum. This may be seen as an incubation time before the actual healing starts.

4.2.2 The Geometrical Aspects of the Wound on the Growth Factor Concentration

In this section the concentration of the growth factor is considered numerically for several wound geometries. The healing process is not modeled analytically here; subsequent work will be devoted to this. The first example is for an elliptic wound whose

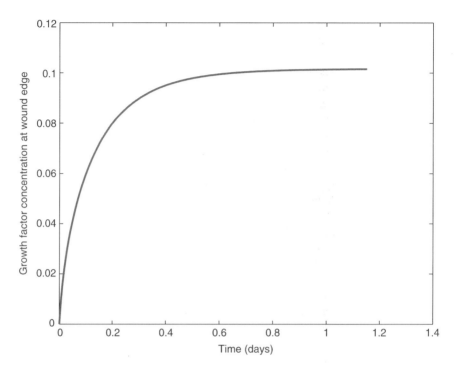

Fig. 9 The accumulation of the growth factor concentration profile at the wound edge for a circular wound. The concentration is expressed in arbitrary units

equation is given by $x^2 + 4y^2 = 1$. In the simulations the thickness of the active layer is chosen constant over all positions of the wound edge. The contour lines of the growth factor concentration are shown in Figure 10.

For the same geometry, the growth factor concentration over the edge of the wound is shown in Figure 11. In Figure 11 the growth factor concentration increases from the position where the curvature of the elliptic wound is maximal up to a maximum value where the curvature of the wound is minimal. There is a small decrease of the growth factor concentration on the wound edge as $x \to 0$, that is the position with the minimum curvature. This is consistent with the results using the analytic solution in [16] plotted in Figure 12. The analytic solution also predicts a slight decrease of the growth factor concentration as the wound radius becomes larger, that is, the wound curvature decreases. This effect takes place if the wound radius exceeds the radius for which the growth factor concentration is maximal. It can be seen in Figure 10 that the maximum concentration is located near the wound edge and near the spot of minimum curvature of the wound edge. This is contrary to the expectations where for a spherical wound there exists a critical size under which the wound is expected to heal within the lifetime of an animal. We remark that the results for the concentration at the wound edge from the analytical solution for a circular wound due to Adam agree perfectly with the results from the finite element method. For the

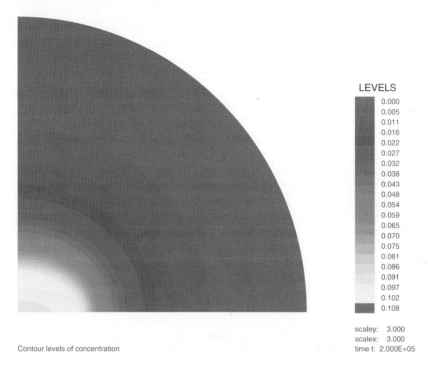

LEVELS

	0.000
	0.005
	0.011
	0.016
	0.022
	0.027
	0.032
	0.038
	0.043
	0.048
	0.054
	0.059
	0.065
	0.070
	0.075
	0.081
	0.086
	0.091
	0.097
	0.102
	0.108

scaley: 3.000
scalex: 3.000
time t: 2.000E+05

Contour levels of concentration

Fig. 10 Contour lines of the growth factor concentration for an elliptic wound. The concentration is expressed in arbitrary units

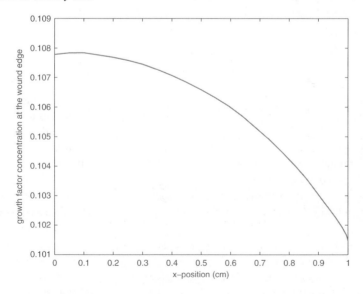

Fig. 11 Line plot of the growth factor concentration over the edge of the elliptic wound. The equation of the elliptic wound edge is given by $x^2 + 4y^2 = 1$. The concentration is expressed in arbitrary units

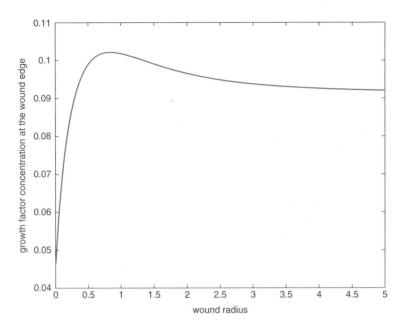

Fig. 12 The growth factor concentration at the wound edge as a function of the wound radius as computed using the analytic solution by Adam [10]. The thickness of the active layer is fixed. The concentration is expressed in arbitrary units

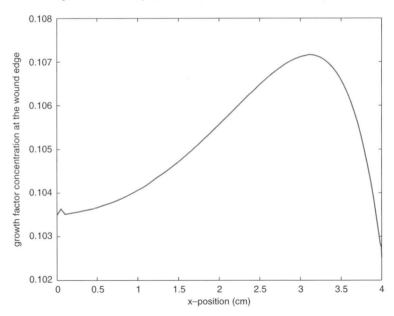

Fig. 13 Line plot of the growth factor concentration over the edge of the elliptic wound. The equation of the elliptic wound edge is given by $\frac{x^2}{16} + y^2 = 1$. The concentration is expressed in arbitrary units

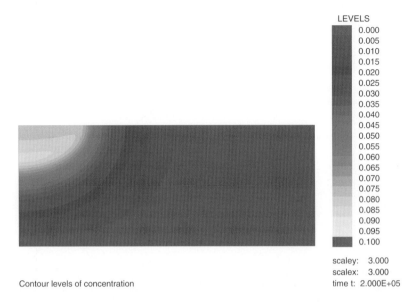

LEVELS

0.000
0.005
0.010
0.015
0.020
0.025
0.030
0.035
0.040
0.045
0.050
0.055
0.060
0.065
0.070
0.075
0.080
0.085
0.090
0.095
0.100

scaley: 3.000
scalex: 3.000
time t: 2.000E+05

Contour levels of concentration

Fig. 14 Contour lines of the growth factor concentration for an elliptic wound in three dimensions with rotational symmetry. The concentration is expressed in arbitrary units

case of contour lines around the elliptic wound, the maximum curvature is such that the radius of curvature is smaller than the critical radius in Figure 11. For a more elongated elliptic wound, $\frac{x^2}{16} + y^2 = 1$, the concentration of the growth factor along the wound edge differs significantly. A line plot is shown in Figure 13. It can be seen that the maximum concentration of the growth factor is obtained at neither of the locations of extremal curvature of the wound edge. At the location at the wound edge where the growth factor concentration is maximal, the wound is most likely to heal if the wound edge concentrations are almost critical. This is in line with the results from the analytic solution due to Arnold and Adam [16].

4.2.3   Incorporation of the Wound Depth

In this application we consider a circular wound with a depth and an active layer. This configuration is modeled by rotational symmetry and hence only two spatial coordinates are needed. The contour lines of the growth factor concentration are shown in Figure 14 and a graph of growth factor concentration over the wound edge is displayed in Figure 15. It can be seen that the concentration of the growth factor is maximal near the symmetry axis. This implies that at this position the concentration is more likely to satisfy the healing condition than at the other spots if the wound edge concentrations are almost critical. Furthermore, due to accumulation of the growth factor concentration at the wound edge during the transient part, the healing process will start a little earlier there than at any other location at the wound edge.

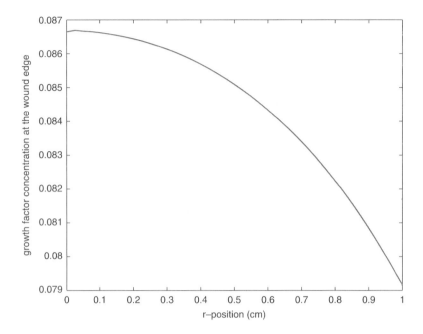

Fig. 15 Line plot of the growth factor concentration over the edge of the elliptic wound with rotational symmetry. The equation of the wound edge in polar coordinates is given by $r^2 + 4(z-2)^2 = 1$. Here r denotes the radial coordinate. The concentration is expressed in arbitrary units

5   Conclusions

Some mathematical models for epidermal wound healing have been presented. Further, some results from simulations have been described. An innovation in the present paper is the assumption to allow overlapping between the processes of neovascularization and wound contraction. The nonlinear impact of the aspect ratio of the wound on the healing time is demonstrated. Further, some features of the model due to Adam are highlighted. An important consequence of the model due to Adam is that along the perimeter of non-circular wounds, there are regions where the growth factor concentration may exceed or may be less than the threshold concentration for wound healing. Further, the models predict an incubation time (waiting time) before actual healing sets in.

References

1. Sherratt J.A., Murray J.D. (1991) Mathematical analysis of a basic model for epidermal wound healing. J Math Biol 29:389–404
2. Filion J., Popel A.P. (2004) A reaction diffusion model of basic fibroblast growth factor interactions with cell surface receptors. Ann Biomed Engng 32(5):645–663

3. Maggelakis S.A. (2003) A mathematical model for tissue replacement during epidermal wound healing. Appl Math Modell 27(3):189–196
4. Gaffney E.A., Pugh K., Maini P.K. (2002) Investigating a simple model for cutaneous wound healing angiogenesis. J Math Biol 45(4):337–374
5. Murray J.D. (2004) Mathematical biology II: spatial models and biomedical applications. Springer, New York
6. Greenspan H.P. (1972) Models for the growth of a solid tumor by diffusion. Stud Appl Math 52:317–340
7. Osher S., Fedkiw R. (2004) Level set methods and dynamic implicit surfaces. Springer, New York
8. Adam J.A. (1999) A simplified model of wound healing (with particular reference to the critical size defect). Math Comput Modell 30:23–32
9. Adam J.A. (2002) The effect of surface curvature on wound healing in bone: II. The critical size defect. Math Comput Modell 35:1085–1094
10. Adam J.A. (2002) Healing times for circular wounds on plane and spherical bone surfaces. Appl Math Lett 15:55–58
11. Maggelakis S.A., Adam J.A. (1990) Diffusion regulated growth characteristics of a prevascular carcinoma. Bull Math Biol 52:549–582
12. Maggelakis S.A. (2004) Modeling the role of angiogenesis in epidermal wound healing. Discr Cont Sys 4:267–273
13. Vermolen F.J., van Baaren E., Adam J.A. (2006) A simplified model for growth factor induced healing of wounds. Math Comput Modell 44:887–989
14. Adam J.A. (2002) The effect of surface curvature on wound healing in bone. Appl Math Lett 15:59–62
15. Adam J.A. (2004) Inside mathematical modeling: building models in the context of wound healing in bone. Discr Cont Dyn Sys B, 4(1):1–24
16. Arnold J.S., Adam J.A. (1999) A simplified model of wound healing II: The critical size defect in two dimensions. Math Comput Modell 30:47–60
17. Arnold J.S. (2001) A simplified model of wound healing III: The critical size defect in three dimensions. Math Comput Modell 34:385–392
18. Javierre E., Vuik C., Vermolen F.J., Segal A. (2005) A level set method for particle dissolution in a binary alloy. Report at the Delft Institute of Applied Mathematics, 05–03
19. Chen S., Merriman B., Osher S., Smereka P. (1997) A simple level set method for solving Stefan problems. J Comput Phys 135:8–29

Appendix: Derivation of the Wound Healing Rate Equation

It is known from basic multivariable calculus that the area of the surface $z = f(x, y)$ over a closed region D in the $x - y$ plane is given by

$$A = \int_D dS = \int_D \sqrt{1 + f_x^2(x, y) + f_y^2(x, y)} dx dy, \tag{33}$$

provided that f and its first-order partial derivatives are continuous on D. For a spherical surface of radius a, we have $f(x, y) = \sqrt{a^2 - x^2 - y^2}$, so

$$dA = \frac{a}{\sqrt{a^2 - x^2 - y^2}} dx dy = \frac{ar}{\sqrt{a^2 - r^2}} dr d\theta, \tag{34}$$

when expressed in polar coordinates (r, θ). Hence for the region $D = \{r \leq R\}$, we have

$$A(R) = \int_0^{2\pi} \int_0^R \frac{ar}{\sqrt{a^2 - r^2}} dr d\theta = 2\pi a \left(a - \sqrt{a^2 - R^2}\right). \tag{35}$$

Note that this reduces to the expected result $A(R) = \pi R^2$ as $\frac{a}{R} \to \infty$. From this expression it follows that if a circular wound shrinks from an initial radius R_0 (at $t = 0$) to $R(t)$ at time t, then, the magnitude of the change in area ΔA is

$$\Delta A = 2\pi a \left(\sqrt{a^2 - R^2(t)} - \sqrt{a^2 - R_0^2}\right). \tag{36}$$

Assuming a constant thickness of the spherical skull of $h \ll a$, in terms of the total volume recovered by the healing process in $[0, t]$, the healed wound volume is approximated by

$$\Delta V = \Delta A h = 2\pi h \int_0^t \int_{R(\tau)}^{R_0} \frac{r \overline{S}(R(\tau), r)}{\sqrt{a^2 - r^2}} dr d\tau. \tag{37}$$

In the above equation the function $\overline{S}(R(t), r)$ represents a measure of the volume rate of the regeneration of bone or tissue per unit area as a function of position and the wound radius. The right-hand side of Equation (37) (when multiplied by the factor $2\pi h$) represents the time-accumulated volume of bone accrued at the (locally) instantaneous rate per unit surface area $\overline{S}(R(t), r)$. Thus Equation (37) is essentially a statement of volume conservation: as the wound heals in a symmetrical manner (in this model) the total volume that has been "healed" (replaced by new bone) as the radius decreases from $R(0)$ to $R(t)$ in time $t \geq 0$ is approximately $h \Delta A$. This will be inaccurate if the condition $h \ll a$ is not satisfied, but an appropriate geometrical adjustment can be made in this case. For the animal models initiating this paper it is not a concern.

The choice of the new bone growth rate per unit area $\overline{S}(R(t), r)$ should be a reasonable description of the nature of the healing in such wounds. Unfortunately, as far as known, there appears to be no useful clinical data available in the current literature (despite an extensive search). Thus, it is necessary to choose a phenomenologically plausible functional form for $\overline{S}$; something that is a fairly sharp maximum at the wound edge, and that therefore falls sharply away from the wound edge. The reason for this is based on the fundamental notion that limited wound damage induces increased cell proliferation in the vicinity of the wound edges, and this rate of proliferation *per unit area* accelerates as the wound heals. After total healing has occurred the tissue returns to its normal state.

Equation (37) is combined with Equation (36) and subsequently the result is differentiated with respect to time, to obtain

$$\frac{dR^2}{dt} = -2\sqrt{a^2 - R^2(t)} \int_{R(t)}^{R_0} \frac{r\overline{S}(R(t), r)}{\sqrt{a^2 - r^2}} dr. \tag{38}$$

If the measure for the rate of bone or tissue regeneration is chosen to be a Riemann integrable function, then, it can be seen that the right-hand side is Lipschitz continuous with respect to R and that $R(t) = R_0$ is the only solution here. This was not noted in [10]. Therefore, to remedy this error we will use a Dirac Delta distribution for this function $\overline{S}$ in order to have a non-zero healing rate at the initial stage. Mathematically, this corresponds to the initiation of the healing process. For this purpose we will use $\overline{S}(R(t), r) = S(R(t))\delta(r - R(t))$ in the integral formulation. This is a correction to the earlier work [10]. Here $\delta(x)$ denotes the Dirac Delta "function" and $S(R)$ represents a continuous rate function depending on R. This form for $\overline{S}$ then represents an inward moving "wave of healing" as the wound shrinks. For these functions, note that $R(t) \leq R_0$ and for any arbitrary function $f(r)$:

$$\lim_{\epsilon \to 0} \int_{R(t)}^{R_0} f(r)\delta_\epsilon(r - R(t))dr = \frac{1}{2}f(R(t)). \tag{39}$$

Then, this gives the following relation:

$$\frac{dR^2}{dt} = -2\sqrt{a^2 - R^2(t)} \int_{R(t)}^{R_0} \frac{rS(R(t))\delta(r - R(t))}{\sqrt{a^2 - r^2}} dr = -R(t)S(R(t)). \tag{40}$$

From the nature of the Dirac Delta distribution, it follows that this choice predicts a nonzero initial healing rate. Noting that the dependence of the radius of the sphere, a, vanishes, S could be made dependent on a; alternatively the dependence of a follows from the solution of the equation for the growth factor. Simplifying Equation (40) yields

$$\frac{dR}{dt} = -\frac{1}{2}S(R(t)), \text{ with } R(0) = R_0. \tag{41}$$

This equation is integrated to obtain R as a function of t. A second assumption that can be introduced is that the wound only heals if the growth factor concentration exceeds this threshold value. Therefore at each time during the healing process, it is necessary to know whether the growth factor concentration at the wound edge exceeds the threshold value.

Numerical Modelling of Self Healing Mechanisms

Joris JC Remmers[1] and René de Borst[2]

[1] *Delft University of Technology, Faculty of Aerospace Engineering, Delft, The Netherlands*
  E-mail: J.J.C.Remmers@TUDelft.nl
[2] *Eindhoven University of Technology, Department of Mechanical Engineering, Eindhoven,*
  The Netherlands E-mail: r.d.borst@TUE.nl

A number of self healing mechanisms for composite materials have been presented in the previous chapters of this book. These methods vary from the classical concept of micro-encapsulating of healing agents in polymer systems to the autonomous healing of concrete. The key feature of these self healing mechanisms is the transport of material to the damaged zone in order to establish the healing process. Generally, this material is a fluid and its motion is driven by capillary action which enables transportation over relatively large distances requiring little or no work. In the micro-encapsulated polymers as developed by White et al. [1], this liquid material is a healing agent, which is enclosed in the material by micro-encapsulation. When the capsule is ruptured by a crack, the healing agent will flow into the crack, driven by capillary action. Polymerisation of this healing agent is triggered by contact with catalysts which are inserted in the material and whose position is fixed. The new polymerised material will rebond the crack surfaces.

In the case of autonomous healing of concrete, the roles of healing agent and catalysts are reversed. Here, the uncarbonised cementitious material can be considered as the healing agent. However, the position of this constituent in the material is fixed and will only react when it is contacted by a mobile catalyst, which is water in this particular case. The water is not stored in the concrete beforehand but will reach the uncarbonised material from outside. The majority of the fluid will reach the uncarbonised material through the cracks, and, as a result of the porous structure of concrete, fluids can also reach these parts by penetration into the bulk material from external boundaries and crack surfaces.

In order to increase the efficiency of these self healing mechanisms, the knowledge of the behaviour of the different constituents and their mutual interaction is of key importance. Indeed, this behaviour can be analysed experimentally, but numerical simulations on the microscopic level of observation can be of assistance. In this chapter, a numerical method is presented that allows to analyse mechanisms that play a role in self healing processes. The method is able to simulate the three main stages of the self healing process: (i) the nucleation and growth of cracks due to external loads, (ii) the transport of fluid material in the cracks and the porous bulk material,

---

[1] Currently at Eindhoven University of Technology, Department of Mechanical Engineering

S. van der Zwaag (ed.), *Self Healing Materials. An Alternative Approach to 20 Centuries of Materials Science*, 365–380.
© 2007 *Springer*.

and (iii) the subsequent rebonding of the crack due to chemical reaction of the healing agent and a catalyst.

The method is composed of state of the art numerical techniques that are based on the finite element method. First, the evolution of damage or crack growth can be simulated by the cohesive segments method [2], which is an extension of the partition of unity approach to cohesive fracture. The cohesive segments method is able to simulate the nucleation, growth and coalescence of multiple interacting cracks, irrespective of the structure of the underlying finite element mesh. The subsequent flow of fluids through the deformable, porous medium is governed by a classical two-phase theory [3]. A partition-of-unity extension of this model accounts for the flow of fluids through the cracks [4, 5]. The final stage of the healing process, the rebonding of the crack surfaces, is simulated by a reversed cohesive constitutive model [6], in which the strength of the cohesive zone is restored as a function of time and stress state at the discontinuity.

The proposed method allows for a truly multiscale analysis since the events that occur on different length scales are captured within a single numerical scheme. In principle, the method can be used to analyse the behaviour of small structural components or coupons with a specific length in the order of tens of millimetres, the typical size of the micro-cracks in the material is of the order of millimetres. Finally, the smallest length scale is the opening of these cracks which serve as a channel for fluid flow. The magnitude of this opening is in the order of a few microns, depending on the material.

1   The Simulation of Fracture

In the cohesive zone approach to fracture, the process zone is lumped in a single plane or line ahead of the existing crack [7, 8]. The relation between the work expended in this cohesive zone and that in the crack tip field is typically such that the stress singularity is cancelled and the near tip stresses in the process zone are finite. Needleman [9] extended the approach by inserting cohesive constitutive relations at specified planes in the materials, whether or not there is a crack. Apart from the fact that this approach can capture crack nucleation, there is an additional technical advantage. By specifying the cohesive relation along a surface, there is no need to determine the length of the cohesive zone.

An independent constitutive relation that describes the separation of the cohesive surface governs the failure characteristics of the material. This cohesive constitutive relation, together with the constitutive relation for the bulk material and the appropriate balance laws and boundary and initial conditions, completely specify the problem. Fracture, if it takes place, emerges as a natural outcome of the deformation process without introducing any additional failure criterion.

The cohesive zone approach is useful, especially when fracture takes place along well-defined interfaces as, for example in a lamellar solid. Here, the cohesive zone can be incorporated in the finite element mesh beforehand by means of interface

 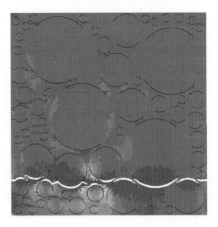

Fig. 1 Simulation of fracture in a cementitious material. The figure on the left shows the finite element mesh of a specimen with circular aggregates. The figure on the right shows the hydrostatic stress and the deformed geometry after total fracture. The displacements are magnified by a factor 10.0. (From Tijssens et al. [12])

elements. However, for a solid that is homogeneous on the scale modelled, the crack path is often not known and the placement of cohesive zones can be problematic. To allow for crack growth in arbitrary directions, interface elements are placed between all continuum elements in the mesh [10]. This approach enables the simulation of complex fracture phenomena such as crack branching [11] and crack initiation away from the crack tip. Tijssens et al. [12] have been using the method to simulate nucleation and coalescence of micro-cracks in the micro-structure of heterogeneous materials, see Figure 1.

Unfortunately, the approach of placing interfaces between all continuum elements is not completely mesh independent. In fact, since the interface elements are aligned with element boundaries, the orientation of cracks is restricted to a limited number of angles. In addition, the initial stiffness of the cohesive zone, which is used to simulate a perfect bond prior to cracking, reduces the overall stiffness of the body, which gives rise to an ill-posed problem. Increasing the dummy stiffness can reduce the magnitude of this error at the expense of possible numerical instabilities [13].

In order to avoid finite element mesh-related problems, the cohesive surface is incorporated in the continuum elements by using the partition-of-unity property of finite element shape functions [14, 15] in conjunction with a discontinuous mode incorporated at the element level. In this approach, the cohesive zone is represented by a jump in the displacement field of the continuum elements [16, 17]. The magnitude of the displacement field is determined by an additional set degrees of freedom, which is added to the existing nodes of the finite element mesh. A crack (or cohesive surface) can be extended during the simulation at any time and in any direction, irrespective of the structure of the underlying finite element mesh as shown in Figure 2. In addition, most numerical problems that are related to the initial dummy stiffness are avoided since the cohesive zone is only inserted upon crack propagation. Finally,

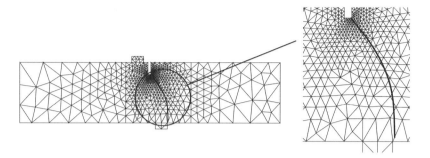

Fig. 2 Crack growth in a single edge notched beam, simulated with the partition of unity approach to cohesive fracture. The cohesive surface (the bold line) crosses the continuum elements. (from Wells and Sluys [16])

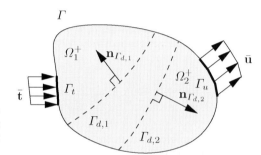

Fig. 3 Domain Ω with two discontinuities, $\Gamma_{d,1}$ and $\Gamma_{d,2}$ (dashed lines). It is assumed that the discontinuities do not cross

an important advantage over traditional remeshing techniques is that the topology of the finite element mesh is not modified: The number of nodes and elements remains the same, as well as their mutual connectivity.

In order to accommodate the simulation of multiple interaction crack patterns, the aforementioned partition of unity approach to cohesive fracture has been extended to the cohesive segments method in which multiple discontinuities are defined [2]. Since each part is supported by a unique additional set of degrees of freedom, the interaction between different cracks can be modelled as well. The cohesive segments method can be used to simulate the interfaces between the different constituents in the material, as well as cohesive fracture in the bulk material.

1.1 Kinematic Relations

A key feature of the partition of unity-based approaches to cohesive fracture is that the discontinuity that represents the cohesive zone is incorporated in the kinematic relations that form the basis of the element formulation. Consider the body Ω in Figure 3 that is crossed by two discontinuities Γ_1 and Γ_2 that do not cross. Each discontinuity splits the body in two parts, denoted by Ω_j^+ and Ω_j^-. The total displacement of a material point $\mathbf{x}$ in the body consists of a regular part $\hat{\mathbf{u}}$ plus an additional part $\tilde{\mathbf{u}}$ that is added on one side of the discontinuity only:

$$\mathbf{u}(\mathbf{x}, t) = \hat{\mathbf{u}}(\mathbf{x}, t) + \sum_{j=1}^{m} \mathcal{H}_{\Gamma_{d,j}}(\mathbf{x})\tilde{\mathbf{u}}_j(\mathbf{x}, t), \tag{1}$$

where $\mathcal{H}_{\Gamma_{d,j}}(\mathbf{x})$ is the Heaviside jump function associated with discontinuity j, which is equal to one when $\mathbf{x}$ resides in the Ω_j^+ part of the domain and equal to zero when this point is in the Ω_j^-. Assuming a small strain formulation, the strain field can be obtained by differentiating Equation (1) with respect to the position $\mathbf{x}$:

$$\epsilon(\mathbf{x}, t) = \nabla^s \hat{\mathbf{u}}(\mathbf{x}, t) + \sum_{j=1}^{m} \mathcal{H}_{\Gamma_{d,j}}(\mathbf{x})\nabla^s \tilde{\mathbf{u}}_j(\mathbf{x}, t) \qquad \mathbf{x} \notin \Gamma_{d,j}, \tag{2}$$

where a superscript s denotes the symmetric part of a differential operator. Note that at the discontinuities $\Gamma_{d,j}$, the strains are not defined. There, the magnitude of the displacement jump

$$[\![\mathbf{u}]\!]_j(\mathbf{x}, t) = \tilde{\mathbf{u}}_j(\mathbf{x}, t), \tag{3}$$

is the relevant kinematic quantity.

1.2   Momentum Balance

The discontinuous displacement and strain fields can be used as an input for the quasi-static equilibrium equations without body forces:

$$\nabla \cdot \boldsymbol{\sigma} = \mathbf{0} \qquad \mathbf{x} \in \Omega, \tag{4}$$

where $\boldsymbol{\sigma}$ is the Cauchy stress in the bulk material. The problem is completely specified with the boundary conditions:

$$\mathbf{n}_t \cdot \boldsymbol{\sigma} = \bar{\mathbf{t}} \qquad \mathbf{x} \in \Gamma_t, \tag{5}$$

$$\mathbf{u} = \bar{\mathbf{u}} \qquad \mathbf{x} \in \Gamma_u, \tag{6}$$

$$\mathbf{n}_{\Gamma_{d,j}} \cdot \boldsymbol{\sigma} = \mathbf{t}_j \qquad \mathbf{x} \in \Gamma_{d,j}, \tag{7}$$

where $\bar{\mathbf{t}}$ are the prescribed tractions on Γ_t with outward normal vector $\mathbf{n}_t$, $\bar{\mathbf{u}}$ are the prescribed displacements on Γ_u and $\mathbf{t}_j$ are the tractions at discontinuity $\Gamma_{d,j}$, which can be regarded as an internal boundary. The normal $\mathbf{n}_{\Gamma_{d,j}}$ at discontinuity $\Gamma_{d,j}$ points into Ω_j^+.

The weak form of the equilibrium equation can be obtained by multiplying Equation (4) by an admissible displacement field $\delta\mathbf{u}$ and integrating over the domain Ω:

$$\int_{\Omega} \delta\mathbf{u} \cdot (\nabla \cdot \boldsymbol{\sigma}) \, d\Omega = 0. \tag{8}$$

By taking the space of the admissible variations to be the same as the actual displacement field, Equation (1), the variations of the displacements can be decomposed as:

$$\delta \mathbf{u} = \delta \hat{\mathbf{u}} + \sum_{j=1}^{m} \mathcal{H}_{\Gamma_{\mathrm{d},j}} \delta \tilde{\mathbf{u}}_j \,. \tag{9}$$

Applying Gauss' theorem, using the symmetry of the Cauchy stress tensor and using the boundary conditions at the external boundary Γ_t and at the discontinuity planes $\Gamma_{\mathrm{d},j}$ gives the following equilibrium equations in weak form:

$$\int_{\Omega} \nabla^s \delta \hat{\mathbf{u}} : \boldsymbol{\sigma} \, \mathrm{d}\Omega + \sum_{j=1}^{m} \int_{\Omega} \mathcal{H}_{\Gamma_{\mathrm{d},j}} \nabla^s \delta \tilde{\mathbf{u}}_j : \boldsymbol{\sigma} \, \mathrm{d}\Omega + \sum_{j=1}^{m} \int_{\Gamma_{\mathrm{d},j}} \delta \tilde{\mathbf{u}}_j \cdot \mathbf{t}_j \mathrm{d}\Gamma$$

$$= \int_{\Gamma_t} \delta \hat{\mathbf{u}} \cdot \bar{\mathbf{t}} \mathrm{d}\Gamma + \sum_{j=1}^{m} \int_{\Gamma_t} \mathcal{H}_{\Gamma_{\mathrm{d},j}} \delta \tilde{\mathbf{u}}_j \cdot \bar{\mathbf{t}} \mathrm{d}\Gamma \,. \tag{10}$$

The first two terms in the equation represent the internal force in the bulk material. The stress $\boldsymbol{\sigma}$ can be obtained from the constitutive relation:

$$\dot{\boldsymbol{\sigma}} = \mathbf{C} \dot{\boldsymbol{\epsilon}} \,, \tag{11}$$

where $\dot{\boldsymbol{\epsilon}}$ is the strain-rate and $\mathbf{C}$ is the consistent tangent matrix of the bulk material. Note that any material law can be used in this formulation. An example of the use of a partition of unity based fracture model in combination with a non-linear constitutive relation for the bulk material can be found in [18]. The third term in Equation (10) contains the contribution of the cohesive surfaces to the internal force of the system. The tractions $\mathbf{t}_j$ are a function of the opening of the discontinuity $\mathbf{v}_j$ and can be expressed in rate form as well:

$$\dot{\mathbf{t}}_j = \mathbf{T} \dot{\mathbf{v}}_j \,, \tag{12}$$

where $\mathbf{T}$ is the consistent tangent of the cohesive law. An example of a cohesive relation is given in the next section.

1.3   Finite Element Discretisation

The discretisation of the equilibrium equations is performed by employing the partition of unity property of finite element shape functions. In standard finite element methods, the continuous displacement field is discretised by finite element shape functions. It was shown by Babuška and Melenk [14] that when a field is not continuous, it can still be discretised using these standard finite element shape functions in combination with an enhanced basis function, according to:

$$f(\mathbf{x}, t) = \sum_{i=1}^{n_{nod}} \phi_i(\mathbf{x}) \left(a_i(t) + \sum_{j=1}^{m} \gamma_j(\mathbf{x}) b_{ij}(t) \right), \tag{13}$$

where $\gamma_j(\mathbf{x}, t)$ is an enhanced basis with m orthogonal terms and b_{ij} are the additional nodal degrees of freedom that support the enhanced basis functions. The number m of enhanced base functions may be different for each node i in the model. However, in order to avoid linear dependency, the enhanced basis γ_j and the shape functions ϕ_i may not originate from the same span of functions. In this case, the enhanced basis functions $\gamma_j(\mathbf{x}, t)$ are equal to the step functions $\mathcal{H}_{\Gamma_{\mathrm{d},j}}$. The total displacement field inside an element can be written as:

$$\mathbf{u} = \mathbf{Na} + \sum_{j=1}^{m} \mathcal{H}_{\Gamma_{\mathrm{d},j}} \mathbf{Nb}_j . \tag{14}$$

where $\mathbf{N}$ is a matrix containing the shape functions ϕ_i for a the nodes that support this element. The strain field as defined in Equation (2) is equal to:

$$\boldsymbol{\epsilon} = \mathbf{Ba} + \sum_{j=1}^{m} \mathcal{H}_{\Gamma_{\mathrm{d},j}} \mathbf{Bb}_j , \tag{15}$$

where the matrix $\mathbf{B}$ is the standard operator that maps the nodal displacement components to strains in the bulk material. Finally, the magnitude of the jump in the displacement field $[\![\mathbf{u}]\!]_j$ as derived in Equation (3) is equal to:

$$[\![\mathbf{u}]\!]_j = \mathbf{Nb}_j . \tag{16}$$

The admissible displacement fields as presented in Equation (9) and their spatial derivatives can be discretised in a similar fashion. Substitution of these fields into (10) gives the discrete equilibrium equations. Details on this operation can be found in [2].

1.4 Implementation

The evolution of micro-cracks can be divided into three stages: (i) the nucleation, (ii) the growth, and finally (iii) the coalescence with other segments. In all cases, the same criterion is used to determine when and in which direction a segment is created or extended. In order to arrive at a consistent numerical model, this fracture criterion is intimately connected to the cohesive constitutive relation that governs the debonding process of a cohesive segment. This fracture criterion and the constitutive relation are discussed at the end of this section.

When the stress state in an integration point in the bulk material exceeds the strength of the material, a new cohesive segment is added. The segment, which is assumed to be straight, crosses this integration point and is extended into the neighbouring elements until it touches the outer boundaries of these elements, see Figure 4. The patch of neighbouring elements consists of all elements that share one of the nodes of the central element that contains the integration point in which the criterion was violated. The nodes that support these outer boundaries are not enhanced in

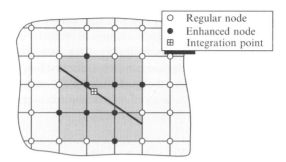

Fig. 4 Creation of a new cohesive segment (bold line) in virgin material. The segment crosses the integration point in the bulk material in which the criterion was violated. The segment is stretched until it touches the boundary of the patch of elements (grey area) that share the nodes of the element in which the yield criterion was violated

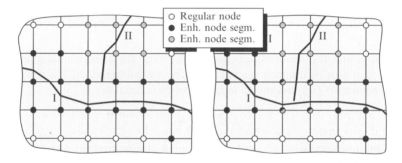

Fig. 5 Merging of two cohesive segments (bold lines). Segment II is extended until it touches segment I (right picture). Segment I can be regarded as a free edge so that segment II will not have a tip in this case. Consequently, all nodes of the element will be enhanced in order to support displacement discontinuity of segment II. Note that the nodes of these elements are enhanced twice

order to guarantee a zero-crack opening at the tips of the new segments. When the stress state in the tip violates the fracture criterion for a specific angle, the segment is extended into the next element in that direction, until it touches the boundary of that element.

A key feature of the method is the possibility of having multiple, interacting cohesive segments. To accommodate this, each cohesive segment is supported by a unique set of additional degrees of freedom. When two segments meet within a single element, the nodes that support the elements are enhanced twice. Consider the situation depicted in Figure 5a, in which segment II is approaching another segment (I), which can either be a material interface or a previously created cohesive crack. When the fracture criterion at the tip of segment II is violated, the segment is extended accordingly. The position of segment I marks a discontinuity and can therefore be considered as a free edge in the material. Hence, segment two is only extended until it touches segment I, see Figure 5b. Since this segment forms a free edge, there is no new tip for segment II and all nodes of the corresponding element are enhanced.

To avoid sudden jumps in stresses when a discontinuity is inserted, the constitutive behaviour of the cohesive segment is related to the stress state in the bulk material that violates the fracture criterion. The initial tractions in this irreversible model are taken to be equal to the normal and shear tractions in the bulk material at the exact moment of nucleation [2, 19].

When the normal traction component $t_{n,0}$ is positive, the cohesive segment is assumed to open as a cleavage crack. The cohesive tractions in the normal and the shear directions, t_n and t_s, decrease monotonically to zero from their initial values $t_{n,0}$ and $t_{s,0}$ as a function of the normal displacement jump:

$$t_n = t_{n,0}\left(1 - \frac{\Delta_n}{\Delta_{cr}}\right) ; \qquad t_s = t_{s,0}\left(1 - \frac{\Delta_n}{\Delta_{cr}}\right) \text{sgn}(\Delta_s), \qquad (17)$$

where Δ_n and Δ_s are the normal and sliding component of the displacement jump $[\![\mathbf{u}]\!]_j$ respectively, and sgn(.) is the signum function:

$$\text{sgn}(x) = \frac{x}{|x|}. \qquad (18)$$

The characteristic length of the cohesive law is determined by the critical normal opening displacement Δ_{cr} at which the crack has fully developed and the tractions have reduced to zero. This parameter is related to the fracture toughness $\mathcal{G}_c$, or the area under the softening curve and can be determined as follows:

$$\Delta_{cr} = \frac{2\mathcal{G}_c}{t_{n,0}}. \qquad (19)$$

A secant unloading algorithm is used for the case that the discontinuity starts to close [19].

2   Fluid Flow in Fractured Porous Materials

The second part of the method involves the simulation of the flow of fluids in the fractured material. Traditionally, the flow of fluids in a deformable, porous medium has been a topic of attention in the geotechnical and petrochemical sciences. Recently, the simulation of the flow of fluids gained interest in the biomedical community as well to model the behaviour of human tissues which are, in fact, porous media and must be analysed as such. Many of the current models are based on the seminal works of Terzaghi [20] and Biot [21], see Lewis and Schrefler [3] for a recent account.

2.1   Governing Equations for a Fluid-Saturated Porous Medium

The bulk is considered as a two-phase medium subjected to the restrictions of small displacement gradients and small variations in the concentrations. Furthermore, the assumptions are made that there is no mass transfer between the constituents, that convective terms and the gravity acceleration can be neglected, and that the processes which we consider, occur isothermally and quasi-statically. With these assumptions, the balances of linear momentum for the solid and the fluid phases read, e.g. Abellan and de Borst [22]:

$$\nabla \cdot \boldsymbol{\sigma}_\pi + \hat{\mathbf{p}}_\pi = \mathbf{0}, \tag{20}$$

with $\boldsymbol{\sigma}_\pi$ the stress tensor of constituent π. As in the remainder of this paper, $\pi = s$, f, with s and f denoting the solid and fluid phases, respectively. Further, $\hat{\mathbf{p}}_\pi$ is the source of momentum for constituent π from the other constituent, which takes into account the possible local drag interaction between the solid and the fluid. Evidently, the latter source terms must satisfy the momentum production constraint $\sum_{\pi=s,f} \hat{\mathbf{p}}_\pi = \mathbf{0}$. Adding both momentum balances, taking into account the momentum production constraint, and defining $\boldsymbol{\sigma} = \boldsymbol{\sigma}_s + \boldsymbol{\sigma}_f$, one obtains the "standard" equilibrium equation for the mixture:

$$\nabla \cdot \boldsymbol{\sigma} = \mathbf{0}, \tag{21}$$

Under the same assumptions as for the balance of momentum, one can write the balance of mass for each phase as:

$$\dot{\rho}_\pi + \rho_\pi \nabla \cdot \mathbf{v}_\pi = 0, \tag{22}$$

with ρ_π the apparent mass density and $\mathbf{v}_\pi$ the absolute velocity of constituent π. We multiply the mass balance for each constituent π by its volume ratio n_π, add them and utilise the constraint $\sum_{\pi=s,f} n_\pi = 1$ to give:

$$\nabla \cdot \mathbf{v}_s + n_f \nabla \cdot (\mathbf{v}_f - \mathbf{v}_s) + n_s \rho_s^{-1} \dot{\rho}_s + n_f \rho_f^{-1} \dot{\rho}_f = 0 \tag{23}$$

The change in the mass density of the solid material is related to its volume change by

$$(\alpha - 1)\nabla \cdot \mathbf{v}_s = n_s \rho_s^{-1} \dot{\rho}_s, \tag{24}$$

with K_s the bulk modulus of the solid material, K_t the overall bulk modulus of the porous medium, and $\alpha = 1 - K_t/K_s$ the Biot coefficient [3]. For the fluid phase, a phenomenological relation is assumed between the incremental changes of the apparent fluid mass density and of the fluid pressure p [3]:

$$Q^{-1}\mathrm{d}p = n_f \rho_f^{-1} \mathrm{d}\rho_f, \tag{25}$$

with $Q^{-1} = (\alpha - n_f)/K_s + n_f/K_f$ the compressibility modulus, and K_f is the bulk modulus of the fluid. Inserting Equations (24) and (25) into the balance of mass of the total medium, Equation (23), gives:

$$\alpha \nabla \cdot \mathbf{v}_s + n_f \nabla \cdot (\mathbf{v}_f - \mathbf{v}_s) + Q^{-1} \dot{p} = 0 \tag{26}$$

2.2   Weak Form and Micro-Macro Coupling

To arrive at the weak form of the balance equations, we multiply the momentum balance of Equations (21) and the mass balance of Equations (26) by kinetically admissible test functions for the displacements of the skeleton, $\boldsymbol{\eta}$, and for the pressure, ζ. Substitution into Equations (21) and (26), using Darcy's relation

$$n_f(\mathbf{v}_f - \mathbf{v}_s) = -k_f \nabla p, \tag{27}$$

with k_f the permeability coefficient of the porous medium, integrating over the domain Ω and using the divergence theorem leads to the corresponding weak forms:

$$\int_\Omega (\nabla \cdot \boldsymbol{\eta}) \cdot \boldsymbol{\sigma} \; d\Omega + \int_{\Gamma_d} [\![\boldsymbol{\eta} \cdot \boldsymbol{\sigma}]\!] \cdot \mathbf{n}_{\Gamma_d} \; d\Omega = \int_\Gamma \boldsymbol{\eta} \cdot \mathbf{t}_p \; d\Omega, \tag{28}$$

and

$$\begin{aligned}
&-\int_\Omega \alpha \zeta \nabla \cdot \mathbf{v}_s \; d\Omega + \int_\Omega k_f \nabla \zeta \cdot \nabla p \; d\Omega - \int_\Omega \zeta Q^{-1} \dot{p} \; d\Omega \\
&+ \int_{\Gamma_d} \mathbf{n}_{\Gamma_d} \cdot [\![\zeta \, n_f (\mathbf{v}_f - \mathbf{v}_s)]\!] \; d\Gamma = \int_\Gamma \zeta \mathbf{n}_\Gamma \cdot \mathbf{q}_p \; d\Gamma
\end{aligned} \tag{29}$$

Because of the presence of a discontinuity inside the domain Ω, the power of the external tractions on Γ_d and the normal flux through the faces of the discontinuity are essential features of the weak formulation. Indeed, these terms enable the momentum and mass couplings between the discontinuity (the small scale) and the surrounding porous medium (the large scale).

The momentum coupling stems from the tractions across the faces of the discontinuity and the pressure applied by the fluid in the discontinuity onto the faces of the discontinuity. We assume stress continuity from the cavity to the bulk, so that we have:

$$\boldsymbol{\sigma} \cdot \mathbf{n}_{\Gamma_d} = \mathbf{t}_d - p \mathbf{n}_{\Gamma_d}, \tag{30}$$

with $\mathbf{t}_d$ the cohesive tractions. Therefore, the weak form of the balance of momentum becomes:

$$\int_\Omega (\nabla \cdot \boldsymbol{\eta}) \cdot \boldsymbol{\sigma} \; d\Omega + \int_{\Gamma_d} [\![\boldsymbol{\eta}]\!] \cdot (\mathbf{t}_d - p \mathbf{n}_{\Gamma_d}) \; d\Gamma = \int_\Gamma \boldsymbol{\eta} \cdot \mathbf{t}_p \; d\Gamma \tag{31}$$

Since the tractions have a unique value across the discontinuity, the pressure p must have the same value at both faces of the discontinuity, and, consequently, this must also hold for the test function for the pressure, ζ. Accordingly, the mass transfer coupling term for the water can be rewritten as follows:

$$\begin{aligned}
&-\int_\Omega \alpha \zeta \nabla \cdot \mathbf{v}_s \; d\Omega + \int_\Omega k_f \nabla \zeta \cdot \nabla p \; d\Omega - \int_\Omega \zeta Q^{-1} \dot{p} \; d\Omega \\
&+ \int_{\Gamma_d} \zeta \mathbf{n}_{\Gamma_d} \cdot \mathbf{q}_d d\Gamma = \int_\Gamma \zeta \mathbf{n}_\Gamma \cdot \mathbf{q}_p \; d\Gamma,
\end{aligned} \tag{32}$$

where $\mathbf{q}_d = n_f [\![\mathbf{v}_f - \mathbf{v}_s]\!]$ represents the fluid flux through the faces of the discontinuity.

To quantify the influence of the "micro"-flow inside the discontinuity on the "macro"-scale, we recall the balance of mass, which, for the "micro"-flow in the cavity reads:

$$\dot{\rho}_f + \rho_f \nabla \cdot \mathbf{v} = 0$$

subject to the assumptions of small changes in the concentrations and that convective terms can be neglected, cf. Equation (22). We assume that the first term can be neglected because the problem is monophasic in the cavity and the velocities are therefore much higher than in the porous medium, although this assumption is not essential. With this assumption, and focusing on a two-dimensional configuration, the mass balance inside the cavity simplifies to:

$$\frac{\partial v}{\partial x} + \frac{\partial w}{\partial y} = 0, \tag{33}$$

with $v = \mathbf{v} \cdot \mathbf{t}_{\Gamma_d}$ and $w = \mathbf{v} \cdot \mathbf{n}_{\Gamma_d}$ the tangential and normal components of the fluid velocity in the discontinuity, respectively, see Figure 6. Accordingly, the difference in the fluid velocity components that are normal to both crack faces is given by:

$$[\![w_f]\!] = -\int_{-h}^{h} \frac{\partial v}{\partial x} dy \tag{34}$$

To proceed, the velocity profile of the fluid flow inside the discontinuity must be known. Different possibilities exist, but here we follow Ref. [5], in which a Newtonian fluid was assumed. Together with the balance of momentum for the fluid in the discontinuity and integrating from $y = -h$ to $y = h$, Figure 6, we obtain the following velocity profile:

$$v(y) = \frac{1}{2\mu} \frac{\partial p}{\partial x} (y^2 - h^2) + v_f, \tag{35}$$

with μ the viscosity of the fluid. The essential boundary $v = v_f$ has been applied at both faces of the cavity, and stems from the relative fluid velocity in the porous medium at $y = \pm h$:

$$v_f = (\mathbf{v}_s - n_f^{-1} k_f \nabla p) \cdot \mathbf{t}_{\Gamma_d}.$$

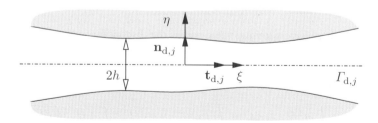

Fig. 6 Geometry of a crack in the domain. The width of the cavity $2h$ is a function of the position on the cavity Γ_d and is equal to the displacement jump $[\![\mathbf{u}_j]\!]$

Substitution of Equation (35) into Equation (34) and again integrating with respect to y then leads to:

$$[\![w_f]\!] = \frac{2}{3\mu}\frac{\partial}{\partial x}\left(\frac{\partial p}{\partial x}h^3\right) - 2h\frac{\partial v_f}{\partial x}, \tag{36}$$

This equation gives the amount of fluid attracted in the tangential fluid flow. It can be included in the weak form of the mass balance of the "macro"-flow to ensure the coupling between the "micro"-flow and the "macro"-flow. Since the difference in the normal velocity of both crack faces is given by $[\![w_s]\!] = 2\dot{h}$, the mass coupling term becomes:

$$\mathbf{n}_{\Gamma_d}\cdot\mathbf{q}_d = n_f[\![w_f - w_s]\!] = n_f\left(\frac{2h^3}{3\mu}\frac{\partial^2 p}{\partial x^2} + \frac{2h^2}{\mu}\frac{\partial p}{\partial x}\frac{\partial h}{\partial x} - 2h\frac{\partial v_f}{\partial x} - 2\dot{h}\right). \tag{37}$$

As in the preceding section, the interpolation of each component of the displacement field $\mathbf{u}$ of the solid phase is enriched with discontinuous functions, cf. Equation (1). With respect to the interpolation of the pressure p we note that the fluid flow normal to the discontinuity is discontinuous. Since the fluid velocity is related to the pressure gradient via Darcy's law, the gradient of the pressure normal to the discontinuity is therefore discontinuous across the discontinuity. Accordingly, the enrichment of the interpolation of the pressure must be such that the pressure itself is continuous, but has a discontinuous first spatial derivative. The distance function $\mathcal{D}_{\Gamma_d}$ defined as

$$\mathbf{n}_{\Gamma_d}\cdot\nabla\mathcal{D}_{\Gamma_d} = \mathcal{H}_{\Gamma_d}, \tag{38}$$

satisfies this requirement, and accordingly, all nodes whose support is cut by the discontinuity hold additional pressure degrees of freedom such that:

$$p = \mathbf{H}\hat{\mathbf{p}} + \mathcal{D}_{\Gamma_d}\mathbf{H}\tilde{\mathbf{p}}, \tag{39}$$

where $\mathbf{H}$ contain the shape functions H_i used as partition of unity for the interpolation of the pressure field p, and $\hat{\mathbf{p}}$ and $\tilde{\mathbf{p}}$ are the nodal arrays assembling the amplitudes that correspond to the standard and enhanced interpolations of the pressure field.

The choices for N_i and H_i are driven by modelling requirements. Indeed, the modelling of the fluid flow inside the cavity needs the second derivatives of the pressure, see Equation (37). Hence, the order of the finite element shape functions H_i has to be sufficiently high, otherwise the coupling between the fluid flow in the cavity and the bulk will not be achieved. Further, the order of the finite element shape functions N_i must be greater than or equal to the order of H_i for consistency in the discrete balance of the momentum equation.

3   The Healing Process

The final part of the numerical tool is the simulation of the rebonding of crack surfaces. The basic idea of the healing model is that the cohesive constitutive relation that governs the opening of the crack is extended with a term in which the strength and stiffness of the interface is increased to simulate crack bonding. The approach has a phenomenological character and avoids the detailed analysis on a smaller level of observation, where the actual chemical reactions of the rebonding process play a role.

The actual rebonding of crack surfaces can be triggered in different ways. In the encapsulated polymer material [1], the healing agent starts to polymerise when it is contacted by a catalyst that is embedded in the material. In the case of self healing concrete, the rebonding is initiated when the amount of water in a part of uncarbonised concrete has reached as certain threshold level. In both situations, the pressure p of the fluid at a certain position in the model is the main parameter that triggers the onset of the healing model. Important parameters that characterise the healing process are the time needed to form the bond and the final strength of the new bond. In this section, an example of a phenomenological mode-I cohesive law with healing will be presented [6].

A simple mode-I cohesive constitutive relation can be written as, cf. Equation (17):

$$t_n = t_{n,0} \left(1 - \frac{\Delta_n}{\Delta_{cr}} \right)$$

In this simple model, the tractions in the shear directions are not defined. When at a certain instant t_r the catalyst and the healing agent start to react, the rebonding of the surfaces is initiated. At this moment the opening of the crack is defined as $\Delta_{n,r}$. Because of the chemical reaction, the stiffness of the interface will be recovered up to a certain value r ($0 \le r \le 1$). A recovery up to $r = 1$ implies that the interface has fully recovered and the interface possesses the same properties as it had prior to cracking. This recovery is a function of time and is governed by a healing parameter $h(t)$. An example of a healing function which can properly mimic the behaviour of rebonding processes, is the arctan function, shown in Figure 7. In this figure, the healing time has been taken equal to 1 s. and the recovery factor r at $t \to \infty$ equals 1. The total traction including the healing effect is then given by

$$t_n = t_{n,0} \left(1 - \frac{\Delta_n}{\Delta_{cr}} \right) + r(h(t)) \cdot (t_{n,0} - t_n(\Delta_{n,r})) \tag{40}$$

where h is the time-dependent healing function and $\Delta_{n,r}$ the normal opening at the instant when the recovery started.

Although this cohesive healing law incorporates characteristics of the healing process, it is far from complete. An important factor in the healing process is the loading state of the interface. When the crack is closed, e.g. in a fatigue-loading situation, the rebonding takes place much faster and the final strength of the bond improves significantly.

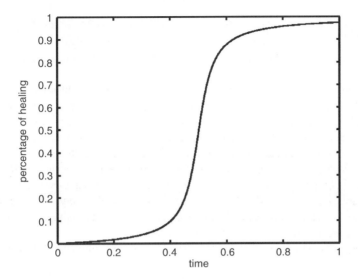

Fig. 7 Healing curve as a function of time

4 Concluding Remark

In this chapter, a numerical strategy for the simulation of self healing mechanisms has been outlined. It consists of three components: (i) a cohesive-crack formulation, (ii) a transport model for the liquid in the cracks, and (iii) a healing function that augments the cohesive crack model. Currently, numerical simulation of self healing processes is just at its infancy, and major steps will have to be taken, especially regarding the implementation of the proper physics in the constitutive model, before reliable simulations can be done.

References

1. White S.R., Sottos N.R., Geubelle P.H., Moore J.S., Kessler M.R., Sriram S.R., Brown E.N., Viswanathan S. (2001) Autonomic healing of polymer composites. Nature 409:794–797
2. Remmers J.J.C. (2006) Discontinuities in materials and structures, a unifying computational approach. Ph.D. dissertation. Delft University of Technology, Delft, The Netherlands
3. Lewis R.W., Schrefler B.A. (1998) The finite element method in the static and dynamic deformation and consolidation of porous media, 2nd edn. Wiley, Chichester
4. de Borst R., Réthoré J., Abellan M.A. (2006) A numerical approach for arbitrary cracks in a fluid-saturated medium. Arch Appl Mech 75:595–606
5. Réthoré J., de Borst R., Abellan M.A. (2007) A two-scale approach for fluid flow in fractured porous media. Int J Numer Methods Eng doi: 10.1002/nme.1962
6. Schimmel E.C., Remmers J.J.C. (2006) Development of a constitutive model for self-healing materials. Report DACS-06-003, ISSN 1574–6992

7. Dugdale D.S. (1960) Yielding of steel sheets containing slits. J Mech Phys Solids 8:100–108
8. Barenblatt G.I. (1962) The mathematical theory of equilibrium cracks in brittle fracture. Adv Appl Mech 7:55–129.
9. Needleman A. (1987) A continuum model for void nucleation by inclusion debonding. J. Appl Mech 54:525–531
10. Xu X.P., Needleman A. (1993) Void nucleation by inclusion debonding in a crystal matrix. Model Simul Mater Sci Eng 1:111–132.
11. Xu X.P., Needleman A. (1994) Numerical simulations of fast crack-growth in brittle solids. J Mech Phys Solids 42:1397–1434
12. Tijssens M.G.A., Sluys L.J., van der Giessen E. (2001) Simulation of fracture in cementitious composites with explicit modeling of microstructural features. Eng Fract Mech 68:1245–1263
13. Schellekens J.C.J., de Borst R. (1992) On the numerical integration of interface elements. Int J Numer Methods Eng 36(1):43–66
14. Babuška I., Melenk J.M. (1997) The partition of unity method. Int J Numer Methods Eng 40(4):727–758
15. Moës N., Dolbow J., Belytschko T. (1999) A finite element method for crack growth without remeshing. Int J Numer Methods Eng 46:131–150
16. Wells G.N., Sluys L.J. (2001) A new method for modelling cohesive cracks using finite elements. Int J Numer Methods Eng 50(12):2667–2682
17. Moës N., Belytschko T. (2002) Extended finite element method for cohesive crack growth. Eng Fract Mech 69:813–833
18. Simone A., Sluys L.J. (2004) The use of displacement discontinuities in a rate-dependent medium. Comp Methods Appl Mech Eng 193(27–29):3015–3033
19. Camacho G.T., Ortiz M. (1996) Computational modelling of impact damage in brittle materials. Int J Solids Struct 33:2899–2938
20. Terzaghi K. (1943) Theoretical soil mechanics. Wiley, New York.
21. Biot M.A. (1965) Mechanics of incremental deformations. Wiley, Chichester
22. Abellan M.A., de Borst R. (2005) Wave propagation and localisation in a softening two-phase medium. Comp Methods Appl Mech Eng 195:5011–5019

About the Authors

John Adam is currently a professor in the Department of Mathematics & Statistics at Old Dominion University in Norfolk, Virginia, USA. He holds a BSc. in Physics and a Ph.D. in Theoretical Astrophysics, both from the University of London. His principal research interests are currently mathematical modelling of cancer, wound healing, as well as theoretical aspects of meteorological optics.

Magnus (H.M.) Andersson is currently a research scientist at the Beckman Institute at the University of Illinois at Urbana-Champaign, USA. He holds an MSc. in Engineering Physics and a Ph.D. in Fluid Mechanics. His principal research interests are in autonomous materials systems, specifically mechanochemical assessment experiments and techniques.

Rolf A.T.M. van Benthem is currently part-time professor in Coatings Technology at the Department of Chemical Engineering and Chemistry of the Eindhoven University of Technology, the Netherlands. At DSM, the Netherlands, he is R&D Competence Manager in Coatings and Thermoset Materials. He holds an MSc. in Organic Chemistry and a Ph.D. in Organic Chemistry and Homogeneous Catalysis. His research interests include polymer chemistry (synthesis and application), network formation, and coatings technology.

Sheba D. Bergman is currently a postdoctoral associate in the Chemistry department at the University of California, Santa Barbara, USA. She holds a Ph.D. in Chemistry from Tel Aviv University. Her principal research interests are the development of novel polymeric materials that exhibit both re-mendability and autonomous healing.

Amit Bhasin is an associate research scientist at the Texas Transportation Institute, Texas A&M University, USA. He holds an MSc. and a Ph.D. in Civil Engineering. His research interests are in the effect of surface properties of asphalt binder and aggregates and the impact of moisture on the fatigue damage process.

Ian P. Bond is currently a reader in Aerospace Materials in the Department of Aerospace Engineering at the University of Bristol, UK. He holds a BSc. in Materials Science and a Ph.D. in Polymer Composite Materials. His principal research interests include the design, fabrication, and application of highly damage tolerant/self healing composites and multifunctional smart materials for self-actuation.

Rene de Borst is Dean of Mechanical Engineering and Distinguished Professor at Eindhoven University of Technology, the Netherlands. He holds an MSc. in Civil Engineering and a Ph.D. in Engineering Sciences, both from Delft University of Technology, the Netherlands. His primary research interests lie in the development of numerical methods for materials, structures, flows, and interaction problems.

Peter Fratzl is director at the Max Planck Institute of Colloids and Interfaces in Potsdam, Germany. He holds an engineering degree from the Ecole Polytechnique, Paris, and a doctorate in Physics from the University of Vienna. His principal research interests are in biological and biomimetic materials.

Simon A. Hayes is currently a lecturer at The University of Sheffield in the UK. He holds a BEng. in Materials Science, a Master of Education (MEd), and a Ph.D.

in Composite Technology. His principal research interests are in smart sensing of cure state and damage development in composite structures as well as self healing technologies. He holds two patents in these areas.

Jeff T. De Hosson is currently a professor of Applied Physics at the Faculty of Mathematics and Natural Sciences of the University of Groningen, the Netherlands. He holds an MSc. degree in theoretical physics and chemical physics from the Utrecht University and a Ph.D. in Physics from the University of Groningen, the Netherlands. His research interests are in dislocation dynamics, interface science, and in situ transmission/scanning electron microscopy.

Peizheng Huang is currently an associate professor at the College of Aerospace Engineering, Nanjing University of Aeronautics and Astronautics, P.R. China. She holds an MSc. in Mechanical Engineering and a Ph.D. in Solid Mechanics. Her principal research interest is in the defect evolution in metallic materials.

Frank R. Jones is Professor of Polymers and Composites in the Department of Engineering Materials, The University of Sheffield, UK. He holds a Ph.D. in Polymer Chemistry after industrial experience of plastics materials, which led to his research interests in linking the micromechanics of composites to the structure of the resin and interphase in the presence of environments. He holds two patents in self healing resins.

Henk Jonkers is currently a postdoctoral research associate at the Faculty of Civil Engineering and Geosciences at the Technical University Delft, the Netherlands. He holds an MSc. in Marine Biology and a Ph.D. in Microbial Ecology. His principal research interests concern the impact of microbial communities on natural- and man-made materials and ecosystems with focus on development and application of bioinspired sustainable materials in civil engineering.

Michael W. Keller is currently a postdoctoral researcher at the University of Illinois at Urbana-Champaign, USA. He holds an MSc. and Ph.D. in Theoretical and Applied Mechanics. His principal research interests are solid mechanics and materials.

Victor C. Li is the E. Benjamin Wylie Collegiate Professor of Civil and Environmental Engineering at the University of Michigan, USA. He holds an MSc. in Applied Mechanics, and a Ph.D. in Solids and Structures. His principal research interests are in micromechanics-based cement-based composite materials design and engineering for robustness, integrated materials and structural design, and structural durability and sustainability.

Zhonghua Li is currently a professor at the Department of Engineering Mechanics of Shanghai Jiaotong University, P.R. China. He holds an MSc. in Engineering Mechanics. His principal research interests are fracture, creep, and the defect evolution in metallic materials.

Dallas N. Little is the Snead Chair Professor in Civil Engineering at Texas A&M University, USA. He holds an MSc. and a Ph.D. in Civil Engineering. His research interests include fatigue damage of asphalt materials and the impact of fillers and surface properties of asphalt binder and aggregate on the microcrack growth and healing processes.

Roger Lumley is currently a senior research scientist with the CSIRO Light Metals Flagship in Melbourne, Australia. He holds a Ph.D. in Metallurgy. His principal research interests are involved with the metallurgy of aluminium alloys, in particular processing–microstructure–property relationships.

W. (Marshall) Ming is currently an assistant professor in Polymer Chemistry & Coatings Technology at Eindhoven University of Technology, the Netherlands. He holds a Ph.D. in Polymer Chemistry & Physics. His current research interests are centred on nature-inspired functional, nanostructured polymeric/hybrid materials and their applications.

Jeffrey S. Moore is currently a professor in the Department of Chemistry and in Materials Science and Engineering at the University of Illinois at Urbana-Champaign, USA. He holds a Ph.D. in materials science and engineering. His principal research interests involve the synthesis and study of large organic molecules and the discovery of new polymeric materials.

Etelvina Javierre Perez is currently a postdoctoral researcher at the Faculty of Aerospace Engineering at the Technical University Delft, the Netherlands. She holds an MSc. in Applied Mathematics and a Ph.D. in Computational Materials Science. Her principal research interests are in the design and implementation of numerical methods for solving moving boundary problems, especially those arising from phase transformations in materials science.

Joris J.C. Remmers is an assistant professor at the Faculty of Aerospace Engineering at Delft University of Technology, the Netherlands. He holds an MSc. in Aerospace Engineering and a Ph.D. in Computational Mechanics. His principal research interests are in the development of numerical methods for the simulation of fracture. As of September 2007, he has taken a position at Eindhoven University of Technology, Department of Mechanical Engineering.

Miranda W.G. van Rossum is a graduate student at the Department of Applied Mathematics at the Technical University Delft, the Netherlands. Her interests are in mathematical modelling of complex phenomena.

Wim G. Sloof is currently an associate professor at the Department of Materials Science and Engineering of the Delft University of Technology, the Netherlands. He holds an MSc. and a Ph.D. in Materials Science. His principal research interests are oxidation of metals and materials, and coating systems for high temperature applications.

Nancy R. Sottos is currently the Donald B. Willett Professor of Materials Science and Engineering at the University of Illinois Urbana-Champaign, USA. She holds a Ph.D. in Mechanical Engineering. Her principal research interests are autonomous materials systems with emphasis on self healing polymers, mechanical behaviour of polymers, and composites, and reliability and adhesion of thin film interfaces.

Richard S. Trask is currently a Research Fellow in the Department of Aerospace Engineering at the University of Bristol, UK. He holds a BSc. in Materials Science and Engineering, MSc. in Advanced Materials Technology, and a Ph.D. in Damage Tolerant Composite Sandwich Structures. His principal research interests include structural damage tolerance using novel concepts/bioinspired and biomimetic self healing composites, and the development of advanced composite manufacturing techniques.

Russell Varley is currently a Senior Research Scientist at CSIRO Manufacturing and Materials Technology, Australia. He holds a Ph.D. in Materials Engineering. His principal research interests are in the understanding of structure–property relationships of high-performance polymers and the effect upon long-term performance and durability.

Fred Vermolen is currently an assistant professor at the Faculty of Applied Mathematics at the Delft University of Technology, the Netherlands. He holds an MSc. in Applied Physics and a Ph.D. in Materials Science/Numerical Mathematics. His principal research interests are models in mathematical biology, materials science, and flow in porous media.

Hua Wang is currently a professor at the Department of Engineering Mechanics of Shanghai Jiaotong University, P.R. China. He holds an MSc. in Mechanical Engineering and a Ph.D. in Solid Mechanics. His principal research interest is in simulation of microstructure evolution of materials.

Richard Weinkamer is currently research group leader at the Max Planck Institute of Colloids and Interfaces in Potsdam, Germany. He holds an MSc. in Mathematics and a Ph.D. in Physics. His principal research interests are in computational mechanobiology to investigate the influence of mechanical forces on maintenance and healing processes in living biological materials.

Scott White is currently the Donald B. Willett Professor of Engineering in the Department of Aerospace Engineering at the University of Illinois, USA. He holds a Ph.D. in Engineering Mechanics from The Pennsylvania State University. His principal research is the development of self-generating and self-regulating materials systems that find use in a wide range of industries from microelectronics to aerospace to biomedicine.

Gareth J. Williams is currently a postgraduate student in the Department of Aerospace Engineering at the University of Bristol, UK. He holds an MEng. in Aeronautical Engineering. His principal research interests are utilizing resin-bearing hollow glass fibres to impart self-healing functionality to carbon fibre-reinforced plastic for aerospace applications.

Hugo R. Williams is currently a postgraduate student in the Department of Aerospace Engineering at the University of Bristol, UK. He holds a MEng. in Aeronautical Engineering. His principal research interests are self healing of composite sandwich structures using vascular networks and the development of biomimetic optimization and design tools.

Gijsbertus (Bert) de With is currently full professor in Materials Science at the Department of Chemical Engineering and Chemistry of the Eindhoven University of Technology, the Netherlands. He holds an MSc. in Chemistry and a Ph.D. in Technological Sciences. His research interests include the chemical and mechanical processing as well as the chemo-mechanical behaviour of multi-phase materials.

Fred Wudl is currently a professor of Chemistry and Materials Science and co-Director of the Center for Polymers and Organic Solids at the University of

California, Santa Barbara, USA. He holds a BSc. in Chemistry and a Ph.D. in Organic Chemistry. His principal research interests are in the design and synthesis of novel polymeric systems with applications in polymeric photovoltaics and polymeric light-emitting diodes. Recently he has initiated research into re-mending polymers using reversible polymerizations.

En-Hua Yang is currently a Ph.D. student at the Department of Civil and Environmental Engineering at the University of Michigan, USA. He holds an MSc. in Materials Science and an MS in Applied Mechanics. His principal research interests are in the design, characterization, and application of ductile cementitious composites.

Hiroyuki Y Yasuda is currently an associate professor of Materials Science at the Research Centre for Ultra-High Voltage Electron Microscopy of Osaka University, Japan. He holds an MSc. and a Ph.D. in Materials Science and Engineering from Osaka University. His research interests are in dislocation dynamics, phase transformations, and transmission electron microscopy.

Wenting Zhang is currently a Ph.D. student in the Department of Engineering Materials at the University of Sheffield, UK. Her principal research interest is in the development and characterization of novel functional healing resin systems and facilitating their adoption in commercial applications.

Sybrand van der Zwaag is currently a professor at the Faculty of Aerospace Engineering at the Technical University Delft, the Netherlands. He holds an MSc. in Metallurgy from the Delft University of Technology and a Ph.D. in Applied Physics from Cambridge University. His principal research interests are in the design and realization of novel metallic and polymeric systems with improved mechanical properties for aerospace and space applications, and self healing materials in particular.

Springer Series in
MATERIALS SCIENCE

Editors: R. Hull   R. M. Osgood, Jr.   J. Parisi   H. Warlimont

Springer Series in
MATERIALS SCIENCE

Editors: R. Hull   R. M. Osgood, Jr.   J. Parisi   H. Warlimont